Lecture Notes in Computer Science 16077

The series Lecture Notes in Computer Science (LNCS), including its subseries Lecture Notes in Artificial Intelligence (LNAI) and Lecture Notes in Bioinformatics (LNBI), has established itself as a medium for the publication of new developments in computer science and information technology research, teaching, and education.

LNCS enjoys close cooperation with the computer science R & D community, the series counts many renowned academics among its volume editors and paper authors, and collaborates with prestigious societies. Its mission is to serve this international community by providing an invaluable service, mainly focused on the publication of conference and workshop proceedings and postproceedings. LNCS commenced publication in 1973.

Jannik Matuschke · José Verschae
Editors

Approximation and Online Algorithms

23rd International Workshop, WAOA 2025
Warsaw, Poland, September 18–19, 2025
Proceedings

Springer

Editors
Jannik Matuschke
KU Leuven
Leuven, Belgium

José Verschae
UC Chile
Santiago, Chile

ISSN 0302-9743 ISSN 1611-3349 (electronic)
Lecture Notes in Computer Science
ISBN 978-3-032-06705-0 ISBN 978-3-032-06706-7 (eBook)
https://doi.org/10.1007/978-3-032-06706-7

This Springer imprint is published by the registered company Springer Nature Switzerland AG
The registered company address is: Gewerbestrasse 11, 6330 Cham, Switzerland

Preface

The 23rd Workshop on Approximation and Online Algorithms (WAOA 2025) focused on the design and analysis of algorithms for online and computationally hard problems. Both types of problems have a large number of applications in a wide variety of fields. The workshop was held in Warsaw, Poland, on September 18–19, 2025. WAOA 2025 was part of ALGO 2025, which also hosted ESA, ALGOCLOUD, ALGOWIN, ATMOS, IPEC, and WABI.

WAOA 2025 invited submissions on topics related to approximation and online algorithms, including algorithmic game theory, coloring and partitioning, computational economics and mechanism design, experimental methods for approximation and online algorithms, FPT-approximation algorithms, geometric problems, graph algorithms and network design, inapproximability results, packing and covering, matroids and submodular functions, new paradigms in approximation and online optimization, online selection problems, resource augmentation, relaxations and tightness of formulations, robust and stochastic problems, and scheduling problems.

In response to the call for papers, we received 43 submissions. Each submission was reviewed by at least three referees in a double-blind review process. Submissions were judged primarily on originality, technical quality, clarity, and relevance to the conference topics. Based on the reviews, the program committee selected 15 papers. This volume contains the final revised versions of these papers, as well as an invited contribution from our plenary speaker, Sahil Singla.

The EasyChair conference system was used to manage the electronic submissions, the review process, and the electronic program committee discussions. We would like to thank all authors who submitted papers to WAOA 2025 and all participants of WAOA 2025, including the presenters of the accepted papers. Special thanks go to the plenary speaker Sahil Singla for accepting our invitation. We would also like to thank all members of the program committee and the external reviewers for their diligent work in evaluating the submissions and their contributions to the discussions. Additionally, we would like to express our gratitude to the members of the WAOA steering committee for their support. Last but not least, we are grateful to all the local organizers of ALGO, especially Marcin Pilipczuk and Paweł Rzążewski, who served as local chairs of the organizing committee.

August 2025

Jannik Matuschke
José Verschae

Organization

Steering Committee

Roberto Solis-Oba	University of Western Ontario, Canada
Evripidis Bampis	Sorbonne Université, France
Thomas Erlebach	Durham University, UK
Christos Kaklamanis	University of Patras, Greece
Nicole Megow	Universität Bremen, Germany
Laura Sanità	Bocconi University, Italy
Martin Skutella	Technische Universität Berlin, Germany

Program Committee

Umang Bhaskar	Tata Institute of Fundamental Research, India
Karthekeyan Chandrasekaran	University of Illinois Urbana-Champaign, USA
Yann Disser	TU Darmstadt, Germany
Franziska Eberle	Technische Universität Berlin, Germany
Yuri Faenza	Columbia University, USA
Cristina Fernandes	University of São Paulo, Brazil
Thekla Hamm	TU Eindhoven, The Netherlands
Lisa Hellerstein	New York University, USA
Sungjin Im	University of California, Santa Cruz, USA
Naonori Kakimura	Keio University, Japan
Kim-Manuel Klein	Christian-Albrechts-University of Kiel, Germany
Jannik Matuschke (Chair)	Katholieke Universiteit Leuven, Belgium
Arturo Merino	Universidad de O'Higgins, Chile
Malin Rau	Chalmers University of Technology, Sweden
Rebecca Reiffenhäuser	University of Amsterdam, The Netherlands
Melanie Schmidt	HHU Düsseldorf, Germany
Seeun William Umboh	The University of Melbourne, Australia
José Verschae (Chair)	Pontifical Catholic University of Chile, Chile
Yu Yokoi	Tokyo Institute of Technology, Japan

Additional Reviewers

Bandopadhyay, Susobhan
Berndt, Sebastian
Bernstein, Aaron
Blauth, Jannis
Buchem, Moritz
Böhm, Martin
Cohen, Ilan
Couto, Yan
Dey, Dipan
Encz, Koppány
Goswami, Mayank
Guo, Longkun
Gálvez, Waldo
Herold, Martin
Hommelsheim, Felix
Huang, Zhengcheng
Jain, Rhea
Jäger, Sven
Kawase, Yasushi
Klimm, Max
Kozma, László
Kucera, Petr
Kumar, Nithish Kumar
Kurita, Kazuhiro
Livanos, Vasilis
Lutz, Annette
Maack, Marten
Marwaha, Kunal
Pandit, Yeshwant
Pedrosa, Lehilton L. C.
Polak, Adam
Pollner, Tristan
Reuter, Janina
Rohwedder, Lars
Rubio, Benjamín
Sadhukhan, Arpan
Sarkar, Sherry
Schewior, Kevin
Schlöter, Jens
Schweichhart, Jonas
Shu, Xinkai
Soma, Tasuku
Szabo, Daniel
Thelen, Linda
Vuorinen, Aapeli
Whittington, Philip
Wienöbst, Marcel
Zhu, Weihao
Zych-Pawlewicz, Anna

Invited Talk

Beyond Competitive Analysis for Online Algorithms

Sahil Singla

Georgia Institute of Technology, USA

Competitive analysis has long been the standard framework for evaluating online algorithms under worst-case inputs. While powerful, it often yields weak or even vacuous bounds for central problems—such such as online matchings, load balancing, and network design. Yet many of these problems become far more tractable once we move beyond the purely worst-case lens. This has motivated a range of beyond worst-case models that exploit additional structure or information to achieve stronger guarantees.

In this talk, I will survey several such models and their algorithmic implications: stochastic arrival processes and prophet-type inequalities; algorithms with sample access to the input; recourse models that allow limited revisions to past decisions; and optimal policy design as a benchmark beyond the offline optimum. I will emphasize unifying ideas across these approaches and highlight recent advances that overcome long-standing worst-case barriers.

Contents

Interval-Constrained Bipartite Matching over Time

Andreas Abels[1], Mariia Anapolska[2(✉)], and Christina Büsing[2]

[1] Heinrich-Heine-Universität Düsseldorf, Düsseldorf, Germany
andres.abels@hhu.de
[2] RWTH Aachen University, Aachen, Germany
{anapolska,buesing}@combi.rwth-aachen.de

Abstract. Interval-constrained online bipartite matching problem frequently occurs in medical appointment scheduling: Unit-time jobs representing patients arrive online and are assigned to a time slot within their given feasible time interval. We consider a variant of this problem where reassignments are allowed and extend it by a notion of time that is decoupled from the job arrival events. As jobs appear, the current point in time gradually advances, and once the time of a slot is passed, the job assigned to it is fixed and cannot be reassigned anymore. We analyze two algorithms for the problem with respect to the resulting matching size and the number of reassignments they make. We show that FirstFit with reassignments according to the shortest augmenting path rule is $\frac{2}{3}$-competitive with respect to the matching cardinality, and that the bound is tight. The competitive ratio remains $\frac{2}{3}$ if we restrict the algorithm to make only a constant number $k \geqslant 1$ of reassignments per job arrival. This fills the gap between the known optimal algorithm that makes no reassignments, which is $\frac{1}{2}$-competitive, on the one hand, and an earliest-deadline-first strategy (EDF), which we prove to obtain a maximum matching in this over-time framework, but which suffers $\Omega(n^2)$ reassignments in the worst case, on the other. In the full paper, we show that FirstFit remains $\frac{2}{3}$-competitive also if the sets of feasible slots per job that are not intervals, and that this is the best possible deterministic competitive ratio, while EDF loses its optimality.

Keywords: Online matching · Bipartite matching · Competitive analysis

1 Introduction

Consider the process of assigning medical appointments of equal duration to patients. Patients request appointments one by one and expect an immediate response. Patient's health condition determines the time interval within which

A. Abels—The research was conducted during the author's employment at RTWH Aachen Unversity.

J. Matuschke and J. Verschae (Eds.): WAOA 2025, LNCS 16077, pp. 1–17, 2026.
https://doi.org/10.1007/978-3-032-06706-7_1

they require an appointment: the urgency of treatment imposes the deadline, and the amount of time needed to prepare the patient for treatment, e.g., additional diagnostics before a surgery, determines the earliest possible appointment day. When urgent requests, that is, those with a short feasible time interval, arrive, previously scheduled patients with looser deadlines may be postponed in favor of urgent patients. As time passes, a growing part of the schedule is in the past and therefore cannot be changed, whereas appointment slots in the future can still be reassigned to different patients. The goal is to schedule as many patients as possible and minimize changes to the schedule.

The described appointment scheduling cannot be modeled precisely as a classical online job scheduling problem, since the decision on whether to reject a job must be made upon job's arrival, not at its latest possible start time. Instead, we model the process as a variant of online bipartite matching with reassignments. In online bipartite matching, vertices of one part of the graph, here called "slots", are known in advance, and vertices of the other part, here called "jobs", which represent patients, are revealed one by one together with their incident edges. Upon arrival, each job is either immediately assigned to an adjacent slot or irrevocably rejected. It is possible for the algorithm to reallocate, i.e., reassign, jobs to different slots to make place for a new job.

In online bipartite matching with reassignments, any reassignment along the graph edges is feasible. In our application, however, if appointment scheduling continues as the appointments take place, reassigning a job to an already passed slot is impossible, and the job assigned to it also cannot be reassigned. Therefore, we extend this well-studied online problem by a notion of time. As time passes, more jobs arrive and slots are processed, so that once a slot is in the past, jobs assigned to it cannot be reassigned anymore. The task is to maximize the number of assigned jobs and minimize the number of reassignments. We call the described problem Bipartite Matching over Time (BMT). Furthermore, our jobs feature interval constraints: a job can be allocated only to a slot between its given earliest and latest slot. This yields a special version of the problem, called *Interval-Constrained Bipartite Matching over Time* (ICBMT), which is the main focus of this work.

We analyze two algorithmic approaches to ICBMT with respect to the number of scheduled jobs and of required reassignments: a greedy FirstFit algorithm that uses shortest augmenting paths to augment the assignment when there is no feasible slot available, which is analogous to an optimal assignment strategy for online bipartite matching, and the earliest-deadline-first strategy, which is motivated by an optimal algorithm for related machine scheduling problems. We analyze both approaches in the ICBMT setting first and then also in the BMT setting.

1.1 Related Work

ICBMT is strongly related to the online bipartite matching problem and the online assignment problem. In these problems, the servers (possibly with capacity greater than one) are known in advance, and clients are revealed online, each with a subset of servers the client can be served by. Typical objectives are to maximize the number of assigned clients [19] or to balance the load of the servers [6].

In the classical maximum online bipartite matching, all matching decisions are irrevocable, and the aim is to maximize the number of matched clients. No deterministic algorithm can have a competitive ratio greater than $\frac{1}{2}$, which is achieved by a greedy algorithm. Karp, Vazirani, and Vazirani present a randomized algorithm RANKING, which is strictly $(1 - \frac{1}{e})$-competitive, and show the tightness of the bound [19]. The proof has later been corrected, refined, and generalized [7,12,14].

In many applications, it is reasonable to allow reassignments; that is, client rejections are irrevocable, but the assigned server of an already assigned client can be changed. In this problem, it is trivial to achieve a maximum matching at the cost of a large number of reassignments. One widely studied strategy for online bipartite matching with reassignments is Shortest Augmenting Paths (SAP), which is also the basis of an optimal algorithm for offline maximum bipartite matching [18]. Grove et al. show a lower bound of $\Omega(n \log n)$ reassignments for SAP [15]. They also show a matching upper bound of $\mathcal{O}(n \log n)$ reassignments for the case of all clients having a degree of two. This upper bound for SAP is later also proven for arrivals in random order [10] and for trees [9]. For general graphs, Bernstein, Holm, and Rotenberg show an amortized upper bound of $\mathcal{O}(n \log^2 n)$ reassignments for SAP [5]. Chaudhuri et al. conjecture that the lower bound is tight for general bipartite graphs [10].

Further exact approaches to online bipartite matching with reassignments, which are different from SAP, have also been studied [8,10] and do not improve the number of reassignments.

Another branch of research concerns approximation algorithms for online bipartite matching. Bosek et al. construct $(1-\epsilon)$-approximate matchings applying $\mathcal{O}((1 - \epsilon)n)$ reassignments [8]. Bernstein et al. approximate the problem for every p-norm simultaneously. Their algorithm is 8-competitive with $O(n)$ reassignments [6]. Gupta, Kumar, and Stein consider the assignment version of the problem, where each server can serve multiple clients, and present a 2-competitive algorithm for maximum load minimization with at most $O(n)$ reassignments [16]. Recently, Shin et al. consider the online bipartite matching problem under a bounded number k of allowed reassignments per arrival [23]. They show that a modified SAP-algorithm is $(1 - \frac{2}{k+2})$-competitive for this setting and that this bound is tight already on paths.

The influence of recourse has also been studied for more general matching problems, such as minimum-cost matching [17,22] and maximum matching on general graphs [2,4,21].

Another closely related problem is unit job scheduling with integer release times and deadlines. The interval constraints directly correspond to release times and deadlines, and the maximum cardinality objective corresponds to throughput maximization. Here, jobs can be buffered until they are executed or their deadline arrives. The offline version of this problem is optimally solved by the Earliest Deadline First strategy, which can be implemented in $O(n)$ time [13,24]. For the online version, Baruah, Haritsa and Sharma show that any algorithm that is never idle is at least 2-competitive [3]. Chrobak et al. give a barely random algorithm, an algorithm that only uses a single random bit, that is $\frac{5}{3}$-competitive [11]. Note that in these scheduling problems, deadlines and release dates are not necessarily integers. Moreover, in contrast to ICBMT, decisions on rejections of jobs do not have to be made immediately upon job's arrival, as jobs can remain pending for several time steps.

1.2 Our Contribution

We introduce a new variant of online bipartite matching: Interval-Constrained Bipartite Matching over Time (ICBMT). In contrast to related problems that use recourse, the over-time nature of our problem eliminates some of the augmenting paths that may be needed to recover an offline optimal solution to the problem.

We first consider a FirstFit-type algorithm for ICBMT that uses a shortest augmenting path in case there is no directly available slot, and assigns the earliest free slot otherwise (Sect. 2). FirstFit does not only mimic the common first-come-first-serve practice in appointment scheduling, but is also a natural choice. Due to the over-time property of the problem, algorithms that leave early slots unoccupied risk missing a possible match or have to augment the matching for the current time slot. For this algorithm, we give a tight lower and upper bound of $\frac{2}{3}$ on the competitive ratio with respect to the cardinality of the final matching. In addition, we give an instance on which the algorithm performs $\Omega(n \log n)$ reassignments. To show this result, we have to specify the algorithm more precisely. For every new assignment, there might be several shortest augmenting paths ending in the same slot, and the choice of one impacts the result. Our algorithm always selects the lexicographically smallest shortest path, where the order is with respect to the sequence of slots in the path. This is a natural extension of the FirstFit strategy, which also favors early slots for the current job. We conjecture that there is no deterministic strategy for selecting the shortest paths that yields strictly better results, but a randomization over different strategies might be promising.

Our results imply that, by allowing only a constant number of reassignments for each new assignment, we retain the competitive ratio of $\frac{2}{3}$ for only a linear number of reassignments in total (Sect. 2.2). At first glance, this result sounds amazing, but one should not mistake this for the superior algorithm. For every number $k \in \mathbb{N}$ of reassignments allowed per single new assignment, there exists an instance on which this restricted algorithm k-FirstFit only achieves a $\frac{k+1}{k+2}$-fraction of the assignment constructed by the unconstrained FirstFit strategy on the same instance.

These properties of FirstFit are in contrast to the performance of the earliest-deadline-first strategy: it guarantees a maximum matching; that is, it is 1-competitive with respect to the cardinality of the final matching (Sect. 3). However, this strategy makes $\Omega(n^2)$ reassignments in the worst case, and restricting the number of reassignments per job arrival makes the algorithm uncompetitive.

For BMT, in which the sets of feasible slots for each job are not necessarily intervals, we show that FirstFit and k-FirstFit remain $\frac{2}{3}$-competitive with the same number of reassignments, and that it is the best possible competitive ratio any deterministic algorithm can attain. In contrast, EDF becomes only $\frac{1}{2}$-competitive ([1], Sec. 5).

Our main results are summarized in Table 1. All competitive ratios are tight. In addition, we analyze the performance of the algorithms in two simpler special cases: uniform interval lengths of jobs and a common arrival time for all jobs.

A full version with missing results and proofs is provided online.

Table 1. Performance bounds for algorithms presented in this work on instances with n jobs. Competitive ratios in bold are the best possible for a deterministic algorithm in the corresponding setting. revFor FirstFit, no matching upper bound on worst-case number of reassignments is known.

		FirstFit	k-FirstFit	EDF
assigned jobs (competitive ratio)	interval-constrained	$\frac{2}{3}$	$\frac{2}{3}$	$\mathbf{1}$
	unconstrained	$\mathbf{\frac{2}{3}}$	$\mathbf{\frac{2}{3}}$	$\frac{1}{2}$
reassignments (both settings)		$\Omega(n \log n)$	$\Theta(n)$	$\Theta(n^2)$

1.3 Problem Definition and Preliminaries

In *Interval-Constrained Bipartite Matching over Time*, we are given processing capacity of one slot per time unit $t \in \mathbb{N}$; we identify slots with time units they belong to. Jobs $j \in \{1, \ldots, n\}$ are revealed one by one together with their arrival times $a_j \in \mathbb{N}_0$ and their intervals $[r_j, d_j]$, $r_j, d_j \in \mathbb{N}$, which is the set of their feasible slots. We assume all numbers to be polynomial in the (unknown) number of jobs. The arrival times are non-decreasing in the order of jobs' arrival, i.e., $a_j \leqslant a_{j+1}$ for any $j < n$. Several jobs can have the same arrival time, i.e., arrive within the same time unit, but not simultaneously; the order in which they arrive is fixed and, upon arrival of the first such job, the remaining jobs with the same arrival time are not yet visible to the algorithm. This assumption models the classical online setting. The case when all requests arriving at one day are scheduled simultaneously is considered in [1], Sect. 4. For all jobs $j \leqslant n$, the *preparation time* $r_j - a_j$ is at least 1, as it is a common in-hospital practice to fix a day schedule the day before. However; all results in this work also hold for nonnegative preparation times. The jobs will be denoted as tuples of their attributes $j = (a_j, r_j, d_j)$, and instances as sequences of jobs.

Upon arrival, job j must be assigned to some slot $t \in [r_j, d_j]$. The already scheduled jobs can be reassigned to different slots to free a slot in the job j's interval, but they cannot be rejected. If no reassignment combination allows to assign job j, the job is irrevocably rejected. The assignments in slots $t \leqslant a_j$ are fixed and cannot be involved in reassignments for job j. This models the fact that the appointment schedule for the days already past, including the current day, cannot be changed. We therefore call the slots up to a_j *fixed.*

The goal is to produce online an assignment $\alpha\colon \{1, \dots, n\} \to \mathbb{N} \cup \{\bot\}$ of jobs to slots, where $\alpha(j) = \bot$ denotes that the job is rejected. The assignment should optimize the following objectives: (1) maximize the number of assigned jobs, and (2) minimize the number of reassignments. We consider the two objectives separately.

Any instance of ICBMT has an *underlying bipartite graph* consisting of the job set on one side and all feasible slots on the other, with edges connecting jobs to their feasible slots. An *offline augmenting path for a solution* $\alpha\colon \mathcal{J} \to \mathbb{N} \cup \{\bot\}$ of ICBMT is an augmenting path with respect to the matching α in the underlying bipartite graph.

Remark 1. An ICBMT solution has an optimal number of assigned jobs if and only if it admits no offline augmenting paths.

In fact, an optimal offline solution for ICBMT is a maximum matching in the underlying bipartite graph.

The main difference between ICBMT and the classical online bipartite matching with reassignments is the temporal component of ICBMT, which leads to a partially fixed assignment as time progresses. However, if all jobs in an instance arrive at the same time, then ICBMT with the objective of minimizing rejections reduces to the online maximum bipartite matching. We call this special case Static ICBMT problem.

Definition 1. Static ICBMT *is a special case of the ICBMT problem, in which all jobs of an instance have the same arrival time, i.e., for any instance* $I = ((a_j, r_j, d_j))_{j=1}^{n}$ *of Static ICBMT holds* $a_j = a$ *for all* $j \in \{1, \dots, n\}$.

Any instance of Static ICBMT is also an instance of online bipartite matching with reassignments: Indeed, since the arrival time of jobs is constant, no part of the assignment becomes fixed, and all reassignments are allowed at any time. Hence, a feasible solution for online bipartite matching is also a feasible solution for Static ICBMT on the same set of jobs.

2 Greedy Approach: FirstFit

One natural and widely used appointment assignment strategy is to greedily assign the earliest possible slot to every patient. If necessary, a few other patients are reassigned.

The algorithm we call FirstFit in this work follows this greedy paradigm: Assign jobs to early slots and minimize the number of reassignments for each

Algorithm 1: FirstFit

Input: job $j' = (a_{j'}, r_{j'}, d_{j'})$, current slot assignment $\alpha \colon \mathcal{J} \to \mathbb{N}$
Output: a new assignment $\alpha' \colon \mathcal{J} \cup \{j'\} \to \mathbb{N} \cup \{\bot\}$

```
F:={t ∈ ℕ | t > a_j' ∧ α^{-1}(t) = ∅};                        /* free slots */
𝒫:={p | p is a shortest augmenting path from j' to F} ;
P:= arg min{t ∈ F | t is the endpoint of p ∈ 𝒫} ;             /* earliest target slot */
if P exists then
  | α':=α Δ {(j,t) | (j,t) ∈ P or (t,j) ∈ P};
else
  | α':=α ∪ (j',⊥);
return α'
```

job arrival. It is a special case of the greedy shortest augmenting path strategy for bipartite matching that additionally considers the special structure of instances of ICBMT, namely the total order on the set of slots. The algorithm is defined as follows.

The input consists of a job $j' = (a_{j'}, r_{j'}, d_{j'})$ and a current slot assignment $\alpha \colon \mathcal{J} \to \mathbb{N}$, where $\mathcal{J} \subseteq \{1, \ldots, n\}$ is the set of previously assigned jobs. FirstFit searches for a shortest sequence of reassignments that frees up one slot in the interval of j' and assigns the last job of the sequence to a free slot t^*. If no such sequence of reassignments exists, job j' is rejected.

The sequence of reassignments results from a *shortest augmenting path* P from job j' to a free slot t^* in a bipartite job-slot graph $G_{\alpha,j'}$. The graph $G_{\alpha,j'} = (V^T \dot\cup V^J, E^\alpha \dot\cup E^{\overline{\alpha}})$ is a digraph containing a vertex for each non-fixed slot and job: $V^T = \{t \in \cup_{j \in V^J} [r_j, d_j] \mid t > a_{j'}\}$ and $V^J = \{j \in \mathcal{J} \mid \alpha(j) > a_{j'}\} \dot\cup \{j'\}$. There are edges $E^{\overline{\alpha}}$ from each job to its feasible slots it is not assigned to, $E^{\overline{\alpha}} = \{(j,t) \mid j \in V^J,\ t \in V^T,\ t \in [r_j, d_j],\ \alpha(j) \neq t\}$, and edges E^α from each filled slot to the job assigned to it, $E^\alpha = \{(t,j) \mid t \in V^T,\ j \in V^J,\ \alpha(j) = t\}$. The path P is augmenting with respect to the current assignment α. Among multiple possible shortest paths, we choose the one with *earliest target slot*, which motivates the name FirstFit. If such a path exists, then the new assignment α' is a symmetric difference between the edges of the old assignment and of the path P, denoted $\alpha \Delta \{e \mid e \in P\}$.

Note that we do not need to enumerate the set of paths $\mathcal{P}$: the earliest-target shortest augmenting path can be found efficiently via breadth-first search on $G_{\alpha,j'}$, if the nodes of one level are processed in the chronological order. Notice that FirstFit can also be used for Bipartite Matching over Time without interval constraints, where sets of feasible slots of jobs are of arbitrary structure (see full version [1]). On interval-constrained instances, however, we additionally have the following property.

Remark 1. When assigning job j', FirstFit for ICBMT always chooses the earliest free slot in the current assignment in or after the interval of job j', i.e., slot $t^* := \min\{t \geqslant r_{j'} \mid \alpha^{-1}(t) = \emptyset\}$, as the target free slot.

The proof of the remark can be found in the full version of the paper [1].

If multiple shortest paths to the earliest possible slot exist, one is selected arbitrarily. We will discuss the influence of the choice of the path on the algorithm's performance at the end of Sect. 2.

2.1 Matching Size Under FirstFit

We start our analysis of FirstFit with the primary objective in ICBMT – the number of assigned jobs, i.e., the cardinality of the resulting matching. In general, FirstFit can make unnecessary jobs rejections, as we will see later. On instances of Static ICBMT, however, FirstFit achieves a maximum number of assigned jobs.

Remark 2. FirstFit yields a maximum matching on instances of Static ICBMT.

In fact, as mentioned above, Static ICBMT is a special case of maximum online bipartite matching. FirstFit, in turn, coincides with the SAP strategy for online bipartite matching complemented with a tie-breaking rule between paths of the same length: the earliest terminal slot is preferred. Hence, the optimality of FirstFit follows from the optimality of SAP [10].

To prove the next optimality result, we introduce the notion of a closed interval.

Definition 2. *A* closed interval *in an assignment* $\alpha\colon \mathcal{J} \to \mathbb{N} \cup \{\perp\}$ *for ICBMT is a time interval* $I \subseteq \mathbb{N}$ *without empty slots such that all jobs* $j \in \alpha^{-1}(I)$ *have intervals* $[r_j, d_j] \subseteq I$.

Observe that any alternating path with respect to the matching α in the underlying bipartite graph that starts in a closed interval can visit only slots of this interval and jobs assigned to them.

Theorem 1. *If all jobs in an instance of ICBMT have intervals of the same length, then FirstFit yields a maximum matching.*

The proof shows that for a FirstFit solution, there are no offline augmenting paths in the underlying bipartite graph (see Remark 1). To this end, we show that the interval of each rejected job lies in a closed interval, so that augmenting paths starting with the rejected job never reach an empty slot. The proof can be found in [1].

Theorem 2. *On instances with at least two different interval lengths, FirstFit is at most $\frac{2}{3}$-competitive for ICBMT with respect to matching cardinality.*

Proof. We define two job types: *Urgent* jobs with preparation time of 1 and interval length δ, and *regular* jobs with preparation time 2 and intervals of length $\Delta \geqslant 3\delta - 2$. Consider now the following instance $\mathcal{I}$ of ICBMT (see Fig. 1). First, $\ell{:=}\delta - 1$ regular jobs j_i for $i \in \{1, \ldots, \ell\}$ arrive at time -1 (we extend the time scale to negative numbers to simplify the notation). Their interval is

$[1, \Delta]$, and FirstFit assigns them to slots $\alpha(i) = i$ for $i \in \{1, \dots, \ell\}$. Next, for each $i \in \{1, \dots, \delta\}$, an urgent job u_i with interval $[i, i + \ell]$ is revealed at time $i - 1$. Upon arrival of job u_i, the first $\ell + i - 1$ slots are occupied, so u_i is assigned to slot $\ell + i$, the last slot of its interval.

Finally, further ℓ urgent jobs v_i, $i \in \{1, \dots, \ell\}$ with interval $[\delta, \delta + \ell]$ are revealed also at time ℓ. All slots in their interval are occupied. The assignment up to the current time point ℓ cannot be modified, so the regular jobs cannot be reassigned.

Hence, jobs v_i, $i \in \{1, \dots, \ell\}$, are all rejected by FirstFit, so FirstFit achieves the objective value $\mathrm{FF}(\mathcal{I}) = 2\delta - 1$. An optimal assignment β assigns all jobs to a feasible slot:

$$\begin{aligned} \beta(j_i) &= \delta + \ell + i, \; 1 \leqslant i \leqslant \ell, \\ \beta(u_i) &= i, \qquad\qquad 1 \leqslant i \leqslant \delta, \\ \beta(v_i) &= \delta + i, \qquad\; 1 \leqslant i \leqslant \ell; \end{aligned}$$

hence, $\mathrm{OPT}(\mathcal{I}) = 3\delta - 2$. From

$$\frac{\mathrm{FF}(\mathcal{I})}{\mathrm{OPT}(\mathcal{I})} = \frac{2\delta - 1}{3\delta - 2} \xrightarrow[\delta \to \infty]{} \frac{2}{3}$$

follows that the competitive ratio of FirstFit is not greater than $\frac{2}{3}$.

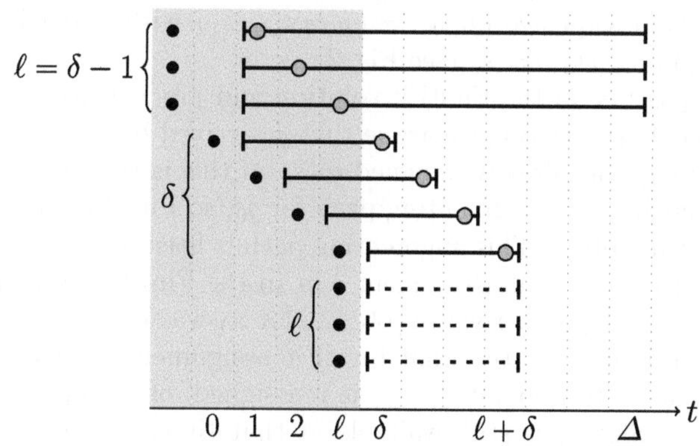

Fig. 1. Upper bound instance for $\delta = 4$, $\ell = 3$. The arrival order of jobs is from top to bottom. The arrival times are marked with dots and the FirstFit assignment with yellow circles. The grayed-out part of the solution is fixed upon arrival of the last ℓ jobs. The dashed jobs are rejected. (Color figure online)

We complement this result with a tight lower bound. Even if we drop the interval constraints and allow for arbitrary sets of feasible slots, FirstFit is at least $\frac{2}{3}$-competitive.

Theorem 3. *FirstFit is at least $\frac{2}{3}$-competitive for ICBMT with respect to the number of assigned jobs.*

Proof. Consider an arbitrary instance of ICBMT. Let $f\colon \mathcal{J}_{\mathrm{FF}} \to \mathbb{N}$ be the FirstFit assignment and $o\colon \mathcal{J}_{\mathrm{OPT}} \to \mathbb{N}$ an optimal assignment. Without loss of generality, we can select assignment o so that $\mathcal{J}_{\mathrm{FF}} \subseteq \mathcal{J}_{\mathrm{OPT}}$. Indeed, observe that the sets $\{\mathcal{J} \subseteq \{1,\dots,n\} \mid \exists$ feasible assignment $\mathcal{J} \to \mathbb{N}\}$ of jobs assigned by some assignment form a matroid. It is clearly an independence system; to see that the augmentation property holds, consider two assignments on job sets $\mathcal{J}$ and $\mathcal{J}'$ with $|\mathcal{J}| < |\mathcal{J}'|$ and with underlying matchings M and M', respectively. The symmetric difference $M \Delta M'$ contains an M-augmenting path [20, Proof of Theorem 10.7]. Augmenting M along this path yields a new matching M'' that assigns all jobs of $\mathcal{J}$ and one additional job from $\mathcal{J}' \setminus \mathcal{J}$.

To compare the cardinalities of the two assignments, we construct alternating job-slot sequences that help to partition the job set. For each job $j_0 \in \mathcal{J}_{\mathrm{OPT}} \setminus \mathcal{J}_{\mathrm{FF}}$ rejected by FirstFit but assigned by the optimal algorithm, we construct a sequence $Q(j_0)$ by first adding job j_0 and its optimal slot $s_0 = o(j_0)$ to it. Next, in each iteration $k \in \mathbb{N}$, we consider the job $j_k = f^{-1}(s_{k-1})$ that is assigned by FirstFit to the optimal slot s_{k-1} of the previous job in the sequence. By our assumption, every job assigned by f also has a slot under assignment o. Job j_k and its optimal slot $s_k = o(j_k)$ are added to the sequence $Q(j_0)$. We repeat these iterations until we reach a slot s_k that is free in f.

We show that each sequence contains at least three jobs. If a sequence $Q(j_0)$ for some rejected job j_0 contains only one job, then $Q(j_0) = (j_0, o(j_0))$, and slot $o(j_0)$ is free in assignment f. So FirstFit would have assigned j_0 to this slot.

Now consider a sequence $Q(j_0) = (j_0, s_0, j_1, s_1)$ with two jobs, with $s_0 = o(j_0)$, $s_1 = o(j_1)$ and $f(j_1) = s_0$ (see Fig. 2).

Since slot s_1 is free in the FirstFit solution and job j_1 is assigned to slot s_0, we have $s_0 < s_1$. Furthermore, we argue that j_1 arrived before j_0 and that upon arrival of j_0, job j_1 was already assigned to s_0. If this is the case, then the path (j_0, s_0, j_1, s_1) has been an augmenting path for j_0, so FirstFit would reassign j_1 instead of rejecting job j_0. This augmenting path is feasible because all its edges are contained in the optimal assignment o or in the FirstFit assignment f.

So suppose that, upon arrival of job j_0, slot s_0 was occupied not by j_1, but by some further job j_2. In the final FirstFit assignment we have $f(j_1) = s_0$; hence, at some later time point, job j_1 is reassigned, or assigned straight upon its arrival, to s_0. Slot s_1 is still available at that time; hence, the augmenting path (re-)assigning j_1 to s_0 can be shortened by assigning j_1 to a free slot s_1. This contradicts slot s_1 remaining empty in the final FirstFit assignment. So j_1 was occupying slot s_0 upon arrival of j_0, and FirstFit would have assigned j_0 by reassigning j_1.

Next, we express the number of jobs assigned or rejected by both algorithms through the number of sequences n_Q. We have shown that each sequence contains at least three jobs, where the first job is in $\mathcal{J}_{\mathrm{OPT}} \setminus \mathcal{J}_{\mathrm{FF}}$ and the others in $|\mathcal{J}_{\mathrm{FF}} \cap \mathcal{J}_{\mathrm{OPT}}|$. This yields a bound of $n_Q \leqslant \frac{1}{2}|\mathcal{J}_{\mathrm{FF}} \cap \mathcal{J}_{\mathrm{OPT}}|$. As each job in $\mathcal{J}_{\mathrm{OPT}} \setminus \mathcal{J}_{\mathrm{FF}}$ occurs at the start of a unique sequence, we have

$$|\mathcal{J}_{\mathrm{OPT}} \setminus \mathcal{J}_{\mathrm{FF}}| = n_Q \leqslant \frac{1}{2}|\mathcal{J}_{\mathrm{FF}} \cap \mathcal{J}_{\mathrm{OPT}}|.$$

Finally, we obtain

$$|\mathcal{J}_{\mathrm{OPT}}| = |\mathcal{J}_{\mathrm{OPT}} \cap \mathcal{J}_{\mathrm{FF}}| + |\mathcal{J}_{\mathrm{OPT}} \setminus \mathcal{J}_{\mathrm{FF}}| \leqslant \frac{3}{2}\,|\mathcal{J}_{\mathrm{OPT}} \cap \mathcal{J}_{\mathrm{FF}}| \leqslant \frac{3}{2}\,|\mathcal{J}_{\mathrm{FF}}|\,.$$

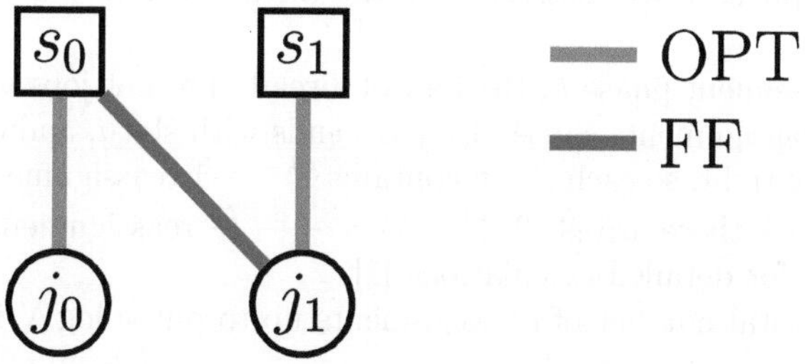

Fig. 2. An alternating sequence with two jobs.

2.2 Reassignments by FirstFit

Another natural performance measure for online algorithms for ICBMT is the number of reassignments made by the algorithm. Before we present our analysis of FirstFit for this measure, note that we did not specify how ties between shortest augmenting paths from the current job to the earliest free slot are resolved. Although each of the shortest paths yields the same number of reassignments in that iteration, we will see that the resulting solutions lead to different numbers of reassignments in later iterations.

All possible shortest augmenting paths can be uniquely represented and compared by the sequence of slots they visit on the way from the new job to the target slot. One possible rule for path selection is to choose the lexicographically smallest sequence of slots. Intuitively, this rule favors early slots for incoming jobs, and the path can be found in polynomial time.

We show in Lemma 1 that with this path choice rule, which we denote *lex-min* for short, FirstFit makes at least $\Omega(n \log n)$ reassignments in the worst case. Observe that this bound coincides with the lower bound on the worst-case number of reassignments for the SAP strategy for online bipartite matching [15].

Lemma 1. *In the worst case, FirstFit makes $\Omega(n \log n)$ reassignments on instances with n jobs, even on instances of Static ICBMT.*

Proof. Here, we provide only a sketch of the proof; for the full proof, see [1].

Consider an upper-triangle instance with $N = 2^n$ jobs arriving at time 0. Jobs are enumerated in the order of arrival, job $i \in \{1, \ldots, N\}$ has interval $[1, N+1-i]$, see Fig. 3.

To describe how FirstFit works on this instance, we divide the set of jobs into $n+1$ subsequent phases: phase $k \leqslant n$ contains $\frac{N}{2^k}$ jobs, and phase $n+1$ contains one job.

In the first phase, the first half of all jobs arrives and is assigned by FirstFit to the first $\frac{N}{2}$ slots. In the second phase, the half of the remaining jobs arrives. As there is no free slot for any of these new jobs, each new job causes a reassignment of one of the jobs of phase one. According to the lex-min rule for choosing augmenting paths, jobs of phase 2 are assigned to the first $\frac{N}{4}$ slots, and the first $\frac{N}{4}$ jobs of phase 1 are reassigned to the third block of $\frac{N}{4}$ slots, i.e., to slots $\frac{N}{2}+1,\ldots,\frac{3N}{4}$.

In each subsequent phase k, the half of already present jobs gets reassigned. The chain of reassignments caused by job i ends with slot i, and each job moves $\frac{N}{2^{k-1}}$ slots to the right, so each chain contains $2^{k-1}-1$ reassignments (see Fig. 3). Hence, in phase k there are $\frac{N}{2^k}(2^{k-1}-1) = \frac{N}{2} - \frac{N}{2^k}$ reassignments in total. See the full version for detailed calculations [1].

Hence, the total number of reassignments up to phase $\log N$ is

$$\sum_{k=2}^{\log N} \left(\frac{N}{2} - \frac{N}{2^k}\right) = \frac{N}{2}(\log(N)-1) - N\sum_{k=2}^{\log N}\frac{1}{2^k}$$
$$= \frac{N}{2}(\log(N)-1) - N\left(\frac{1}{2} - \frac{1}{N}\right) \in \Omega(N\log N)\,.$$

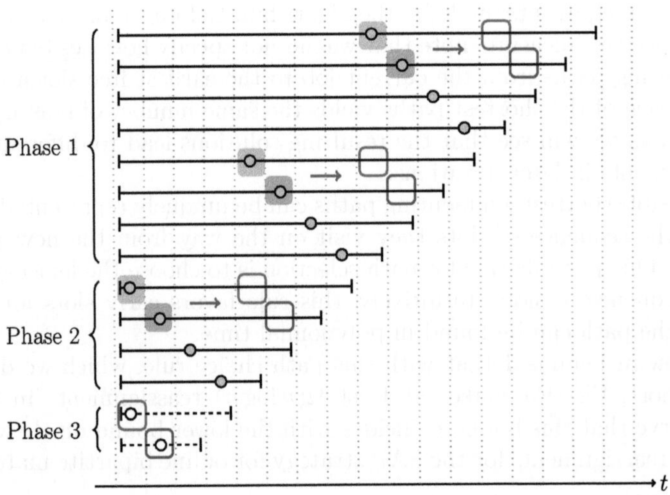

Fig. 3. Example instance with the reassignments in phase 3. Under lex-min rule, jobs of the current phase occupy the first slots in their order of arrival, and all other jobs are assigned in parallel diagonal blocks. The reassignment chains propagate this block structure through the instance.

Observe that on the ICBMT instance in Lemma 1, a different path choice rule, e.g., the *lex-max* rule that chooses the lexicographically maximal sequence of visited slots, would lead to only $\frac{N}{2}$ reassignments in total.

However, the lex-max rule is also sub-optimal: Consider an instance similar to that in Fig. 3, but in which the second half of the jobs have right-aligned intervals $[i - \frac{N}{2}, \frac{N}{2}]$ for $i \in \{\frac{N}{2}+1, \ldots, N\}$. By reasoning analogously to the proof of Lemma 1, the lex-max path choice rule yields $\mathcal{O}(n \log n)$ reassignments on this instance, whereas lex-min requires only $\frac{N}{2}$ reassignments.

As for the upper bound on the number of reassignments, there is a known upper bound of $\mathcal{O}(n \log^2 n)$ reassignments for online bipartite matching under the Shortest Augmenting Path strategy. This bound holds in particular for Static ICBMT as a special case of online bipartite matching, but not necessarily for ICBMT in general. Unfortunately, we were unable to show an analogous upper bound for ICBMT. Our problem differs from online bipartite matching in two aspects: The interval structure of feasible sets and the temporal component that fixes parts of the graph during the online arrival process. While the interval constraints possibly reduce the complexity of the problem by restricting possible adversarial instances, the temporal aspect of ICBMT instances makes any SAP-based algorithm less potent, since there are fewer augmenting paths available, as Fig. 4 demonstrates.

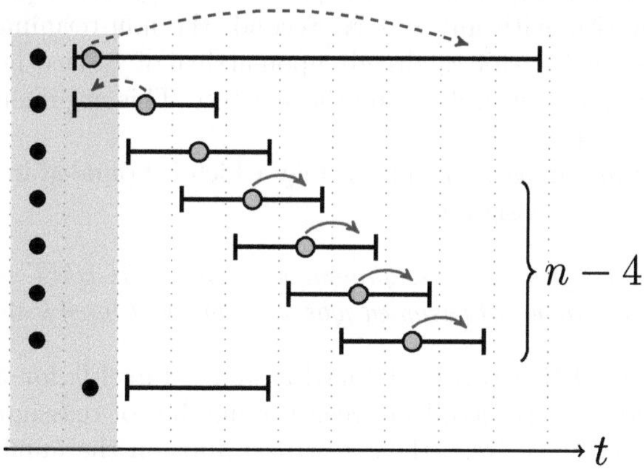

Fig. 4. An instance with $n = 8$ jobs with the FirstFit assignment, identical to an SAP assignment, upon arrival of the last job. The reassignments by FirstFit in ICBMT are shown in red. Online bipartite matching admits a shorter augmenting path of length 2, shown in blue dashed. (Color figure online)

Finally, we again consider a special case of instances with uniform interval length, assuming that these instances allow for better bounds also with respect to the number of reassignments.

Lemma 2. *On instances of ICBMT with uniform interval length* δ*, FirstFit makes* $\Theta(\delta n)$ *reassignments in the worst case.*

The complete proof is provided in [1]. An exemplary instance implies a lower bound of $\mathcal{O}(n)$. To obtain an upper bound, we show that on the described instances, FirstFit never reassigns a job to an earlier slot. The total number of reassignments per job to later slots is then clearly bounded by the length of the interval δ.

Limited Number of Reassignments. Next, we introduce a modification of FirstFit which makes a limited number of reassignments per job arrival, analogously to the modified SAP strategy considered by Shin et al. [23]. *FirstFit with path limit k*, or k-FirstFit for short, is only allowed to use augmenting paths whose length does not exceed $2k + 1$ for a fixed $k \in \mathbb{N}$, that is, paths which contain at most k already scheduled jobs. Thus, k-FirstFit reassigns at most a constant number of jobs per new assignment.

Clearly, k-FirstFit makes at most $kn = \mathcal{O}(n)$ reassignments. Interestingly, both Theorem 3 and the example in Theorem 2 also hold for FirstFit with path limit. To see why Theorem 3 holds for k-FirstFit, recall that the only nontrivial part of the proof is showing that alternating sequences of length 4 do not exist. The argument is based on two main statements. First, a path encoded by an alternating sequence (j_0, s_0, j_1, s_1) is an augmenting path for the presumably rejected job j_0. Since the augmenting path uses only one reassignment, it is feasible for k-FirstFit with any $k \in \mathbb{N}$. Second, the non-terminal job j_1 of the sequence was assigned to slot s_0 already upon arrival of j_0, as otherwise the path used to later reassign j_1 would not be the shortest. This argument clearly holds for k-FirstFit as well.

Hence, we obtain an assignment strategy which is bounded in both the number of rejections and reassignments.

Corollary 1. *The k-FirstFit assignment strategy for ICBMT is $\frac{2}{3}$-competitive with respect to the number of assigned jobs and makes at most $k \cdot n$ reassignments.*

Since the competitive ratios of FirstFit and k-FirstFit for any $k \in \mathbb{N}$ are equal, there is no clear trade-off between the number of reassignments and the number of accepted jobs. Nevertheless, with a limit on the number of reassignments per new assignment, we expect to obtain worse solutions on average. For instance, for any $k \in \mathbb{N}$, there are instances on which $(k + 1)$-FirstFit yields a maximum matching, while k-FirstFit assigns only a $\frac{k+1}{k+2}$-fraction of jobs, see [1]. In particular, k-FirstFit is not optimal for Static ICBMT. Therefore, although 1-FirstFit has the best worst-case guarantee on the number of reassignments, choosing a bigger k or using unlimited FirstFit might still be preferable.

3 Proactive Reassignments: Earliest Deadline First

So far, we have assumed that reassignments are made only if it is otherwise impossible to assign a newly arriving job. Now we investigate an assignment algorithm that strategically reassigns jobs upon arrival of new requests.

Algorithm 2: EDF

```
Input: job j' = (a_{j'}, r_{j'}, d_{j'}), current slot assignment α: 𝒥 → ℕ
Output: new assignment α': 𝒥 ∪ {j'} → ℕ ∪ {⊥}
S:={j'} ∪ {j ∈ 𝒥 | α(j) > a_{j'} and d_j > d_{j'}};
sort S by deadlines;
initialize α': j ↦ { ⊥,     j ∈ S,
                   { α(j),  j ∈ 𝒥 \ S;
foreach j ∈ S do
    if no free slot in [r_j, d_j] then
        α':=α ∪ (j', ⊥);                      /* reject j', keep old assignment */
        return α';
    else
        α'(j):= min{t ∈ [r_j, d_j] | (α')^{-1}(t) = ∅};      /* earliest free slot */
return α'
```

The *EDF* algorithm (Algorithm 2) mimics the optimal offline assignment strategy Earliest Deadline First. It maintains a maximal matching in the future part of the schedule and rejects jobs that cannot be matched in addition to the already scheduled ones. Ties in the ordering of jobs in the schedule are broken according to non-decreasing deadlines.

In contrast to machine scheduling, where jobs with earlier deadlines are prioritized and may push out other jobs from the schedule, our EDF for ICBMT respects the arrival order of jobs and rejects later jobs with an earlier deadline, if they cannot be assigned without discarding some already accepted jobs.

Theorem 4. *EDF maximizes the number of assigned jobs.*

The full proof is given in [1]. It shows that a final EDF solution admits no offline augmenting paths. To this end, we show for each rejected job that its interval is contained in a closed interval, which we explicitly construct.

Clearly, EDF is inefficient in terms of the number of reassignments. We show that it achieves the trivial upper bound of $\Omega(n^2)$ reassignments per n jobs.

Lemma 3. *EDF makes $\Omega(n^2)$ reassignments in the worst case, even on instances of Static ICBMT.*

Proof. Consider again an instance consisting of n jobs with monotonously decreasing deadlines, as in the proof of Lemma 3 (see Fig. 3). Upon each job arrival, all already present jobs will be reassigned one slot further, in order to assign the new job to the first slot. Hence, the algorithm makes $\sum_{i=1}^{n}(i-1) = \frac{n(n-1)}{2}$ reassignments in total.

To complete the comparison of FirstFit and EDF, we consider the number of reassignments of EDF on instances with uniform interval lengths.

Lemma 4. *If all n jobs have the same interval length δ, then EDF makes $\Theta(\delta n)$ reassignments.*

We observe that EDF makes only forward reassignments, which yields the upper bound on the number of reassignments. The lower bound follows from an example instance provided in the full version [1]. If the interval length δ is considered fixed, then, as for FirstFit as well, we obtain a worst-case bound of $\Theta(n)$ reassignments.

Finally, we consider a version of EDF with a limited number of allowed reassignments per iteration, analogously to k-FirstFit: κ-EDF works identically to EDF if an arriving job causes at most $\kappa = \kappa(n)$ reassignments, and rejects the job otherwise. In contrast to FirstFit, limiting the number of reassignments of EDF to asymptotically less than n reduces its competitivity. Consider instances as in Fig. 3, where the i-th job in the order of arrival has interval $[1, \ n - i + 1]$ for $i \in \{1, \ldots, n\}$ and causes reassignments of all previously accepted jobs. Here, κ-EDF accepts only $\kappa + 1$ first jobs and rejects the rest. Hence, for any $\kappa \in o(n)$, κ-EDF has no constant competitivity ratio.

Acknowledgments. The authors thank anonymous reviewers for their constructive feedback, which helped to greatly improve the presentation of this contribution. This work was partially supported by DFG (German Research Foundation) - RTG 2236/2.

Disclosure of Interests. The authors have no competing interests to declare.

References

1. Abels, A., Anapolska, M., Büsing, C.: Interval-constrained bipartite matching over time (2025). https://arxiv.org/abs/2402.18469
2. Angelopoulos, S., Dürr, C., Jin, S.: Online maximum matching with recourse. J. Comb. Optim. **40**(4), 974–1007 (2020). https://doi.org/10.1007/s10878-020-00641-w
3. Baruah, S.K., Haritsa, J.R., Sharma, N.: On-line scheduling to maximize task completions. In: 15th Real-Time Systems Symposium, RTSS, pp. 228–236 (1994)
4. Bernstein, A., Dudeja, A.: Online matching with recourse: random edge arrivals. In: 40th Annual Conference on Foundations of Software Technology and Theoretical Computer Science, FSTTCS, pp. 11:1–11:16 (2020)
5. Bernstein, A., Holm, J., Rotenberg, E.: Online bipartite matching with amortized O $(\log^2 n)$ replacements. J. ACM **66**(5), 1–23 (2019)
6. Bernstein, A., Kopelowitz, T., Pettie, S., Porat, E., Stein, C.: Simultaneously load balancing for every p-norm, with reassignments. In: 8th Innovations in Theoretical Computer Science Conference, ITCS, vol. 67, pp. 51:1–51:14 (2017)

7. Birnbaum, B.E., Mathieu, C.: On-line bipartite matching made simple. SIGACT News **39**(1), 80–87 (2008)
8. Bosek, B., Leniowski, D., Sankowski, P., Zych, A.: Online bipartite matching in offline time. In: 55th Annual Symposium on Foundations of Computer Science, FOCS, pp. 384–393 (2014)
9. Bosek, B., Leniowski, D., Sankowski, P., Zych-Pawlewicz, A.: A tight bound for shortest augmenting paths on trees. Theor. Comput. Sci. **901**, 45–61 (2022)
10. Chaudhuri, K., Daskalakis, C., Kleinberg, R.D., Lin, H.: Online bipartite perfect matching with augmentations. In: 28th International Conference on Computer Communications, INFOCOM. pp. 1044–1052 (2009)
11. Chrobak, M., Jawor, W., Sgall, J., Tichý, T.: Online scheduling of equal-length jobs: randomization and restarts help. SIAM J. Comput. **36**(6), 1709–1728 (2007)
12. Devanur, N.R., Jain, K., Kleinberg, R.D.: Randomized primal-dual analysis of RANKING for online bipartite matching. In: 24th Annual Symposium on Discrete Algorithms, SODA, pp. 101–107 (2013)
13. Frederickson, G.N.: Scheduling unit-time tasks with integer release times and deadlines. Inf. Process. Lett. **16**(4), 171–173 (1983)
14. Goel, G., Mehta, A.: Online budgeted matching in random input models with applications to adwords. In: 19th Annual Symposium on Discrete Algorithms, SODA, pp. 982–991 (2008)
15. Grove, E.F., Kao, M.-Y., Krishnan, P., Vitter, J.S.: Online perfect matching and mobile computing. In: Akl, S.G., Dehne, F., Sack, J.-R., Santoro, N. (eds.) WADS 1995. LNCS, vol. 955, pp. 194–205. Springer, Heidelberg (1995). https://doi.org/10.1007/3-540-60220-8_62
16. Gupta, A., Kumar, A., Stein, C.: Maintaining assignments online: matching, scheduling, and flows. In: 25th Annual Symposium on Discrete Algorithms, SODA, pp. 468–479 (2014)
17. Gupta, V., Krishnaswamy, R., Sandeep, S.: Permutation strikes back: the power of recourse in online metric matching. In: Approximation, Randomization, and Combinatorial Optimization, APPROX/RANDOM.,pp. 40:1–40:20 (2020)
18. Hopcroft, J.E., Karp, R.M.: An $n^{5/2}$ algorithm for maximum matchings in bipartite graphs. SIAM J. Comput. **2**(4), 225–231 (1973)
19. Karp, R.M., Vazirani, U.V., Vazirani, V.V.: An optimal algorithm for on-line bipartite matching. In: 22nd Annual Symposium on Theory of Computing, STOC, pp. 352–358 (1990)
20. Korte, B., Vygen, J.: Combinatorial Optimization. AC, vol. 21. Springer, Heidelberg (2018). https://doi.org/10.1007/978-3-662-56039-6
21. Liu, A.H., Toole-Charignon, J.: The power of amortized recourse for online graph problems. In: In: Chalermsook, P., Laekhanukit, B. (eds) Approximation and Online Algorithms. WAOA 2022. LNCS, vol. 13538. Springer, Cham (2022). https://doi.org/10.1007/978-3-031-18367-6_7
22. Megow, N., Nölke, L.: Online minimum cost matching with recourse on the line. In: Approximation, Randomization, and Combinatorial Optimization, APPROX/RANDOM, pp. 37:1–37:16 (2020)
23. Shin, Y., Kim, K., Lee, S., An, H.: Online graph matching problems with a worst-case reassignment budget. CoRR abs/ arXiv: 2003.05175 (2020)
24. Steiner, G., Yeomans, S.: A note on "scheduling unit-time tasks with integer release times and deadlines". Inf. Process. Lett. **47**, 165–166 (1993)

The Online Piercing Set Problem with Recourse

Riju Bindua[1], Minati De[1], Naveen Garg[2(✉)], and Kanav Singla[1]

[1] Department of Mathematics, Indian Institute of Technology Delhi, New Delhi, India
riju.bindua@mail.mcgill.ca, minati@maths.iitd.ac.in, singlakanav2012@gmail.com

[2] Department of Computer Science and Engineering, Indian Institute of Technology Delhi, New Delhi, India
naveen@cse.iitd.ac.in

Abstract. Given a set of geometric objects, A, in $\mathbb{R}^d$, a set of points P is a piercing set if every object in A includes a point in P. We consider the online version of the piercing set problem where recourse is allowed. In the traditional online setting, with irrevocable decisions, no algorithm can guarantee a competitive ratio better than $\Omega(n)$ for piercing intervals, where n is the length of the input sequence. In this paper, we show a 2-competitive algorithm for the problem by allowing at most one recourse per input.

We prove that to maintain an optimum piercing set for intervals, any online algorithm needs $\Omega(n)$ amortised recourse. We also design an algorithm for piercing intervals which provides a trade-off between recourse and competitive ratio; the algorithm is $(1+\epsilon)$-competitive when allowed $(1+2/\epsilon)$ recourse at each step. We extend our results to higher dimensions by obtaining a 2^{d+1}-competitive algorithm with amortised two recourse for piercing axis-aligned hypercubes in $\mathbb{R}^d$.

Keywords: Piercing Set · Online Algorithm · Recourse · Hypercubes · Hitting Set

1 Introduction

Given a collection of sets $S_1, S_2, \ldots, S_n$, $S_i \subseteq U$, a *hitting set* is a set, $H \subseteq U$, such that $S_i \cap H \neq \emptyset, 1 \leq i \leq n$. Finding the minimum hitting set is equivalent to finding the minimum set cover. Both these problems are NP-hard, and they are also hard to approximate within a $\ln n$ factor.

In the geometric version of the hitting set problem, U is a set of points in $\mathbb{R}^d$ and A is a set of geometric objects, like hypercubes, spheres, intervals, etc. The sets S_i are the points of U contained in $a_i \in A$; hence $H \subseteq U$ is a hitting set if every object in A includes a point from H. A closely related problem in the geometric setting is the *piercing set problem.* A piercing set for a set of objects A is a set of points P in $\mathbb{R}^d$ such that every object in A includes a point from P. The objective of the piercing set problem is to find a set P of minimum cardinality. Both these problems are widely studied in

J. Matuschke and J. Verschae (Eds.): WAOA 2025, LNCS 16077, pp. 18–32, 2026.
https://doi.org/10.1007/978-3-032-06706-7_2

computational geometry due to their applications in wireless networks, VLSI design, and resource allocation [1,6,9,15,23,27]. Although these problems are polynomial-time solvable for intervals [23,27], they remain NP-hard for simple objects such as unit disks or axis-aligned unit squares in $\mathbb{R}^2$ [18,19].

In the online hitting set problem, the sets arrive online, and the algorithm should include elements into H so that at every step, H hits all sets which have arrived so far. An $O(\log|U|\log n)$-competitive algorithm and almost matching lower bound of $\Omega\left(\frac{\log|U|\log n}{\log\log|U|+\log\log n}\right)$ is known due to a well-known result of Alon et al. [3]. Designing competitive algorithms that beat this lower bound for various finite geometric set systems is an active area of research. Even and Smorodinsky [17] obtained $\Theta(\log|U|)$-competitive algorithms when objects are intervals in $\mathbb{R}$, or half-planes or unit disks in $\mathbb{R}^2$. For more recent results in this direction, we refer to [2,12,13,24].

Often, relaxing the irrevocability constraint for an online algorithm leads to substantial performance gains. Designing online algorithms which permit small changes in decisions made in earlier steps but have significantly better competitive ratios has been a relatively new area of research. In this paper we consider the problem of hitting and piercing a set of d-dimensional hypercubes when one is allowed bounded recourse.

Surprisingly, no results are known for the online piercing and hitting set problems with recourse despite the $\Omega(n)$ and $\Omega(\log|U|)$ lower bounds on the competitive ratio for even simple objects like intervals [11,17]. In this paper, we show that a little recourse allows us to obtain a constant competitive algorithm for intervals and d-dimensional axis-aligned hypercubes, for fixed d.

1.1 Previous and Related Work

For the online piercing set problem, the result of Alon et al. [3] gives no bound since $|U|$ is not finite. De et al. [10,11] first observed that the competitive ratio of every deterministic online algorithm for piercing intervals in $\mathbb{R}$ is at least $\Omega(n)$, where n is the length of the input sequence. For unit balls in $\mathbb{R}^d$, the best known upper bound of $O(1.321^d)$ and a lower bound of $\Omega(d+1)$ on competitive ratio is known due to Dumitrescu et al. [14]. For translates of an axis-aligned hypercube in $\mathbb{R}^d$, Dumitrescu and Tóth [16] proved that the competitive ratio of any deterministic algorithm is at least 2^d. Later, De et al. [11] designed a deterministic algorithm with competitive ratio $O\left((2+\frac{2}{\alpha})^d\log M\right)$ for α-fat objects in $\mathbb{R}^d$ with diameters in $[1,M]$. They gave a deterministic algorithm with competitive ratio at most $3^d\lceil\log_2 M\rceil+2^d$ for piercing axis-aligned homothetic hypercubes with side lengths in $[1,M]$, and proved a lower bound of $\Omega(d\log M+2^d)$ for the problem.

We now discuss some problems where allowing recourse has led to a significant improvement in the competitive ratio. For the variant of the online Steiner Tree problem where each request either adds or removes a terminal, an $\Omega(\log n)$ lower bound, where n is the number of terminals, was known. Imase and Waxman [22] gave an 8-competitive greedy algorithm with an amortised recourse of $O(k^{1/2})$, where k is the number of requests. Gupta and Kumar [21] showed that this greedy algorithm requires only constant amortised recourse per request.

[22] also gave a 2−competitive algorithm for the online minimum spanning tree problem with an amortised recourse $O(k^{1/2})$, which was a significant improvement to the best achievable competitive ratio of $\Theta(\log(k))$ without recourse. Megow et al. [26] presented a $(1+\epsilon)$-competitive algorithm with an amortised recourse $(1/\epsilon)\log(1/\epsilon)$, for the online minimum spanning tree problem with node arrivals only. They also showed that the lower bound on the amortised recourse used by any $(1+\epsilon)$-competitive algorithm is $\Omega(1/\epsilon)$.

Buchbinder et al. [8] showed that for the online maximum cardinality matching problem, the lower bound on the competitive ratio of a randomised algorithm is $(3+1/\varphi^2)/2$, where φ is the golden ratio. Avitabile et al. [5] gave a $(1+O(\log(k)/k))$-competitive algorithm for the k-edge bounded online maximum matching problem, where at most k changes are allowed per edge. Their work also implied that the lower bound on competitive ratio for any deterministic online algorithm is $1+1/k$ and for any randomised algorithm, it is $1+1/(9k-1)$. Boyar et al. [7] later showed that the greedy algorithm achieves the optimum competitive ratio of 2, when $k=1$ and 3/2, when $k=2$. Angelopoulos et al. [4] improved the lower bound given by [5] for any deterministic online algorithm to $1+1/(k-1)$.

For the online hitting set problem Gupta et.al. [20] show that constant amortised recourse allows for an $O(\log n)$-competitive algorithm. They also obtain similar bounds for the fully dynamic setting when sets that are to be hit arrive and leave.

1.2 Our Contribution

We consider the online version of the piercing and hitting set problems when recourse is allowed. In the traditional online setting, with irrevocable decisions, no algorithm can guarantee a competitive ratio better than $\Omega(n)$ for piercing intervals where n is the number of intervals in the input sequence. Our key results are

- a 2-competitive algorithm with at most one recourse per input for the hitting set problem on intervals.
- a lower bound construction to show that any online algorithm that maintains an optimum piercing set needs amortised $\Omega(n)$ recourse where n is the number of intervals in the input sequence.
- an algorithm for the hitting set problem on intervals exhibiting a trade-off between competitive ratio and recourse. For any fixed integer $k>1$, our algorithm is $\left(1+\frac{1}{k-1}\right)$-competitive with at most $2k-1$ recourse.
- a 2^{d+1}-competitive algorithm for piercing hypercubes in $\mathbb{R}^d$ that has an amortised recourse of two units per input.

Our algorithms are combinatorial and use the number of disjoint intervals/ hypercubes as a lower bound to analyse the competitive ratio.

2 Preliminaries

In an online problem, after each input item arrives, the algorithm needs to make an irrevocable decision, i.e., it cannot be changed upon seeing any future items. The goal is to minimise or maximise an objective function, which is a function of all decisions for a given instance. Allowing *recourse* in an online algorithm permits a relaxation of the irrevocability constraint. This means we can modify decisions taken by the algorithm in earlier steps. Of course, one would like to minimise the number of modifications the algorithm makes for each input. An online algorithm has worst-case recourse α if it makes no more than α modifications for each input. The algorithm has amortised recourse β if the number of modifications made for a sequence of k inputs does not exceed $k\beta$. In this paper we will be counting recourse as the number of *deletions* that the algorithm makes from the hitting/piercing set it maintains.

A hitting set for a collection of geometric objects is also a piercing set; however the converse is not true. Finding a piercing set seems to be an easier problem than finding a hitting set. For intervals in $\mathbb{R}$ an algorithm for finding a piercing set can be used to obtain a hitting set. Let P be a piercing set for a set A of intervals in $\mathbb{R}$. It is no loss of generality to assume that every point in P is an endpoint of an interval in A. If U is the set of points from which the hitting set H is to be picked, one can shrink each interval $a_i \in A$ to an interval $a_i' \subseteq a_i$ such that both endpoints of a_i' are in U and every point in U which hits a_i also hits a_i'. Let A' be this set of shrunk intervals. A piercing set of A' corresponds to a hitting set of A and conversely, every hitting set of A is a piercing set of A'. In this paper our algorithms for intervals are presented for the hitting set problem, assuming that the input intervals have been shrunk so that both endpoints are in U.

There is a simple greedy algorithm for finding a minimum hitting set for a set, A, of intervals in $\mathbb{R}$. The algorithm maintains a set of disjoint intervals $D \subseteq A$ and a hitting set $H \subseteq U$; initially, D and H are empty. Among the intervals in A that are not hit by H, the algorithm picks an interval $a_i = [x_i, y_i]$ with the smallest right endpoint, and adds a_i to D and y_i to H. This step is repeated until H is a hitting set for A. It is easy to prove [25] that the intervals in D are disjoint and that $|H| = |D|$. Since the optimum hitting set has size at least $|D|$, the algorithm computes the optimum hitting set. We refer to this algorithm as the "earliest-deadline-first" algorithm.

Although it is easy to compute the optimum offline hitting set for intervals, for instance, using the Earliest-Deadline-First algorithm, the best competitive ratio achievable by any online algorithm for this problem is $\Omega(\log |U|)$ [17]. The argument illustrates how recourse can be used to improve the competitive ratio, and so we present it here.

Suppose that ALG is an optimum online algorithm that maintains a hitting set $H \subseteq U$ for a set of intervals. To simplify presentation, we assume that ALG adds at most one point to H for every input and that H is initially empty. The first interval a_1 we give to ALG spans all points in U. For any point $u \in U$ that ALG includes in H to hit a_1, we can find an interval a_2 contained in a_1 that spans at least $(|U|-1)/2$ points and is not hit by u; this is the next interval we give to ALG. In general, at the ith step, we give ALG an interval a_i, that is contained in a_{i-1} and is not hit by any point in H and spans at least $|U|/2^{i-1}-1$ points of U. Thus, after $k = \log|U|$ inputs, we would have $|H| = \log|U|$. Since input intervals are nested, any point in a_k hits all intervals and hence OPT is 1. This implies that ALG has competitive ratio at least $\log|U|$. Note that if we were allowed a recourse at step i, we could move the point we picked to hit a_{i-1} to also hit a_i. This is the idea we use to provide a 2-competitive algorithm with one recourse at each step.

3 A 2-Competitive Algorithm with One Recourse for Intervals

We now provide a 2-competitive online algorithm for maintaining a hitting set for intervals when we are allowed at most one recourse with each input.

3.1 The Algorithm

Our algorithm maintains a set $S = \{s_1, s_2, \ldots\}$ of disjoint segments in $\mathbb{R}$ where segment $s_j = [l_j, r_j], l_j \leq r_j$. The set S is initially empty, and during the algorithm, we add segments to S but never delete a segment from S. However, we modify segments in S by moving their right endpoint; the left endpoint of each segment remains fixed.

Let A be the set of intervals that have arrived so far. We maintain the invariant that the set $H = \cup_{i:s_i \in S}\{l_i, r_i\}$ is a hitting set for A. Since we never delete a segment and only modify one segment at each step, it follows that the algorithm has recourse at most 1. We now describe the algorithm by specifying the modifications made to S when an interval $a_i = [x_i, y_i]$ arrives.

Case 1: If a_i is hit by a point in H, we do nothing.
Case 2: If a_i is disjoint with all segments in S, we add a new segment $[x_i, y_i]$ to S, and the points x_i, y_i to the hitting set H.
Case 3: If a_i is properly contained in a segment $s_j = [l_j, r_j]$, we make the following changes.

1. Update the segment s_j to $[l_j, y_i]$. This means replacing r_j by y_i in the hitting set H. Note that $y_i < r_j$.
2. Let $A' \subset A$ be the intervals that are not hit due to this modification in H. Let $a_l = [x_l, y_l]$ be an interval of the shortest length in A'. We add segment $[x_l, y_l]$ to S, and the points x_l, y_l to the hitting set H.

3.2 Analysis of the Algorithm

Lemma 1. *The algorithm maintains the following invariants:*

1. *The segments in S are disjoint.*

Algorithm 1. 2-competitive Algorithm with Worst-Case Recourse 1 to Hit Intervals

```
Initialise A ← ∅, S ← ∅, H ← ∅
while an interval a_i = [x_i, y_i] arrives do
    A ← A ∪ {a_i}
    if a_i is hit by a point in H then
        Do nothing
    else if a_i ∩ s_j = ∅ ∀ s_j ∈ S then
        S ← S ∪ {[x_i, y_i]}, H ← H ∪ {x_i, y_i}
    else
        Let s_j = [l_j, r_j] ∈ S such that s_j ∩ a_i ≠ ∅
        Update s_j ← [l_j, y_i], H ← H \ {r_j} ∪ {y_i}
        Let A' ⊂ A be the intervals that are not hit by any point in H.
        if A' ≠ ∅ then
            Let a_l = [x_l, y_l] be an interval of the shortest length in A'.
            S ← S ∪ {[x_l, y_l]}, H ← H ∪ {x_l, y_l}
        end if
    end if
end while
```

2. $H = \cup_{j=1}^{|S|}\{l_j, r_j\}$ *is a hitting set for intervals in* A.
3. *For each segment* $s_j \in S$*, there exists an interval* $a_i \in A$ *such that* $l_j \leq x_i$ *and* $r_j = y_i$.

Proof. We prove the lemma by induction on $|A|$. Initially, the sets A, S, H are empty, and the invariants are trivially true. Assume that the invariants hold after $k-1$ steps of the algorithm and the interval $a_k = [x_k, y_k]$ arrives. Note that $|A| = k$ and let $|S| = p-1$. We consider how the 3 cases in the algorithm description change the sets S and H and maintain the invariants.

Case 1: If a_k is hit by a point in H, then invariant (2) holds. Since we do not change sets H, S, invariants (1) and (3) continue to hold.

Case 2: a_k is disjoint from all segments in S.

1. We add the segment $s_p = [x_k, y_k]$ to S and since this is disjoint from segments $s_1, \ldots s_{p-1}$, invariant (1) holds.

2. Since x_k, y_k are added to H, H now hits a_k and invariant (2) holds.

3. For the newly added segment s_p, a_k is the interval in A which satisfies the conditions of invariant (3).

Case 3: a_k is properly contained in a segment $s_j \in S$. The segment s_j is shrunk to $[l_j, y_k]$ and possibly a new segment $s_p = [x_l, y_l], l < k$ is added.

1. The shrinking of segment s_j maintains the invariant (1). The addition of a segment $[x_l, y_l]$ to S implies that the interval a_l was not hit by any point in H (after r_j was replaced by y_k). This implies that a_l is disjoint from all segments in S and hence after the addition of segment $[x_l, y_l]$ to S, invariant (1) continues to hold.

2. By replacing r_j with y_k, we ensure that the interval a_k is hit. Recall that $A' \subset A$ is the set of intervals that are not hit due to this modification, and hence all intervals in A' include r_j. Since a_l has the shortest length in A', therefore all intervals in A' must include at least one of x_l and y_l. Hence, adding x_l, y_l to H ensures that invariant (2) holds.

3. For the shrunk segment $s_j = [l_j, y_k]$ and the newly added segment $s_p = [x_l, y_l]$, a_k and a_l respectively are the intervals in A which satisfy the conditions of invariant (3). □

Theorem 1. *There is a deterministic online algorithm for the hitting set problem for intervals that is 2-competitive with worst-case recourse 1.*

Proof. Let OPT be the size of the optimum hitting set for A. By invariant (2) of Lemma 1, the size of the hitting set, H, produced by Algorithm 1 is at most $2|S|$. The inequality arises since for a segment s_j, the endpoints l_j and r_j need not be distinct.

Let $D \subseteq A$ be the collection of intervals given by invariant (3) of Lemma 1. Note that D includes exactly one interval contained in each segment, and since the segments are disjoint (invariant (1), Lemma 1), the intervals in D are disjoint. This implies $\mathsf{OPT} \geq |D|$. Since $|D| = |S|$, we obtain $|H| \leq 2|S| = 2|D| \leq 2\mathsf{OPT}$ establishing a 2-competitive ratio for the algorithm.

The algorithm deletes a point from H when it shrinks a segment by replacing its right endpoint. Since with each arriving interval, at most one segment is shrunk, the algorithm has worst-case recourse 1. □

4 Lower Bound on Recourse for Maintaining Optimum

In this section, we give a lower bound construction to show that any online algorithm needs an amortised recourse $\Omega(n)$ to maintain an optimum piercing set for intervals, even when all the intervals are of the same length. Let $A = \{a_1, a_2, \ldots, a_n\}$, where n is even, be a set of intervals, each of length 5, that arrive in index order. Interval $a_j \in A$ is defined as follows (refer to Fig. 1):

$$a_j = \begin{cases} [-1 - \frac{3}{2}(j+1),\ 4 - \frac{3}{2}(j+1)] & \text{if } j \text{ is odd,} \\ [-4 + \frac{3}{2}j,\ 1 + \frac{3}{2}j] & \text{if } j \text{ is even.} \end{cases}$$

Note that the odd-indexed intervals in A (except a_1) are to the left of 0 and the even-indexed intervals (except a_2) are to the right of 0. Every interval (other than a_{n-1}, a_n) in A intersects exactly two other intervals, one to its left and another to its right. In particular, for any odd $j \neq 1$, the interval a_j intersects a_{j+2} on its left and a_{j-2} on its right. Similarly, for any even $j \neq 2$, the interval a_j intersects a_{j-2} on its left and a_{j+2} on its right.

Let $S = \cup_{j=1-n/2}^{n/2-1} \{[3j-1, 3j+1]\}$ be disjoint segments of the real line where two intervals intersect. We assign each segment in S a red or blue colour as follows: The segment containing 0 is colored blue, and consecutive segments are assigned different colours (see Fig. 1).

Assume $n = 2k, k \in \mathbb{N}$. A minimum piercing set for intervals in A must contain exactly k points, each of which lies in a distinct segment in S. In particular,

- If k is even, the k points of the minimum piercing set must lie in distinct red segments.
- If k is odd, the k points of the minimum piercing set must lie in distinct blue segments.

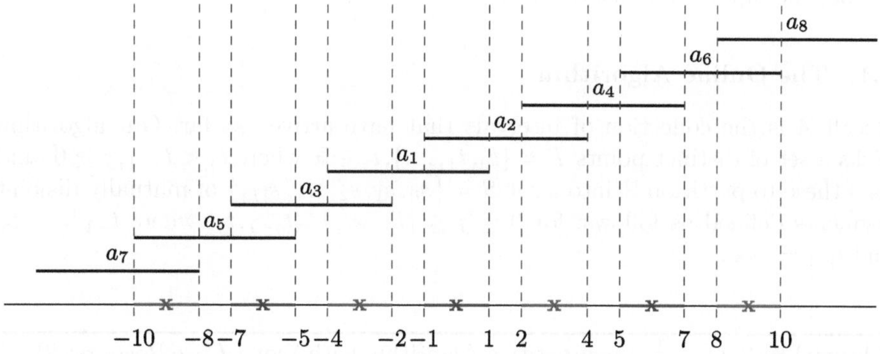

Fig. 1. When $k = 3$, there are three points in the optimum piercing set, each of which must belong to a distinct blue segment. After a_7 and a_8 arrive, the optimum piercing set contains four points, each of which must belong to a distinct red segment. (Color figure online)

The above construction allows us to prove the following theorem.

Theorem 2. *For the piercing set problem for intervals, any algorithm that maintains an optimum piercing set uses $\Omega(n)$ amortised recourse, where n is the number of intervals in the input.*

Proof. Consider an online algorithm that maintains an optimum piercing set, H. After the first $2k$, k odd, intervals from A have appeared as input, the k points in H should lie in k distinct blue segments. After the arrival of intervals a_{2k+1}, a_{2k+2}, the optimum piercing set H should contain $k+1$ points which lie in $k+1$ distinct red segments (Fig. 1).

Since the red and blue segments are disjoint, the algorithm must delete all k points of H in the blue segments and add $k+1$ points from the red segments to H. The arrival of intervals a_{2k+1}, a_{2k+2} leads to a recourse of k and hence the total recourse required to maintain an optimum piercing set for a sequence of n input intervals is $(n/2)(n/2-1)/2$. Hence, the amortised recourse of the algorithm for maintaining an optimum piercing set is $\Omega(n)$. □

In the above instance, the requirement for $\Omega(n)$ recourse can be avoided by allowing for an additive one approximation to the size of the minimum piercing set. However, by making c copies of the instance, it is easy to argue that any additive approximation less than c would require $\Omega(n/c)$ recourse.

5 A $(1 + \epsilon)$-Competitive Algorithm for Intervals

We now propose a $\left(1 + \frac{1}{k-1}\right)$-competitive algorithm for hitting intervals with at most $2k - 1$ recourse, where $k > 1$ is a fixed integer. Our algorithm uses the well-known optimum offline algorithm, the earliest-deadline-first algorithm, for hitting the input intervals, described in Sect. 2.

5.1 The Online Algorithm

Recall A is the collection of intervals that have arrived so far. Our algorithm picks a set of distinct points $T = \{t_0, t_1, \ldots\}, t_i \in \mathbb{R}$ where $t_j < t_{j+1}, j \geq 0$, and uses these to partition $\mathbb{R}$ into a set $S = \left\{s_0, s_1, s_2, \ldots, s_{|T|}\right\}$ of mutually disjoint segments defined as follows: for $0 \leq j \leq |T|$, $s_j = (t_{j-1}, t_j]$ where $t_{-1} = -\infty$ and $t_{|T|} = +\infty$.

Algorithm 2. $(1+\frac{1}{k-1})$-competitive Algorithm with Worst-Case Recourse $2k-1$ to Hit Intervals

Initialise $t_{-1} \leftarrow -\infty, t_0 \leftarrow \infty, s_0 \leftarrow (-\infty, \infty)$,
Initialise $A \leftarrow \emptyset,\ T \leftarrow \emptyset,\ S \leftarrow \{s_0\},\ H \leftarrow \emptyset,\ A_0 \leftarrow \emptyset$
while an interval a_i arrives **do**
    $A \leftarrow A \cup \{a_i\}$
    Let $s_j \in S$ contains the right endpoint of a_i
    $A_j \leftarrow A_j \cup \{a_i\}$
    Apply earliest-deadline-first algorithm to intervals in A_j to obtain H_j.
    if $|H_j| \geq 2k$ **then**
        Let $H_j = \{y_{j1}, y_{j2}, \ldots, y_{j2k}\}$ where $y_{jl} < y_{jl+1}$ for $1 \leq l \leq 2k - 1$
        for $|T| \geq l \geq j$ **do**
            Set $t_{l+1} \leftarrow t_l$
            if $l \geq j + 1$ **then**
                Set $s_{l+1} \leftarrow s_l,\ A_{l+1} \leftarrow A_l,\ H_{l+1} \leftarrow H_l$
            end if
        end for
        Update $t_j \leftarrow y_{j_k}, s_j \leftarrow (t_{j-1}, t_j], s_{j+1} \leftarrow (t_j, t_{j+1}], T \leftarrow T \cup t_{|T|}, S \leftarrow S \cup s_{|T|}$
        Update $A_j \leftarrow \{a_l \in A \mid t_{j-1} < y_l \leq t_j\}$, $A_{j+1} \leftarrow \{a_l \in A \mid t_j < y_l \leq t_{j+1}\}$
        Apply earliest-deadline-first algorithm to intervals in A_j to obtain H_j
        Apply earliest-deadline-first algorithm to intervals in A_{j+1} to obtain H_{j+1}
    end if
    $H = \bigcup_{l=0}^{|T|} H_l$
end while

Initially, $T = \emptyset$ and $S = \{(-\infty, \infty)\}$, and we refine the partition as the algorithm progresses. Let $A_j \subseteq A, j \geq 0$ be the input intervals with right endpoint in segment s_j, i.e. $A_j = \{a_l \mid t_{j-1} < y_l \leq t_j\}$. Let H_j be the hitting set obtained when we run the earliest-deadline-first algorithm on intervals in A_j. Since the sets A_j form a partition of A, the set $H = \cup_j H_j$ is a hitting set for A. We will refer to sets A_j as the intervals and H_j as the hitting set associated with segment s_j.

On arrival of an interval a_i, the algorithm first determines j such that the segment $s_j \in S$ contains the right endpoint of a_i, and adds a_i to the set A_j. It then recomputes H_j by applying the earliest-deadline-first algorithm to A_j. The size of H_j determines the next step of the algorithm.

$|H_j| < 2k$: We leave sets S, T unchanged.

$|H_j| \geq 2k$: Let $H_j = \{y_{j1}, y_{j2}, \ldots, y_{j2k}\}$[1] where $y_{jl} < y_{jl+1}$, $1 \leq l \leq 2k-1$. We split the segment s_j at y_{jk} by including y_{jk} in T. Note that this involves setting t_j to y_{jk} and for $l \geq j$, setting t_{l+1} to t_l.
For $l \geq j+1$, the segment s_l and sets A_l, H_l are renamed to $s_{l+1}, A_{l+1}, H_{l+1}$, respectively, in keeping with our notation. s_j, s_{j+1} are new segments obtained by splitting segment s_j. Input intervals with right endpoints in s_j, s_{j+1} are assigned to sets A_j, A_{j+1} respectively and hitting sets, H_j, H_{j+1}, for these sets of intervals are obtained by running the earliest-deadline-first algorithm.

In both the above cases, the hitting set is $H = \cup_{l \geq 0} H_l$.

5.2 Analysing the Competitive Ratio and Recourse

The size of the hitting set associated with a segment increases monotonically until the segment is split.

Lemma 2. *A segment $s_j, j \geq 0$ is split when $|H_j| = 2k$.*

Proof. Suppose a_i was the interval that arrived when s_j was split into segments s_j, s_{j+1}. The size of the set H_j increased from less than $2k$ to at least $2k$ with the arrival of a_i. Since H_j is the optimum hitting set for a set of intervals, adding an interval can increase $|H_j|$ by almost 1. Hence, the addition of a_i increased the size of the hitting set to $2k$. □

Lemma 3. *The hitting sets associated with the two segments created by splitting a segment have size either k or $k+1$.*

Proof. Suppose a_i was the interval that arrived when s_j was split into segments s'_j, s'_{j+1} (we use the prime to avoid confusion between the new and old segments) and the element t_j added to T. Let H'_j, H'_{j+1} be the optimum hitting sets associated with segments s'_j, s'_{j+1} respectively.

Note that t_j is chosen so that the $2k$ points of the hitting set H_j are equally divided between the new segments s'_j, s'_{j+1}; let H^l_j, H^r_j be these two sets. Since H_j was picked by the earliest-deadline-first rule, H^l_j is the optimum hitting set for the intervals associated with s'_j. Thus $|H'_j| = |H^l_j| = k$.

The k points of H^r_j together with t_j hit all the intervals associated with s'_{j+1} and hence $|H'_{j+1}| \leq k+1$. If $|H'_{j+1}| < k$, then $H'_j \cup H'_{j+1}$, which hits all intervals associated with s_j, has size at most $2k-1$, which implies (by Lemma 2) we would not have split segment s_j. Thus, $k \leq |H'_j|, |H'_{j+1}| \leq k+1$. □

[1] we show later in Lemma 2 that $|H_j| = 2k$.

Recall that running the earliest-deadline-first algorithm on a set of intervals A_j identifies a hitting set H_j and a set $D_j \subseteq A_j$ of disjoint intervals such that $|D_j| = |H_j|$. Let a_{j_1} be the interval in A_j whose right endpoint is the smallest; then a_{j_1} also belongs to D_j.

Claim 1. *For any $j, k, 0 \leq j < k \leq |T|$, intervals in D_j are disjoint with intervals in $D_k \setminus \{a_{k_1}\}$.*

Proof. Suppose an interval $a_i \in D_k \setminus \{a_{k_1}\}$ intersects some interval in D_j. Since a_i has its right endpoint after a_{k_1}, a_i intersects a_{k_1} which contradicts the fact that intervals in D_k are disjoint. □

Lemma 4. *Algorithm 2 has competitive ratio $(1 + \frac{1}{k-1})$.*

Proof. Claim 1 implies that $D = \cup_{j=0}^{|T|} D_j \setminus \{a_{j_1}\}$ is a set of disjoint intervals. Further,

$$|D| = \sum_{j=0}^{|T|}(|D_j| - 1) = \sum_{j=0}^{|T|} |H_j| - |T| - 1.$$

Let OPT denote the size of the minimum hitting set; then $\mathsf{OPT} \geq |D|$. Thus,

$$|H| = \sum_{j=0}^{|T|} |H_j| \leq \mathsf{OPT} + |T| + 1.$$

Since $|H_j| \geq k$, we have $|H| \geq k(|T|+1)$. This implies $|H| \leq \mathsf{OPT}(1+1/(k-1))$ proving the lemma. □

Lemma 5. *Algorithm 2 has recourse at most $2k - 1$.*

Proof. When an interval a_i arrives, the algorithm recomputes the hitting set for A_j, where s_j is the segment containing the right endpoint of a_i. Since $|H_j| \leq 2k - 1$, the algorithm deletes almost $2k - 1$ points from the current hitting set, H, and replaces them with a new set of points. This implies that the algorithm has a worst-case recourse $2k - 1$. □

We capture the results of this section in the following theorem.

Theorem 3. *For any fixed integer $k > 1$, there is a deterministic online algorithm that is $\left(1 + \frac{1}{k-1}\right)$- competitive with at most $2k-1$ recourse for the hitting set problem where objects are intervals in $\mathbb{R}$.*

6 An Amortised 2 Recourse Algorithm for *d*-Dimensional Hypercubes

In this section, we present a 2^{d+1}- competitive algorithm for the piercing set problem with amortised 2 recourse for axis-aligned hypercubes in $\mathbb{R}^d$. Throughout this section, we use hypercube to mean an axis-aligned hypercube.

6.1 A 2^d-Approximation Algorithm

We begin by describing a simple 2^d-approximation algorithm for finding a piercing set for d-dimensional hypercubes. This is an offline algorithm, and let A be the set of hypercubes. Let P be the piercing set; this is initially empty.

We consider the hypercubes in A by increasing side length breaking ties arbitrarily. If the hypercube $a_i \in A$ is hit by a point in P, we do nothing; else, we add a_i to a set D and all the 2^d corner points of a_i to P. We next argue that the hypercubes in D are disjoint.

Claim 2. *If $a_i, a_j \in D$, then a_i, a_j are disjoint.*

Proof. Suppose a_i was considered before a_j. None of the 2^d corners of a_i hit a_j and hence either a_i, a_j are disjoint or a_j is contained in a_i. In the latter case, a_j would have side-length smaller than a_i, which contradicts the ordering in which these hypercubes were considered. □

Hence, an optimum piercing set for the hypercubes in A has size at least $|D|$, i.e. $\mathsf{OPT} \geq |D|$. The set $|P|$ has size exactly $2^d|D|$, which is at most $2^d\mathsf{OPT}$, and this gives us a 2^d-approximation. We refer to this algorithm as the greedy algorithm.

6.2 The Online Algorithm

Our algorithm proceeds in phases. Phase 0 ends with the arrival of the first hypercube. Let A_i, $i \geq 0$ be the hypercubes that have arrived by the end of Phase i. In Phase $i+1$, $i \geq 0$, the algorithm first uses the greedy algorithm to compute a piercing set P_i for the hypercubes in A_i and sets P to P_i. If an arriving hypercube a_j is not hit by a point in P, the algorithm adds an arbitrary corner of a_j to P. Phase $i+1$ ends when $|P| = 2|P_i|$.

Let A be the input hypercubes, and suppose the algorithm is in phase $i+1$ after processing A. Let P be the piercing set after receiving A; it follows from the construction that P hits all hypercubes in A.

Claim 3. *Algorithm 3 has a competitive ratio* 2^{d+1}.

Proof. Since the algorithm is in phase $i+1$, $|P| < 2|P_i|$. Let D_i be the set of disjoint hypercubes found by the greedy algorithm when it was run at the start of phase $i+1$ on the set of hypercubes A_i. Then $|P_i| = 2^d|D_i|$ and $\mathsf{OPT} \geq |D_i|$. This implies

$$|P| < 2|P_i| = 2^{d+1}|D_i| \leq 2^{d+1}\mathsf{OPT},$$

which proves the claim. □

Claim 4. *Algorithm 3 has amortised recourse 2.*

Proof. At the end of phase $j, 1 \leq j \leq i$, $|P| = 2|P_{j-1}|$. All these $2|P_{j-1}|$ in P are deleted at the start of phase $j+1$. In phase j, at least $|P_{j-1}|$ hypercubes were added, and we can charge the points deleted at the start of phase $j+1$ to these hypercubes; this results in an amortised recourse of 2 per input. □

Algorithm 3. 2^{d+1}-competitive Algorithm with Amortised Recourse 2 to Pierce d-dimensional Hypercubes

Initialize $A \leftarrow \emptyset$, $P \leftarrow \emptyset$, $P_{-1} \leftarrow \emptyset$, $i \leftarrow 0$
while a hypercube a_j arrives **do**
  $A \leftarrow A \cup \{a_j\}$
  if a_j is hit by a point in P **then**
    Do nothing
  else
    Let c_j be any arbitrary corner of a_j
    $P \leftarrow P \cup \{c_j\}$
    if $|A| = 1$ or $|P| = 2|P_{i-1}|$ **then** ▷ End of Phase i
      Apply the greedy algorithm on A to obtain the hitting set P_i
      $P \leftarrow P_i$
      $i \leftarrow i + 1$ ▷ Start of Phase $i + 1$
    end if
  end if
end while

We summarise the results of this section in the following theorem.

Theorem 4. *There is a deterministic 2^{d+1}-competitive online algorithm that has amortised recourse of 2 for the piercing set problem in $\mathbb{R}^d$ when the objects are axis-aligned hypercubes.*

7 Conclusion

To the best of our knowledge, our paper provides the first results for the online piercing/hitting of geometric objects with recourse. We believe, with a little work, our results for both intervals and hypercubes can be extended to the fully-dynamic setting when objects arrive and also leave.

We also believe that our results are the best possible for intervals. However, for d-dimensional hypercubes, our recourse guarantees are amortised, and we leave open the problem of obtaining a constant (for fixed d) competitive algorithm with constant recourse for each input. Obtaining constant competitive algorithms with constant recourse for other geometric objects like d-dimensional hyper-rectangles, d-dimensional balls, and α-fat objects is also an interesting direction for future work.

References

1. Agarwal, P.K., Pan, J.: Near-linear algorithms for geometric hitting sets and set covers. Discret. Comput. Geom. **63**(2), 460–482 (2020)
2. Alefkhani, S., Khodaveisi, N., Mari, M.: Online hitting set of d-dimensional fat objects. In: Byrka, J., Wiese, A. (eds.) WAOA 2023. LNCS, vol. 14297. Springer, Cham (2023). https://doi.org/10.1007/978-3-031-49815-2_10

3. Alon, N., Awerbuch, B., Azar, Y., Buchbinder, N., Naor, J.: The online set cover problem. SIAM J. Comput. **39**(2), 361–370 (2009)
4. Angelopoulos, S., Dürr, C., Jin, S.: Online maximum matching with recourse. J. Comb. Optim. **40**(4), 974–1007 (2020)
5. Avitabile, T., Mathieu, C., Parkinson, L.: Online constrained optimization with recourse. Inf. Process. Lett. **113**(3), 81–86 (2013)
6. Bhore, S., Chan, T.M.: Fast static and dynamic approximation algorithms for geometric optimization problems: Piercing, independent set, vertex cover, and matching. In: Proceedings of the 2025 Annual ACM-SIAM Symposium on Discrete Algorithms, SODA 2025, 12-15 January 2025, pp. 2357–2386. SIAM (2025)
7. Boyar, J., Favrholdt, L.M., Kotrbčík, M., Larsen, K.S.: Relaxing the irrevocability requirement for online graph algorithms. Algorithmica **84**(7), 1916–1951 (2022)
8. Buchbinder, N., Segev, D., Tkach, Y.: Online algorithms for maximum cardinality matching with edge arrivals. In: Pruhs, K., Sohler, C. (eds.) 25th Annual European Symposium on Algorithms (ESA 2017), vol. 87 of Leibniz International Proceedings in Informatics (LIPIcs), Dagstuhl, Germany, pp. 22:1–22:14. Schloss Dagstuhl – Leibniz-Zentrum für Informatik (2017)
9. Chan, T.M.: Polynomial-time approximation schemes for packing and piercing fat objects. J. Algorithms **46**(2), 178–189 (2003)
10. De, M., Jain, S., Kallepalli, S.V., Singh, S.: Online piercing of geometric objects. In: 42nd IARCS Annual Conference on Foundations of Software Technology and Theoretical Computer Science, FSTTCS 2022, 18-20 December 2022, vo. 250 of LIPIcs, pp. 17:1–17:16. Schloss Dagstuhl - Leibniz-Zentrum für Informatik (2022)
11. De, M., Jain, S.: Sarat Varma Kallepalli, and Satyam Singh. Online geometric covering and piercing. Algorithmica **86**(9), 2739–2765 (2024)
12. De, M., Mandal, R., Singh, S.: New lower bound and algorithms for online geometric hitting set problem. In: 31st International Computing and Combinatorics Conference, COCOON 2025, pp. 234–247 (2025)
13. De, M., Singh, S., Tóth, C.D.: Online hitting sets for disks of bounded radii. In: 33rd Annual European Symposium on Algorithms, ESA 2025 (2025), page to appear
14. Dumitrescu, A., Ghosh, A., Tóth, C.D.: Online unit covering in Euclidean space. Theor. Comput. Sci. **809**, 218–230 (2020)
15. Dumitrescu, A., Jiang, M.: Piercing translates and homothets of a convex body. Algorithmica **61**(1), 94–115 (2011)
16. Dumitrescu, A., Tóth, C.D.: Online unit clustering and unit covering in higher dimensions. Algorithmica **84**(5), 1213–1231 (2022)
17. Even, G., Smorodinsky, S.: Hitting sets online and unique-max coloring. Discret. Appl. Math. **178**, 71–82 (2014)
18. Fowler, R.J., Paterson, M., Tanimoto, S.L.: Optimal packing and covering in the plane are np-complete. Inf. Process. Lett. **12**(3), 133–137 (1981)
19. Garey, M.R., Johnson, D.S.: Computers and Intractability: A Guide to the Theory of NP-Completeness. W. H Freeman (1979)
20. Gupta, A., Krishnaswamy, R., Kumar, A., Panigrahi, D.: Online and dynamic algorithms for set cover. In: Proceedings of the 49th Annual ACM SIGACT Symposium on Theory of Computing, STOC 2017, pp. 537–550 (2017)
21. Gupta, A., Kumar., A.: Online steiner tree with deletions. In: Chekuri, C (ed.) Proceedings of the Twenty-Fifth Annual ACM-SIAM Symposium on Discrete Algorithms, SODA 2014, pp. 455–467. SIAM (2014)
22. Imase, M., Waxman, B.M.: Dynamic steiner tree problem. SIAM J. Discret. Math. **4**(3), 369–384 (1991)

23. Katz, M.J., Nielsen, F., Segal, M.: Maintenance of a piercing set for intervals with applications. Algorithmica **36**(1), 59–73 (2003)
24. Khan, A., Lonkar, A., Rahul, S., Subramanian, A., Wiese, A.: Online and dynamic algorithms for geometric set cover and hitting set. In: 39th International Symposium on Computational Geometry, SoCG 2023, 12-15 June 2023, vol. 258 of LIPIcs, pp. 46:1–46:17. Schloss Dagstuhl - Leibniz-Zentrum für Informatik (2023)
25. Kleinberg, J., Terdos, E.: Algorithm Design. Pearson (2005)
26. Megow, N., Skutella, M., Verschae, J., Wiese, A.: The power of recourse for online mst and tsp. SIAM J. Comput. **45**(3), 859–880 (2016)
27. Nielsen, F.: Fast stabbing of boxes in high dimensions. Theor. Comput. Sci. **246**(1–2), 53–72 (2000)

Online General Knapsack with Reservation Costs

Elisabet Burjons[1](✉) and Matthias Gehnen[2](✉)

[1] Universitat Politècnica de Catalunya, Barcelona, Spain
elisabet.burjons@upc.edu
[2] RWTH Aachen University, Aachen, Germany
gehnen@cs.rwth-aachen.de

Abstract. In the online general knapsack problem, an algorithm is presented with an item $x = (s, v)$ of size s and value v and must irrevocably choose to pack such an item into the knapsack or reject it before the next item appears. The goal is to maximize the total value of the packed items without overflowing the knapsack's capacit.

As this classical setting is way too harsh for many real-life applications, we will analyze the online general knapsack problem under the reservation model. Here, instead of accepting or rejecting an item immediately, an algorithm can delay the decision of whether to pack the item by paying a fraction $0 \leq \alpha$ of the size or the value of the item. This models many practical applications, where, for example, decisions can be delayed for some costs e.g. cancellation fees. We present results for both variants: First, for costs depending on the value of the items and then for costs depending on the size of the items.

If the reservation costs depend on the value of the items, we find that no algorithm is competitive for reservation costs larger than 1/2 of the item value, and we find upper and lower bounds for the rest of the reservation factor range $0 \leq \alpha < 1/2$. On the other hand, if the reservation costs depend on the size of the items, we find a matching upper and lower bound of 2 for every reservation factor α.

Keywords: online algorithm · knapsack · competitive ratio · reservation

1 Introduction

Assume you are looking for profitable projects to work on next year. Over time, you will receive offers, each offering some profit, but also needing time for completion. With knowledge of the limited amount of hours you can work within a year, you now need to decide whether to accept an offer or not. Naturally (and sadly), you cannot wait for every offer, before you need to decide about accepting one. This is the classical setting of an online knapsack problem, where you need to make a selection of items/projects that promise a maximum combined profit but can be completed within the capacity restriction (number of working hours/knapsack size). While it is known that one cannot hope for any competitive ratio in this setting in theory [18], in a real-life setting the situation is not

J. Matuschke and J. Verschae (Eds.): WAOA 2025, LNCS 16077, pp. 33–47, 2026.
https://doi.org/10.1007/978-3-032-06706-7_3

hopeless: For each contract, you might be able to cancel it after accepting – but likely this will not be possible without some contractual penalty or cancellation fee.

The same situation will also occur if you are looking for concerts to visit in the next months: At some point, and usually, only within a limited time frame, you can buy tickets for events you want to attend. Of course, you want to see as many concerts as possible (each weighted by your preference), but you only have a limited budget for buying tickets. Again, you might run into situations, where you need to decide whether to buy a particular ticket or not, without knowing what other options will arise. Again, it is possible to return or sell already purchased tickets, but this also might come with some cancellation or resell fee.

All those problems can be modeled as a general knapsack problem, but the classical variant is often not flexible enough. When adding the option to postpone a decision for some part of its profit, not only it gets more realistic in many real-life settings but, as we will see in this paper, it will also allow us to achieve bounded competitive ratios. In those settings, it will make no difference whether we model it with some cancellation fee, or some fee to postpone the decision about accepting, as, as it will turn out, there will not be a single item that would be packed without the risk of being removed later on. Therefore, for each item, either it will be rejected immediately, or reserved for later. The setting of postponing decisions for some fee was introduced as the reservation model for the online simple knapsack problem [1], where the value of an item equals its size, and later applied to many other online problems [6–8]. In such problems, an online algorithm has an option, beyond accepting or rejecting an item, to reserve an item to postpone the decision on it, by paying a cost proportional to the value of the item.

In the *knapsack problem*, given a request sequence S, a size function $s\colon S \to \mathbf{R}$ and a value function $v\colon S \to \mathbf{R}$, an algorithm has the task of finding a subset $K \subseteq S$ such that the sum of sizes of K is less or equal to the so-called *knapsack capacity* (which we assume to be normalized to 1 in this paper) and the sum of the values of the items, i.e., the total gain, is maximized. The online variant of this problem reveals the items of S piece by piece, with an algorithm having to immediately decide whether to pack a revealed item or discard it. In the online knapsack with reservation, the items are revealed piece by piece as well, however, an online algorithm now has three available options, namely, pack the revealed item, discard the item, or reserve it. After the full request sequence S is revealed an algorithm can decide which reserved items to pack, as long as they fit into the knapsack. Items already packed cannot be discarded at this stage. The reservation costs must then be paid for each reserved item regardless of whether it is used in the final packing or not.

When we apply the reservation model to the general version knapsack problem, there are two natural ways to do it, either the reservation costs depend on the size or the value of the items. In this paper, we present upper and lower bounds for both cases.

Online algorithms are generally analyzed using competitive analysis, introduced by Sleator and Tarjan [19]. In simple terms, one compares the performance of an online algorithm to the performance of an optimal offline algorithm on the same instance. In the case of maximization problems, like the knapsack problem, one can define the (strict) competitive ratio of an algorithm as follows.

We formally say that an algorithm A is *c-competitive* if there exists a constant $\beta \geq 0$ such that for every request sequence S, if the gain of an algorithm OPT solving the problem optimally on the same request sequence S, is $\text{gain}_{\texttt{OPT}}(S)$, the following inequality holds,

$$\text{gain}_{\texttt{OPT}}(S) \leq c \cdot \text{gain}_{\texttt{A}}(S) + \beta \ .$$

The *strict competitive ratio* of an online algorithm A is the highest ratio of any request sequence S between the gain of A on S and the gain of an algorithm OPT solving the problem optimally on the same request sequence S,

$$\text{CR}(\texttt{A}) = \sup_S \left\{ \frac{\text{gain}_{\texttt{OPT}}(S)}{\text{gain}_{\texttt{A}}(S)} \right\} \ .$$

The strict and non-strict versions of the competitive ratio coincide if β is set to 0.

In the reservation model, the reservation costs for an item are a fixed percentage $0 \leq \alpha$ of the size or the value of the item. Every reserved item has such a cost, regardless of whether the item is ultimately used in the knapsack. Such costs are accounted for in the gain portion of the competitive ratio, where they have to be subtracted from the total value packed into the knapsack by algorithm A.

A first analysis shows that, if the reservation is proportional to the item size, the online general knapsack with reservation cannot yield a bounded strict competitive ratio. This is because an adversary can build instances where the value of every item is exactly α times its size. Thus, no algorithm gains anything from reserving an item, and the lower bounds for the general knapsack problem without reservation hold, which means that the problem is not competitive using the strict definition [18].

The key to this model not being strictly competitive stems from the fact that one can adversarially artificially limit the total gain achievable by an algorithm. The general notion of competitiveness is thus, more natural when the reservation is proportional to the item size.

The non-strict competitiveness is only reasonable in problems where the achievable gain is unbounded, such as the general knapsack problem. In problems where the gain is bounded, every algorithm would be 1-competitive according to such a notion. For instance, in the online simple knapsack, where the value of an item equals its size, with an additive constant of 1 one can claim that every online algorithm is optimal, even without packing anything.

In a setting where the reservation costs are proportional to the item value, we show that algorithms can be competitive also in the strict sense.

1.1 Related Work

The knapsack problem in its offline version is one of the classical optimization problems. It is known to be NP-hard, even the restricted simple knapsack problem [17]. However, it allows fully polynomial approximation scheme algorithms, also for the general case [13]. As is the case for many classical problems, many variants of it exist, in particular, online variants.

When considering the online knapsack problem with irrevocable decisions and without any knowledge about the future, no deterministic algorithm can achieve a bounded competitive ratio, even for the simple knapsack [18]. If randomization is allowed, then there exists a simple 2-competitive algorithm for the simple knapsack using only one randomization bit [4], while the general problem still does not allow a bounded competitive ratio [21].

If one relaxes the blindness of an algorithm, a better competitive ratio can be achieved: One variant is the advice model, where an algorithm can ask an oracle for truthful advice bits [4]. A recent trend investigates the so-called predictions or machine-learned advice, where the given hints no longer need to be truthful [5, 14,20]. Another variant assumes rough item sizes are given in advance, but the actual sizes get revealed in an online manner [9].

Another option is to relax the irrevocability of an algorithm's packing decisions. The setting, when an algorithm is allowed to discard already packed items is also known as knapsack with removability, which is Φ-competitive for the simple knapsack [15], where Φ is the golden ratio. Some models also allow repacking a limited amount of removed items [3]. A simple knapsack variant where one is allowed to remove items with costs is also studied [10].

Furthermore, there is research combining relaxations of blindness and irrevocability. Recently, the knapsack problem with removability was studied in the advice setting [2].

Some variants also relax the packing restriction for an online algorithm, by e.g. allowing the algorithm to exceed the knapsack size [11,16] or by allowing it to cut items into parts [12].

1.2 Our Contributions

In this paper, we present upper and lower bounds for the online general knapsack with reservation costs, when the reservation costs depend both on the value and size of the items.

For the reservation costs depending on the value of the items, we get the upper and lower bounds depicted in Fig. 1. The upper bound is

$$\frac{2(2\alpha + \sqrt{\alpha(1+2\alpha)})(1 + 2\alpha + 2\sqrt{\alpha(1+2\alpha)})}{(1-2\alpha)^2\sqrt{\alpha(1+2\alpha)}},$$

whereas the lower bound is $\frac{2(1+\sqrt{\alpha(2-3\alpha)}-\alpha)}{(1-2\alpha)^2}$. Those constructions work for both the strict- and non-strict definition of the competitive ratio.

In Fig. 1, we can see that the competitive ratio is 2 when the reservation costs go to 0 and the problem becomes not competitive for reservation costs of 0.5 times the value of the item. In between those scenarios, we have upper and lower bounds that do not match but that are asymptotically similar.

When the reservation costs depend on the size of the items, there is an upper and lower bound of 2 for the non-strict competitive ratio regardless of the reservation factor. Here no algorithm can be competitive in the strict sense.

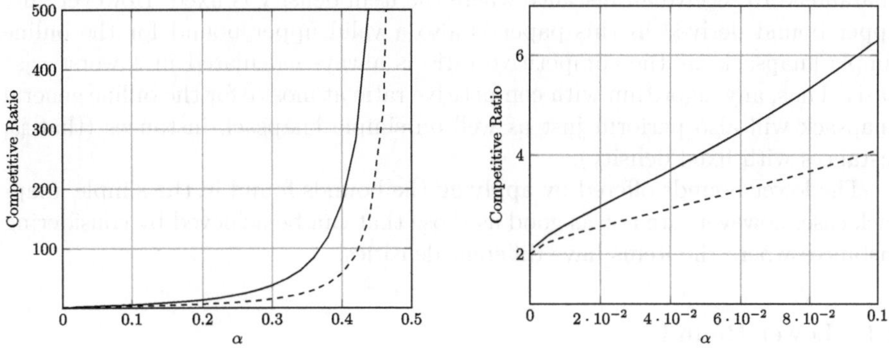

Fig. 1. A schematic plot of the upper and lower bounds for reservation depending on item value. The dashed line represents the general lower bound proven in Theorem 1 and the continuous line is the upper bound achieved in Corollary 1 by analyzing Algorithm 1.

2 Reservation Costs Proportional to Item Value

In this section, we present upper and lower bounds that depend on the value of the item. We present the results using the strict version of the competitive ratio in this setting, which of course, yields also non-strict results in the upper bounds as well for any positive value of β. This is not as straightforward when we consider the lower bounds, but we will see that even the lower bound results will extend to non-strict competitive ratios.

In general, given an item $x{:=}(s, v)$ that has size s and value v, we talk about the density of an item as the ratio between its value and size, namely $d(x) = v/s$. We will sometimes omit the depency on x through the paper if it is clear by context. In a request sequence, we will often mark each item and its size and value by its index in the sequence, namely, the i-th item is $x_i{:=}(s_i, v_i)$.

First, we see that, when the reservation costs depend on the item value, one can translate the lower bounds from the online simple knapsack:

If the reservation costs depend on the value, an adversary can present an instance with fixed density 1, i.e., the items are, for every i, $x_i = (s_i, s_i)$ and mimic a simple knapsack instance. Thus, no algorithm for online general knapsack with reservation costs depending on the value can achieve a better competitive ratio than the bounds for the online simple knapsack with reservation costs.

This observation is valid for both the strict and non-strict versions of the competitive ratio, because, if instead of choosing density 1, we choose an arbitrary density d, one can offset any additive constant β when computing the competitive ratio. For this reason, all of the lower bounds in this section will also be calculated as strict ratios, as one can always scale the density of all of the offered items by an arbitrary value to overcome any constant β.

We cannot make a similar claim for the upper bound, as no algorithm is guaranteed to receive an instance where the item density is fixed. However, any upper bound derived in this paper is also a valid upper bound for the online simple knapsack, as the competitive ratio is always calculated in a worst case basis. Thus, any algorithm with competitive ratio at most c for the online general knapsack will also perform just as well on simple knapsack instances (that is, instances with fixed density).

The lower bounds offered by applying the bounds found in the simple knapsack case, however, are not as good as those that can be achieved by considering instances where the items have different densities.

2.1 Lower Bound

For the lower bound, we build a set of adversarial instances such that no algorithm can achieve a better ratio than the proposed bound. A few observations will help to analyze the behavior of these algorithms:

Proposition 1. *No algorithm packing items before the end of the instance is competitive.*

If an algorithm decides to pack an item before the end of the request sequence has been announced, the adversary can always craft an item x of size 1 and arbitrary value which will render such an algorithm uncompetitive. Because the value of x is arbitrary, this holds true for both the strict and non-strict versions of the competitive ratio.

Thus, we do not need to consider algorithms that preemptively pack during our lower bound analysis. The same holds for algorithms that do not reserve the first item:

Proposition 2. *Any algorithm not reserving any item is uncompetitive.*

If an algorithm decides not to reserve the first item, the adversary can stop the input sequence there, which will render such an algorithm strictly uncompetitive. For the non-strict version, an adversary can offer increasingly denser items of the same size until an arbitrary value is achieved. If an algorithm does not reserve any of these items the algorithm cannot be competitive for any fixed value of β in the non-strict sense. Thus, by scaling the density of the first reserved item, we can assume that the first item offered by the adversary is reserved.

With these considerations, we can construct a lower bound for all other algorithms:

Theorem 1. *No algorithm for the online general knapsack can achieve a competitive ratio smaller than* $\frac{2(1-\alpha+\sqrt{(2-3\alpha)\alpha})}{(1-2\alpha)^2}$ *for a reservation factor of* $0 < \alpha < 0.5$.

Proof. Let $\varepsilon_1, \varepsilon_2 > 0$ and N be a natural number. The adversarial instances consist of items $x_k := (1 - k\varepsilon_1, v_k)$ for $k \geq 0$, starting with $x_0 := (1, 1)$. Each value v_k is larger than its predecessor by a decreasing factor f_k, so $v_k := f_1 f_2 \dots f_k$. Here f_1 is chosen sufficiently large (e.g., by a factor of $(1 + \varepsilon_2)$ larger than the aimed competitive ratio $\frac{2(1-\alpha+\sqrt{(2-3\alpha)\alpha})}{(1-2\alpha)^2}$). Furthermore, f_k decreases slowly towards a factor of 1, such that at any point $k > N$ the factor f_k is at least $(1 - \varepsilon_2) f_{k-N}$. If an algorithm decides to not reserve an item x_k, the adversary stops presenting further items of this form.

If the decision to not reserve the item x_k was made for $k = 0$, the instance ends and the algorithm is not competitive as described above.

If the algorithm decides to not reserve an item x_k for $0 < k < N$, the instance ends as well. The optimal solution can consist of x_k, which is at least $\frac{2(1-\alpha+\sqrt{(2-3\alpha)\alpha})}{(1-2\alpha)^2}$ times larger than the largest reserved item from the algorithm x_{k-1}. Therefore, the competitive ratio in this case also exceeds the claimed one. This case uses the construction that each factor f_i for $i \leq N$ is still larger than $\frac{2(1-\alpha+\sqrt{(2-3\alpha)\alpha})}{(1-2\alpha)^2}$ times it predecessor, which is also the ratio between x_k and x_{k-1}.

If an algorithm decides to not reserve an item x_k for $k \geq N$, the algorithm presents an item $y := (k\varepsilon_1, v_{k-1})$. If the algorithm decides to reserve y, the instance stops; otherwise, copies of y will be presented as long as either a copy gets reserved or more than $\frac{2(1-\alpha+\sqrt{(2-3\alpha)\alpha})}{(1-2\alpha)^2}$ copies of y got presented. If no copies of y got reserved, the optimal solution can pack all of them together, resulting in a packing worth $\frac{2(1-\alpha+\sqrt{(2-3\alpha)\alpha})}{(1-2\alpha)^2} v_{k-1}$ while the algorithm only achieves v_{k-1}. Even ignoring the reservation costs the algorithm has paid so far, the algorithm can not achieve the desired competitive ratio. Note that this assumes ε_1 to be sufficiently small, which can be achieved by choosing ε_1 accordingly for fixed N and ε_2. Therefore, we assume that the algorithm will reserve one copy of y.

In this case, the optimal solution consists of the item x_k together with y, achieving $v_k + v_{k-1}$. The algorithm can pack either y or x_{k-1}, both with value v_{k-1}. Furthermore, the algorithm will have to pay for reservation costs of αv_{k-1} for y plus

$$\alpha \sum_{i=0}^{k-1} v_i \geq \alpha \sum_{i=k-N}^{k-1} v_i \geq \alpha v_{k-N} \sum_{i=0}^{N-1} f_k^i = \alpha v_{k-N} \frac{1 - f_k^N}{1 - f_k}$$

for the reserved items x_i.

Therefore, the algorithm achieves a competitive ratio of

$$\frac{v_k + v_{k-1}}{v_{k-1} - \alpha v_{k-1} - \alpha v_{k-N} \frac{1-f_k^N}{1-f_k}} = \frac{f_k + 1}{1 - \alpha - \alpha \frac{v_{k-N}}{v_{k-1}} \frac{1-f_k^N}{1-f_k}}$$

For every ε_3, the geometric sum can be bounded as a series for a sufficiently large chosen N, resulting in a competitive ratio of

$$\frac{f_k + 1}{1 - \alpha - \alpha(\frac{1}{1-1/f_k} - \varepsilon_3)} .$$

A factor $f_k = \frac{-1+\alpha-\sqrt{2\alpha-3\alpha^2}}{-1+2\alpha}$ promises the minimal competitive ratio among all possible factors. The resulting competitive ratio for this x_k turns out to be exactly

$$\frac{2(1 - \alpha + \sqrt{(2 - 3\alpha)\alpha})}{(1 - 2\alpha)^2} .$$

Note that this minimum exists as it is defined by a trade-off between high total reservation costs (in case of a smaller factor f) and of course the ratio between the algorithm's item and the large item the optimal solution can use (in case of larger factor f).

Finally, note that if the algorithm does reserve all items x_k, at some point it will reserve some item x_y with $f_y < 1 + \varepsilon_2$. Then the instance ends. The algorithm as well as an optimal solution can pack at most the value v_k of the last presented item, however, from the algorithm's value we need to deduct the reservation costs of at least $\alpha N(v_k - 2\varepsilon)$ for the last N items. With sufficiently large N, the algorithm does not achieve a bounded ratio.

2.2 Upper Bounds

In this section we present Algorithm 1, which provides an upper bound that goes to 2 for small values of α, and diverges at $\alpha = \frac{1}{2}$ at the same asymptotical rate as the lower bound.

We already defined the item density for a given item. For this bound, we want to group subsets of items together. For a given set of items X we define its value $v(X) = \sum_{x_i \in X} v_i$ and size $s(X) = \sum_{x_i \in X} s_i$ and accordingly its density as the ratio between them $d(X) = v(X)/s(X)$. To build an algorithm for the upper bound we will consider at all times the densest knapsack R_s, i.e., from the reserved item set take the subset containing the densest items that add up to 1 or more. Observe that R_s might not be packable as is, namely, the least dense item in R_s might not fully fit inside the knapsack. We also define $x_\delta^{R_s}$ or to abbreviate x_δ to be the least dense item in R_s, and $d_\delta^{R_s}$ or d_δ its density.

Let us consider Algorithm 1: This algorithm reserves items as long as they are "dense enough". On a first stage, this algorithm reserves every item until the total size of the reserved items would overflow the knapsack, as shown in line 4. Then, in line 11, the algorithm reserves items as long as they are c times more dense than d_δ, otherwise the offered item is rejected. At every stage, the algorithm maintains R_s containing the densest subset of reserved items with total size 1 or larger as we can see in the loops starting in lines 6 and 13.

Algorithm 1. Algorithm for an upper bound in the setting of reservation costs depending on the value for a given $c \geq 1$. R_s contains the densest reserved items of combined size at least 1

$R_s{:=}\emptyset$; ▷ Densest reserved items up to total size 1
$d_\delta^{R_s}{:=}0$; ▷ Smallest density of all items in R_s
for $k = 1, \ldots, n$ **do**
  if $s(R_s) < 1$ **then** reserve x_k;
    $R_s{:=}R_s \cup x_k$;
    repeat
      Define x_δ as least dense item in R_s;
      if $s(R_s \setminus x_\delta) \geq 1$ **then** $R_s{:=}R_s \setminus x_\delta$;
    until $s(R_s \setminus x_\delta) < 1$;
    Set $d_\delta^{R_s}$ as density of least dense item in R_s;
  else if $d(x_k) \geq c \cdot d_\delta^{R_s}$ **then** reserve x_k;
    $R_s{:=}R_s \cup x_k$;
    repeat
      Define x_δ as least dense item in R_s;
      if $s(R_s \setminus x_\delta) \geq 1$ **then** $R_s{:=}R_s \setminus x_\delta$;
    until $s(R_s \setminus x_\delta) < 1$;
    Set $d_\delta^{R_s}$ as density of least dense item in R_s;
  else reject x_k;
pack the reserved items optimally

Theorem 2. *Algorithm 1 achieves a strict competitive ratio of at most*

$$\frac{2c}{1 - \alpha(\frac{4}{1-\frac{1}{c}} - 2)}.$$

for a given $c > 1$ and $0 < \alpha < \frac{1}{2}$ when the reservation costs depend on the item value.

Proof. Without loss of generality, assume that $d_\delta = 1$ at the end of the algorithm. All of the items with a density larger than c have been reserved and fit in R_v. Let us say all items with a density larger than c have a combined value of v and a combined size of $s < 1$. There can be items with density in the interval $(1, c]$ that are also in R_v and have a combined value of w and a size of t, with $t + s < 1$.

Algorithm 1 is able to pack half of the value of R_v. This is because, by construction, only the last item in R_v might not fit into the knapsack with the rest of them, thus, either the last item alone contains more than half the value of R_v or the rest of the items do. Choosing the appropriate subset of items in R_v, thus, results in a packing with value at least $\frac{v+w+(1-s-t)}{2}$, but we can also be sure that the solution is at least $v + w$ by only taking the items of density larger than 1. Additionally, if the algorithm reserves an amount of x of items with density 1, the algorithm is able to pack a value of at least $\frac{x}{2}$.

The reservation costs are the sum of the reservation costs of v and w and the reservation costs of the less dense items. The costs for the dense items are simply $\alpha(v+w)$. For items of density exactly 1 we can have at most value $x \leq 2-t$, which takes into account the overflowing of R_v at a time when the items that make up v might not have arrived. The reservation costs for the smaller items can be calculated by a geometric sum; and, as at most a size of 2 can be reserved until the marker was increased by a factor of c, the costs are at most $\alpha 2 \sum_{i=1}^{k} \frac{1}{c^i} \leq \alpha(\frac{2}{1-\frac{1}{c}}-2)$. This is also tight as items of combined size $1-\varepsilon$ can be reserved without increasing the marker; followed by one item which then increases the density at the marker and can have size of at most 1.

The optimal solution can consist of reserved and unreserved items. With unreserved items having density at most c, the optimal solution can not be larger than $v+(1-s)c$.

Together, the algorithm achieves a competitive ratio of at least

$$\frac{v+(1-s)c}{\max\left\{\frac{v+w+1-s-t}{2}, v+w, \frac{x}{2}\right\} - \alpha(v+w+x+\frac{2}{1-\frac{1}{c}}-2)}.$$

For an upper bound analysis, and given a factor c and α, we need to minimize this ratio. On the other hand, v, w, s, and t are chosen adversarially to maximize the ratio.

First, note that the ratio is maximized when $w=t=0$, as w only increases the gain of the algorithm and a higher t only contributes even negatively to the reservation costs due to the restriction of $x \leq 2-t$. For this, observe that the algorithm packs at least $v+w$ into the knapsack, which means that by subtracting the reservation costs, the w contribution is $(1-\alpha)w$, which is always positive. Therefore, the ratio looks like this:

$$\frac{v+(1-s)c}{\max\left\{\frac{v+1-s}{2}, v, \frac{x}{2}\right\} - \alpha(v+x+\frac{2}{1-\frac{1}{c}}-2)},$$

where we integrate the 2α term into the geometric sum.

Observe that if $\max\left\{\frac{v+1-s}{2}, v\right\} = v$, $v \geq 1-s$ whereas $\max\left\{\frac{v+1-s}{2}, v\right\} = \frac{v+1-s}{2}$ when $v \leq 1-s$. If we take the derivative with respect to v of both of these ratios we see that it is always negative for any $c>1$ and $0 \leq \alpha \leq \frac{1}{2}$. Thus, to maximize the ratio, every adversarial instance will not present items denser than c. One can also see this in the following way. Even when comparing just the packing of the algorithm with the optimal solution, one easily sees that having dense items favors the algorithm: Not only is the ratio between v and $v/2-\alpha v$ smaller than the ratio between $(1-s)c$ and $(1-s)/2-\alpha(1-s)$, which alone would be a reason to set v (and therefore s) to zero when trying to achieve a worst possible ratio, it also gets more supported by the fact that the algorithm still has to pay further reservation costs (which therefore lessens the ratio even more, when we cut out profit for both the algorithm and the optimal solution). Therefore, the ratio is largest when no items denser than c appear in the request sequence (namely $v=0$ and $s=0$):

$$\frac{c}{\max\left\{\frac{1}{2}, \frac{x}{2}\right\} - \alpha(x+\frac{2}{1-\frac{1}{c}}-2)}.$$

It is easy to see, that the worst case for an algorithm and its competitive ratio is an instance in which the algorithm reserved items with density 1 just to overpack the knapsack slightly (for example by two items of size $\frac{1}{2}+\varepsilon$ each). This is, as the guaranteed gain of an algorithm increases when allowing it to reserve more items of density 1, as the algorithm can use at least half of it for its packing. This holds even if the reservation costs are deducted for twice as much as the algorithm can use in its final packing (as we only consider $\alpha < 0.5$ here). Note that x must be at least one as 1 is the smallest density in R_s at the end of the request sequence, thus if $s = w = 0$ it must mean R_s only contains items of density 1.

Therefore, given a fixed c, the competitive ratio is

$$\frac{c}{\frac{1}{2} - \alpha(1 + \frac{2}{1-\frac{1}{c}} - 2)} = \frac{2c}{1 - \alpha(\frac{4}{1-\frac{1}{c}} - 2)} .$$

Corollary 1. *Algorithm 1 achieves a strict competitive ratio of at most*

$$\frac{2(2\alpha + \sqrt{\alpha(1+2\alpha)})(1 + 2\alpha + 2\sqrt{\alpha(1+2\alpha)})}{(1-2\alpha)^2\sqrt{\alpha(1+2\alpha)}} .$$

for all $0 < \alpha < \frac{1}{2}$ *when* $c = \frac{2\sqrt{2\alpha^2+\alpha}+2\alpha+1}{1-2\alpha}$.

Proof. By Theorem 2, we know that Algorithm 1 achieves a competitive ratio of

$$\frac{2c}{1 - \alpha(\frac{4}{1-\frac{1}{c}} - 2)} .$$

for a given $c > 1$. We can fix an optimal version of Algorithm 1 by choosing the best factor c for each reservation cost α. We take the derivative with respect to c and obtain

$$\frac{2}{1 - \alpha\left(\frac{4}{1-\frac{1}{c}} - 2\right)} - \frac{8\alpha}{\left(1 - \alpha\left(\frac{4}{1-\frac{1}{c}} - 2\right)\right)^2 \left(1 - \frac{1}{c}\right)^2 c}$$

which becomes 0 for $c = \frac{2\sqrt{2\alpha^2+\alpha}+2\alpha+1}{1-2\alpha}$, thus achieving a competitive ratio of

$$\frac{2(2\alpha + \sqrt{\alpha(1+2\alpha)})(1 + 2\alpha + 2\sqrt{\alpha(1+2\alpha)})}{(1-2\alpha)^2\sqrt{\alpha(1+2\alpha)}} .$$

3 Reservation Costs Proportional to Item Size

As we already mentioned in the introduction, when the reservation costs depend on the size of an item, every online algorithm is not competitive in the strict sense:

If items of the type $x = (s, \alpha s)$ for any size s between 0 and 1 are presented, every reservation results in a net gain of 0 if the item gets reserved and then used, and in a negative gain if the item is not used. Thus, no reasonable algorithm reserves any item.

However, any items that directly get taken into the knapsack make the algorithm strictly not competitive. This is because the lower bound approach for the online general knapsack [18] can be applied here. Namely, if the first offered item is not taken, no more items will appear, rendering such algorithms uncompetitive. However, if the first offered item is placed into the knapsack, an item $x = (1, V)$ will be offered of size 1 and arbitrary value V, which would be taken by the optimal solution but cannot be taken by an algorithm with a partially occupied knapsack, rendering all such algorithms uncompetitive as well.

In this section we prove that the general knapsack problem with reservation has a non-strict competitive ratio of 2 for any $\alpha > 0$.

3.1 Lower Bound

We can achieve a lower bound of 2 by constructing an adversarial instance as follows:

Theorem 3. *Given a reservation cost of* $0 < \alpha \leq 1$ *for the item size, no online algorithm solving the online knapsack problem can be c-competitive even in the non-strict meaning for any constant* $c < 2$.

Proof. Given a reservation cost α, let us consider the following adversarial request sequence.

The first item is $x_1 = (\frac{1}{2} + \varepsilon, C)$ for some $\frac{1}{2} > \varepsilon > 0$ and some large value of C. If the item is taken, an item $x_2 = (1, 3C)$ is presented and the request sequence ends. Otherwise, if the item is rejected no other item arrives and the request sequence ends. However, the item can also be reserved, if this is the case, items $x_i = (\frac{1}{2} + \varepsilon^i, C)$ are presented until either x_i is taken for some value of i and an item $(1, 3C)$ is presented, or x_i is rejected for some i, and a complementary item $x'_i = (\frac{1}{2} - \varepsilon^i, C)$ arrives. The adversary will present items x_i until one is taken or rejected or until the sum of the total reservation costs adds up to C.

If the request sequence terminates after an item $(1, 3C)$, the algorithm achieves a gain of at most C, but the optimal gain is at least $3C$, this means that for any constant $\beta > 0$, we can choose C such that the competitive ratio is larger than 2. Here, take into account that C can be an arbitrarily large value that can overcome any chosen constant β.

If the request sequence terminates after an item x_i is rejected, the optimal takes x_i and x'_i into the knapsack, with an total value of $2C$, where the algorithm can only pack a total value of C, achieving again a competitive ratio as close to 2 as necessary for an arbitrary choice of C that overcomes the value of any constant β.

If the request sequence terminates because the first item is rejected, we can choose $C > 2\beta$ and get a competitive ratio larger than 2.

Algorithm 2. Algorithm for reservation costs depending on the size of an item, where D and c are constants greater than 0.

```
R_s := ∅;                                  ▷ Densest reserved items up to total size 1
d_δ^{R_s} := 0;                            ▷ Smallest density of all items in R_s
for k = 1, ..., n do
    if d(x_k) ≤ D then reject x_k;
    else follow Algorithm 1 from line 4 to 18
pack the reserved items optimally
```

Otherwise, if the algorithm keeps reserving every offered item, at some point the reservation costs become than C, at which point the request sequence ends, the gain of such an algorithm is negative and the desired competitive ratio is also achieved by choosing a $C > 2\beta$.

3.2 Upper Bound

Let us consider Algorithm 2 for online general knapsack. This algorithm rejects any item that is less dense than a fixed density D. For the denser items, the algorithm operates just like Algorithm 1, first reserving items until R_s is complete and then reserving items that are a factor c denser than the least dense item in R_s. Observe that in this algorithm c and D are constants and can be chosen. Thus, we prove the following theorem.

Theorem 4. *Given a reservation factor α and an $\varepsilon > 0$, we can choose c and D in Algorithm 2 such that the competitive ratio of Algorithm 2 is less than $2 + \varepsilon$.*

Proof. Given a reservation factor α and an $\varepsilon > 0$, we want to choose c and D in Algorithm 2 such that the competitive ratio of Algorithm 2 is smaller than $2 + \varepsilon$. This means that for every request sequence S

$$\text{gain}_{\texttt{OPT}}(S) \leq (2 + \varepsilon) \cdot \text{gain}_{\texttt{A}}(S) + \beta \, .$$

This also means, by extension, that we need to choose a constant β to satisfy this equation for every request sequence. First, we choose $\beta = D$. Assume D is a constant only depending on ε and α, then β is also a constant fulfilling the requirements. Moreover D is the maximum value of an optimal solution if the offered items have density at most D. Thus, if we reject all items of density smaller than D, this β makes sure that Algorithm 2 is still 1-competitive if no denser items appear.

Now let us choose a reservation factor α' such that the strict competitive ratio in 1 for Algorithm 1 is less than $2 + \varepsilon$ for our given ε. This can be done because the competitive ratio in 1 goes to 2 when α' goes to 0. We set first the competitive ratio obtained in 1 to achieve the desired ratio

$$2 + \varepsilon = \frac{2(2\alpha' + \sqrt{\alpha'(1 + 2\alpha')})(1 + 2\alpha' + 2\sqrt{\alpha'(1 + 2\alpha')})}{(1 - 2\alpha')^2\sqrt{\alpha'(1 + 2\alpha')}} \, .$$

and obtain the appropriate value of α'

$$\alpha' = \frac{\varepsilon^2 - 4\sqrt{\varepsilon^3 + 8\varepsilon^2 + 20\varepsilon + 16} + 10\varepsilon + 16}{2(\varepsilon^2 + 4\varepsilon + 4)}.$$

This reservation factor α' is a constant that only depends on ε. The c associated with α' is $c = \frac{2\sqrt{2\alpha'^2+\alpha'}+2\alpha'+1}{1-2\alpha'}$, which will be our chosen c, is thus, also a constant.

We choose now the threshold density $D = \alpha/\alpha'$, which also a constant depending only on ε and α. Any item of size s with density $d > D$ will have a reservation cost by size of $\alpha \cdot s$, when the reservation factor is α, and a reservation cost by value $\alpha' \cdot d \cdot s > \alpha' \cdot D \cdot s$, when the reservation factor is α'. Thus, if $D = \alpha/\alpha'$, the reservation costs by value exceed the reservation costs by size. That is, $\alpha \cdot s = \alpha' \cdot (\alpha/\alpha') \cdot s = \alpha' \cdot D \cdot s < \alpha' \cdot d \cdot s$.

Assume now that all offered items have density larger than D. Thus, Algorithm 2 reserves exacly the same items as Algorithm 1 would reserve when applied to a reservation factor α'. Thus, all of the reserved items have a reservation cost by size with reservation factor α that is smaller than the reservation cost by value with a reservation factor α'. This means that, by 1 the strict competitive ratio of such a sequence is at most $2 + \varepsilon$.

If Algorithm 2 is offered a request sequence with a mix of items of density larger and smaller than D, we know that the gain of Algorithm 2 has at most a factor $2 + \varepsilon$ over an optimal algorithm choosing only elements denser than D. If the optimal algorithm also considers less dense items, it can gain at most an extra D, which is offset by the value of β. Thus the chosen values of c, D and β are constants depending only on ε and α and, when applied as described to Algorithm 2 , achieve the desired competitive ratio.

4 Conclusions

The general knapsack with reservation costs allows us to model situations where bookings can be canceled for some fee, in a more flexible setting than that of the online simple knapsack. We have presented upper and lower bounds for reservation costs depending on the size and value of the items. In the case of reservation costs depending on size, the bounds are tight, but some gaps remain in the case of reservations depending on item value.

Acknowledgments. We want to thank Henri Lotze, Daniel Mock and Peter Rossmanith for initial discussions on this topic.

References

1. Böckenhauer, H., Burjons, E., Hromkovic, J., Lotze, H., Rossmanith, P.: Online simple knapsack with reservation costs. In: STACS 2021, pp. 16:1–16:18 (2021)

2. Böckenhauer, H.J., Frei, F., Rossmanith, P.: Removable online knapsack and advice. In: STACS 2024, pp. 18:1–18:17 (2024)
3. Böckenhauer, H.J., Klasing, R., Mömke, T., Rossmanith, P., Stocker, M., Wehner, D.: Online knapsack with removal and recourse. In: Combinatorial Algorithms (2023)
4. Böckenhauer, H., Komm, D., Královič, R., Rossmanith, P.: The online knapsack problem: advice and randomization. Theor. Comput. Sci. **527**, 61–72 (2014)
5. Boyar, J., Favrholdt, L.M., Larsen, K.S.: Online unit profit knapsack with untrusted predictions. In: SWAT 2022, pp. 20:1–20:17 (2022)
6. Burjons, E., Frei, F., Gehnen, M., Lotze, H., Mock, D., Rossmanith, P.: Delaying decisions and reservation costs. In: COCOON 2023 (2023)
7. Burjons, E., Gehnen, M., Lotze, H., Mock, D., Rossmanith, P.: The secretary problem with reservation costs. In: Chen, C.-Y., Hon, W.-K., Hung, L.-J., Lee, C.-W. (eds.) COCOON 2021. LNCS, vol. 13025, pp. 553–564. Springer, Cham (2021). https://doi.org/10.1007/978-3-030-89543-3_46
8. Burjons, E., Gehnen, M., Lotze, H., Mock, D., Rossmanith, P.: The online simple knapsack problem with reservation and removability. In: MFCS 2023, pp. 29:1–29:12 (2023)
9. Gehnen, M., Lotze, H., Rossmanith, P.: Online simple knapsack with bounded predictions. In: STACS 2024, pp. 37:1–37:20 (2024)
10. Han, X., Kawase, Y., Makino, K.: Online unweighted knapsack problem with removal cost. Algorithmica **70**(1), 76–91 (2014)
11. Han, X., Kawase, Y., Makino, K., Yokomaku, H.: Online knapsack problems with a resource buffer. In: ISAAC 2019, pp. 28:1–28:14 (2019)
12. Han, X., Makino, K.: Online removable knapsack with limited cuts. Theoret. Comput. Sci. **411**(44), 3956–3964 (2010)
13. Ibarra, O.H., Kim, C.E.: Fast approximation algorithms for the knapsack and sum of subset problems. J. ACM **22**(4), 463–468 (1975)
14. Im, S., Kumar, R., Montazer Qaem, M., Purohit, M.: Online knapsack with frequency predictions. In: NeurIPS 2021, pp. 2733–2743 (2021)
15. Iwama, K., Taketomi, S.: Removable online knapsack problems. In: Widmayer, P., Eidenbenz, S., Triguero, F., Morales, R., Conejo, R., Hennessy, M. (eds.) ICALP 2002. LNCS, vol. 2380, pp. 293–305. Springer, Heidelberg (2002). https://doi.org/10.1007/3-540-45465-9_26
16. Iwama, K., Zhang, G.: Online knapsack with resource augmentation. Inf. Process. Lett. **110**(22), 1016–1020 (2010)
17. Karp, R.M.: Reducibility among combinatorial problems. In: Proceedings of a symposium on the Complexity of Computer Computations, pp. 85–103. The IBM Research Symposia Series (1972)
18. Marchetti-Spaccamela, A., Vercellis, C.: Stochastic on-line knapsack problems. Math. Program. **68**, 73–104 (1995)
19. Sleator, D.D., Tarjan, R.E.: Amortized efficiency of list update rules. In: STOC 1984, pp. 488–492 (1984)
20. Zeynali, A., Sun, B., Hajiesmaili, M.H., Wierman, A.: Data-driven competitive algorithms for online knapsack and set cover, pp. 10833–10841 (2021)
21. Zhou, Y., Chakrabarty, D., Lukose, R.: Budget constrained bidding in keyword auctions and online knapsack problems. In: Internet and Network Economics (2008)

Simple Approximations for General Spanner Problems

Fritz Bökler, Markus Chimani, and Henning Jasper(✉)

Institute of Computer Science, Osnabrück University, Osnabrück, Germany
{fritz.boekler,markus.chimani,henning.jasper}@uos.de

Abstract. Consider a (possibly directed) graph with n nodes and m edges, independent edge weights and lengths, and arbitrary distance demands for node pairs. The *spanner problem* asks for a minimum-weight subgraph that satisfies these demands via sufficiently short paths w.r.t. the edge lengths. For *multiplicative α-spanners* (where demands equal α times the original distances) and assuming that each edge's weight equals its length, the simple GREEDY heuristic by Althöfer et al. (1993) is known to yield strong solutions, both in theory and practice. To obtain guarantees in more general settings, recent approximations typically abandon this simplicity and practicality. Still, so far, there is no known non-trivial approximation algorithm for the spanner problem in its most general form.

We provide two surprisingly simple approximations algorithms. In general, our AUGMENTEDGREEDY achieves the first unconditional approximation ratio of m, which is non-trivial due to the independence of weights and lengths. Crucially, it maintains all size and weight guarantees GREEDY is known for, i.e., in the aforementioned multiplicative α-spanner scenario and even for *additive $+\beta$-spanners*. Further, it generalizes some of these size guarantees to derive new weight guarantees. Our second approach, RANDOMIZEDROUNDING, establishes a graph transformation that allows a simple rounding scheme over a standard multicommodity flow LP. It yields an $\mathcal{O}(n \log n)$-approximation, assuming integer lengths and polynomially bounded distance demands. The only other known approximation guarantee in this general setting requires several complex subalgorithms and analyses, yet we match it up to a factor of $\mathcal{O}(n^{(1/5-\varepsilon})$) using standard tools. Further, on bounded-degree graphs, we yield the first $\mathcal{O}(\log n)$ approximation ratio for constant-bounded distance demands (beyond undirected multiplicative 2-spanners in unit-length graphs).

Keywords: graph spanners · approximation · linear programming

1 Introduction

Let $G = (V, E)$ be a graph with n nodes and m edges. For $u, v \in V$, the distance $d_G^\ell(u, v)$ is the length of a shortest path from u to v in G, subject to lengths

A full version with additional details and references is available on arXiv [5].

J. Matuschke and J. Verschae (Eds.): WAOA 2025, LNCS 16077, pp. 48–63, 2026.
https://doi.org/10.1007/978-3-032-06706-7_4

$\ell \colon E \to \mathbb{Q}_{>0}$. A *spanner* is a subgraph $H = (V, E')$, with $E' \subseteq E$, that preserves these distances up to some error; see Ahmed et al. [1] for an overview. Spanner variations differ in their *distance demand* function $\delta \colon K \to \mathbb{Q}_{>0}$ that maps *terminal pairs* $K \subseteq V \times V$ to maximum allowed distances in H; H is feasible if and only if $d^\ell_H(u, v) \leq \delta(u, v)$ for all $(u, v) \in K$. Most generally, we have:

Definition 1 ((Decoupled Freeform) Spanner Problem). *Given a connected, simple directed or undirected graph $G = (V, E)$, terminal pairs $K \subseteq V \times V$, non-negative weights $w \colon E \to \mathbb{Q}_{\geq 0}$, positive lengths $\ell \colon E \to \mathbb{Q}_{>0}$, and positive distance demands $\delta \colon K \to \mathbb{Q}_{>0}$. Find a feasible spanner $H = (V, E')$ of G with minimum total weight $w(H) {:=} w(E') {:=} \sum_{e \in E'} w(e)$.*

Spanners were first introduced to study communication networks [27]. They since found application in, e.g., network design with bounded distances [8,14], approximate distance oracles [26], graph drawing [3], or passenger assignment [21].

Arguably, Definition 1 is the broadest interpretation of the classic spanner problem. In many applications, the weight and length of an edge are completely independent, we say *decoupled.* However, the literature often considers simplifying assumptions: In *basic* spanner problems, all edges have unit weight and length, i.e., $w(e) = \ell(e) = 1$, for all $e \in E$. In *coupled* spanner problems, each edge's weight equals its length, i.e., $w(e) = \ell(e) > 0$, for all $e \in E$. *Unit-length* or *unit-weight* spanner problems assume $\ell(e) = 1$, or $w(e) = 1$, for all $e \in E$, respectively. Unit-weight variations thus minimize the *size* $|E'|$ of spanners.

In general, every terminal pair $(u, v) \in K$ has an arbitrary distance demand $\delta(u, v)$. We call this *freeform* spanner problems. However, many spanner variations assume a consistent global form: In *linear* (α, β)*-spanners*, for $\alpha \geq 1$ and $\beta \geq 0$ (typically small constants), the distance demands are $\delta(u, v) = \alpha \cdot d^\ell_G(u, v) + \beta$, for all $(u, v) \in K = V \times V$. Notable special cases include *multiplicative* α*-spanners* (or $(\alpha, 0)$-spanners), *additive* $+\beta$*-spanners* (or $(1, \beta)$-spanners) and *distance preservers* (or $(1, 0)$-spanners).

Unfortunately, already severely restricted variations are hard to solve, or at least approximate, in polynomial time (under mild complexity assumptions). The *undirected basic multiplicative* α*-spanner* problem is **NP**-hard [19,27]. Further, the *undirected* decoupled *multiplicative* α*-spanner* problem allows no $2^{\log^{1-\varepsilon} n}$-approximation, for any $1 < \alpha \in \mathcal{O}(n^{1-\varepsilon'})$ and $\varepsilon, \varepsilon' \in (0, 1)$ [16]. See also [5].

Still there are several known approximation algorithms for various special cases, see, e.g., [5]. Many spanner algorithms form a union over several well-chosen paths between terminal pairs. Arguably, the most straightforward approach is the Greedy heuristic, which incrementally adds shortest paths to an initially empty spanner until it is feasible. Famously, Althöfer et al. [2] apply this technique to the *undirected* coupled*multiplicative* α*-spanner* problem, yielding spanners of lightness $\mathcal{O}(n/k)$ and size $\mathcal{O}(n^{1+1/k})$, for all $\alpha = 2k - 1$ $(k \in \mathbb{N}_{\geq 1})$. The *lightness* $\frac{w(E')}{w(MST(G,w))}$ upper bounds the approximation ratio, as it puts the spanner's weight in relation to the weight of a minimum-weight spanning tree $MST(G, w)$ of G (a simple lower bound to the optimum). By violating distance demands by a factor of $(1+\varepsilon)$, $\varepsilon > 0$, Greedy yields $(1+\varepsilon)(2k-1)$-spanners of lightness $\mathcal{O}(n^{1/k}(1+k/(\varepsilon^{1+1/k}\log k)))$ [15]. In practice, Greedy typically outperforms other non-exact approaches [9], often achieving near-optimum weight [4]. By Knudsen [22], Greedy also yields basic additive +2-spanners of size $\mathcal{O}(n^{3/2})$.

To achieve approximation guarantees for broader problem variations, Greedy does not suffice. Corresponding algorithms, e.g., [8,10,13,20], select paths more intricately, sacrificing simplicity and speed. Further, they are restricted to integer lengths[1]. The algorithms are randomized both w.r.t. approximation ratio and feasibility. They tackle the LP relaxation of the *Path-ILP* [28] for spanners as a subroutine. Already solving the LP is notoriously difficult, as variables must be dynamically generated by repeatedly solving (**NP**-hard) *constrained shortest path* problems [17]. The latter is achieved by using an FPTAS, e.g. [25], yielding a $(1+\varepsilon)$-approximation for the relaxation. The spanner approximations then cover some of the terminal pairs by rounding the approximate fractional LP solution. Covering the remaining terminal pairs using shortest path trees, *junction trees* [7], or *shallow-light Steiner trees* [17] yields randomized approximations for directed unit-weight multiplicative α-spanners with ratio $\widetilde{\mathcal{O}}(n^{2/3})$ [13], basic freeform spanners with ratio $\mathcal{O}(n^{3/5+\varepsilon})$ [10], or decoupled multiplicative $(\alpha^2+\alpha\varepsilon)$-spanners with at most $\mathcal{O}(n^{1/2+\varepsilon})$ times the weight of an optimum decoupled α-spanner [8], for $\varepsilon > 0$, respectively. Grigorescu, Kumar and Lin [20] combine several of these ideas: For directed decoupled distance preservers with integer lengths and polynomially bounded distance demands, they present a randomized $\mathcal{O}(n^{1/2+\varepsilon})$-approximation (albeit implicitly, a similar guarantee was achieved before in [8]). Most importantly, they yield the first randomized $\widetilde{\mathcal{O}}(n^{4/5+\varepsilon})$-approximation for the directed decoupled freeform spanner problem with integer lengths and polynomially bounded distance demands. To attain a sublinear ratio for a problem of this generality, the algorithm employs several further complex and practically never-implemented subroutines. It iterates over guesses for the optimum solution value; every iteration not only requires numerous approximations of Path-LP adaptations, but also several constrained shortest path trees [25], and approximations of *minimum-weight distance-preserving junction trees* [10] via *shallow trees* [18]. The running time is not provided in $\mathcal{O}$-notation, but only argued to be generally polynomial; the expected weight guarantee is given in $\widetilde{\mathcal{O}}$-notation, neglecting polylogarithmic factors.

Currently, there are no practical approximations for decoupled, let alone decoupled freeform, spanner problems. New spanner approximations tend to focus on improving weight guarantees in $\overset{(\sim)}{\mathcal{O}}$-notation rather than considering running times or practicality. Yet, despite these efforts, an approximation for general decoupled freeform spanner problems that does not require polynomially bounded integer lengths has remained elusive so far. Consequently, practitioners are prompted to force decoupled instances into a coupled format by discarding

[1] Newer publications say so explicitly; older ones may gloss over this restriction, arising from using an FPTAS to solve constrained shortest path (CSP) problems over integer lengths. Indeed, CSP with rational lengths is strongly **NP**-hard [29].

or combining weights and lengths (e.g. [3,21]). Instances are then often solved with Greedy, which is easy to implement and understood to yield satisfactory results [4,9]. Still, this practice poses significant issues, as it does not guarantee that solutions effectively address the original decoupled problem.

Contribution. We present two simple approximation algorithms for both directed and undirected decoupled freeform spanner problems. Tables 1 and 2 summarize our contributions. See [5] for the full version of this paper.

In Sect. 3, we augment Greedy to become the first unconditional (i.e., no upper bounds or domain restrictions for weights or lengths, etc.) m-approximation for all decoupled freeform spanner problems. The ratio m is non-trivial due to the independence of weights and lengths. Our AugmentedGreedy not only maintains Greedy's running time up to a factor of $\mathcal{O}(\log m)$, but crucially retains all size and weight guarantees typically associated with standard Greedy. Further, it allows us to lift known size guarantees from several coupled settings into decoupled weight guarantees.

The best known approximation algorithm [20] for (un)directed decoupled freeform spanners assumes integer lengths and polynomially bounded maximum distance demand $\bar{\delta}$. It achieves ratio $\widetilde{\mathcal{O}}(n^{4/5+\varepsilon})$, but relies on several complex subalgorithms with complicated analyses. In Sect. 4, for these instances, we provide a simple graph transformation enabling a decoupled freeform spanner approximation via randomized rounding of a standard multicommodity flow LP. Let $\bar{\mu} = (\bar{\delta} + 2)^{n-2}$; our ratio of $\ln(n\bar{\mu}|K|)$ ($\in \mathcal{O}(n \log n)$) accounts for all constants and matches the best known ratio up to a factor of $\mathcal{O}(n^{1/5-\varepsilon})$. Our approximation requires no specialized subroutines, and offers significantly stronger time guarantees as well as easy implementation and analysis. Also, this simple analysis allows us to prove the first $\mathcal{O}(\log n)$-approximation ratio for (un)directed decoupled freeform spanner problems with constant-bounded distance demands and (out)degree (which are **NP**-hard [19]). Previously, for constant-bounded demands only the special case of undirected unit-length multiplicative 2-spanners (i.e., $\delta(u, v) = 2$ for all $\{u, v\} \in K = E$) was known to allow this strong ratio.

2 Preliminaries

Generally, *shortest* means minimum length. In undirected settings, unordered terminal pairs $K \subseteq \binom{V}{2}$ suffice. With OPT we denote the optimum solution value of the considered instance. We define the maximum length $\bar{\ell} := \max_{e \in E} \ell(e)$, weight $\bar{w} := \max_{e \in E} w(e)$, distance demand $\bar{\delta} := \max_{(u,v) \in K} \delta(u, v)$, and (out)degree $\Delta := \max_{q \in V} |\{(q, t) \mid (q, t) \in E\}|$. We can assume w.l.o.g. $\bar{\ell} \le \bar{\delta} \le n\bar{\ell}$, as longer edges cannot be used to satisfy any distance demands and larger demands are trivially satisfied in connected graphs. For a function $f \colon E \to \mathbb{Q}$, we use the shorthands $f(u, v) := f((u, v))$ for $(u, v) \in E$, $f(X) := \sum_{e \in X} f(e)$ for $X \subseteq E$, and $f(G') := f(E')$ for $G' = (V', E') \subseteq G$. Let $\mathrm{LP}(x, y, f)$ denote the running time to solve an LP with $\Theta(x)$ variables and $\Theta(y)$ constraints, where the maximum encoding-length of any coefficient does not exceed that of the largest rational in the image of function f.

Table 1. Guarantees for the (un)directed decoupled freeform spanner problem. $\mathrm{LP}(x, y, f)$ denotes the running time for solving LPs with $\Theta(x)$ variables and $\Theta(y)$ constraints, whose largest numbers arise in the image of function f (cf. Section 2). "—" marks inapplicability due to exponential running times. "\$" denotes the solution is feasible w.h.p and requires integer lengths. Let $\bar{w}, \bar{\ell}, \bar{\delta}, \Delta$ denote the maximum weight, length, distance demand, and (out)degree of an instance, respectively. Let $\varepsilon > 0$, and $\bar{\mu} = (\bar{\delta} + 2)^{n-2}$.

		best known (Grigorescu et al. [20])	our contribution: RandomizedRounding	our contribution: AugmentedGreedy
running time (in $\mathcal{O}$)		$\mathrm{poly}(n, \vert K\vert, \log \bar{w}, \bar{\ell}, \varepsilon^{-1})$, $\Omega(n \log(m\bar{w}) \cdot \mathrm{LP}(\vert K\vert^5 m^8 \bar{\ell}^8, \vert K\vert^3 m^4 \bar{\ell}^4, \max\{\ell, w\}))$	$\mathrm{LP}(\vert K\vert m\bar{\delta}, \vert K\vert(m + n\bar{\delta}), w)$	$\vert K\vert(m + n \log n) \log m$
approx. ratio	general	—	—	m
	$\bar{\delta} \in \mathcal{O}(\mathrm{poly(n)})$	$\widetilde{\mathcal{O}}(n^{4/5+\varepsilon})$ \$	$\ln(n\bar{\mu}\vert K\vert) \in \mathcal{O}(n \log n)$ \$	m
	$\bar{\delta} \in \mathcal{O}(1)$	$\widetilde{\mathcal{O}}(n^{4/5+\varepsilon})$ \$	$\mathcal{O}(n)$ \$	m
	$\bar{\delta}, \Delta \in \mathcal{O}(1)$	$\widetilde{\mathcal{O}}(n^{4/5+\varepsilon})$ \$	$\mathcal{O}(\log n)$ \$	$m \in \mathcal{O}(n)$

Table 2. Overview of Greedy vs. AugmentedGreedy. For coupled multiplicative and basic additive spanners, they obtain the same weight and size guarantees (Corollaries 7 and 8). Let $k \in \mathbb{N}_{\geq 1}$. GEO denotes the geometric setting; $E^{>} \subseteq E$ denotes "high-weight" edges; see Sect. 3 for details.

spanner problem	setting	Greedy: weight ratio guarantee	AugmentedGreedy: weight ratio guarantee	AugmentedGreedy: size
(un)directed decoupled freeform	—	—	m	m (trivial)
undirected decoupled multiplicative	$\alpha = 2k - 1$, $\vert E^{>}\vert \in \mathcal{O}(\sqrt{n} \cdot n^{1/k})$	—	$\mathcal{O}(n^{1+1/k})$	same as Greedy
	GEO, $\alpha > 1$, $\vert E^{>}\vert \in \mathcal{O}(\sqrt{n})$	—	$\mathcal{O}(n)$	same as Greedy
undirected unit-length additive	$2 \leq \beta \in \mathcal{O}(1)$	—	$\mathcal{O}(n^{3/2})$	same as Greedy

Enforcing distance demands $\delta(u, v) = \alpha \cdot d_G^{\ell}(u, v)$, for all *adjacent* node pairs $(u, v) \in E = K$ guarantees feasible multiplicative α-spanners [27]. Further, it is folklore that non-metric edges can be removed from *coupled* α-spanner instances, as they cannot be part of an optimum solution. However, in decoupled settings, non-metric edges *cannot* be removed: Consider an undirected triangle, where two of the edges have unit weight and length, while the third has weight 0.5 and (non-metric) length 3. The optimum 4-spanner has weight 1.5, while removing the third edge leaves us with a minimal solution value of 2. Yet, the following Observation allows us to disregard non-metric distance demands in decoupled freeform spanner problems.

In the *demand graph* $D = (V, K)$, two nodes are adjacent if there is a distance demand between them. Then $d_D^{\delta}(u, v)$ is the distance between nodes u and v in D w.r.t. the distance demands δ. We define the *metric* terminal pairs as those whose distance demands (strictly) satisfy the triangle inequality, i.e., $K' := \{(u, v) \in K \mid d_{D-(u,v)}^{\delta}(u, v) > \delta(u, v)\}$.

Algorithm 1. GREEDY.

Input: graph $G = (V, E)$, terminal pairs K, lengths ℓ, distance demands δ

```
H ← (V, ∅)
for (u, v) ∈ K in non-decreasing order of distance d^ℓ_G(u, v) do
    if d^ℓ_H(u, v) > δ(u, v) then
        add shortest uv-path in G (w.r.t. lengths ℓ) to H
return H
```

Observation 2. *A subgraph $H = (V, E')$ is a feasible (un)directed decoupled freeform spanner of $G = (V, E)$, if and only if $d^\ell_H(u, v) \leq \delta(u, v)$ for all metric terminal pairs (u, v).*

Proof. Necessity follows from $K' \subseteq K$. Conversely, assume $d^\ell_H(u, v) \leq \delta(u, v)$ for all $(u, v) \in K'$. Let $(s, t) \in K \setminus K'$. Let $P \subseteq K$ be a minimum st-path in D w.r.t. δ; w.l.o.g. $P \subseteq K'$. Consequently, $d^\ell_H(s, t) \leq \sum_{(i,j)\in P} d^\ell_H(i, j) \leq \sum_{(i,j)\in P} \delta(i, j) \leq \delta(s, t)$.

3 Augmented Greedy

Arguably, the easiest spanner algorithm is the GREEDY heuristic, shown in Algorithm 1, which disregards weights: Start with an empty spanner $H = (V, \emptyset)$. Until H is feasible, process terminal pairs $(u, v) \in K$ in non-decreasing order of distance in G. If the current distance in H is too long, add a shortest uv-path in G to H. Thereby, we break ties consistently to ensure unique shortest paths, e.g., via well-chosen perturbation or lexicographic ordering. These shortest path computations for all $(u, v) \in K$ dominate the running time of $\mathcal{O}(|K|(m + n \log n))$, where some cases allow refined analyses [2,6]. Despite its simplicity, GREEDY offers strong guarantees for certain spanner problems. For undirected coupled multiplicative α-spanners, it guarantees lightness $\mathcal{O}(n/k)$ and size $\mathcal{O}(n^{1+1/k})$ [2], for $\alpha = 2k-1$ ($k \in \mathbb{N}_{\geq 1}$). For $\alpha \geq \log n$ and $\alpha \geq (\log n)^2$ the lightness improves to $\mathcal{O}(\log n)$ and $\mathcal{O}(1)$, respectively [6]. In the undirected coupled *geometric* setting, i.e., complete graphs with Euclidean lengths, it guarantees lightness $\mathcal{O}(\log n)$ and size $\mathcal{O}(n)$, for all $\alpha > 1$ [6] (observe that this changes the co-domains of w, ℓ, δ to $\mathbb{R}_{>0}$). For spanner problems, where feasible solutions are undirected connected subgraphs that span all nodes, the weight of a minimum-weight spanning tree $w(MST(G, w))$ lower bounds the weight of *all* feasible spanners. Hence, as guarantees [2,6] are absolute or w.r.t. $w(MST(G, w))$, they analogously hold for undirected coupled linear (α, β)-spanners, for any $\beta \geq 0$. Further, for undirected basic additive +2-spanners, GREEDY has (best possible [30]) size guarantee $\mathcal{O}(n^{3/2})$ [22]. However, in general, GREEDY has no known approximation guarantees for spanner problems with decoupled weights and lengths or if distance demands do not take linear form.

Our AUGMENTEDGREEDY overcomes these limitations, see Algorithm 2. It runs in two consecutive phases: Phase 1 achieves the first unconditional approximation ratio for all freeform spanner problems, including decoupled weights

Algorithm 2. AugmentedGreedy. Line 2 is only executed if the problem definition provides $w(MST(G, w)) \leq \text{OPT}$ (see text).

Input: graph $G = (V, E)$, terminal pairs K, weights w, lengths ℓ, distance demands δ
1: binary search for $\hat{w} = \min\{w' \in \mathcal{W} : G[w'] \text{ is a feasible spanner}\}$ // phase 1
2: **[** $\hat{w} \leftarrow \max\{\hat{w}, w(MST(G, w))\}$ **]**
3: construct $G[\hat{w}]$
4: **return** Greedy$(G[\hat{w}], K, \ell, \delta)$ // phase 2

and lengths. For this, it establishes a feasible spanner G' that already attains an initial weight guarantee. Phase 2 uses Greedy to further sparsify G' to obtain the final spanner H. While without phase 1 Greedy does not achieve any weight guarantee in general, executing Greedy after phase 1 ensures it still finds a feasible solution, maintains the weight ratio of the first phase, and achieves Greedy's known guarantees at the same time. Crucially, AugmentedGreedy's total running time closely matches that of Greedy.

The key idea in phase 1 is to determine an adequate lower bound $\hat{w}$ to the optimum solution value OPT. Let $\mathcal{W}$ be the set of distinct edge weights in the instance. For any $w' \in \mathcal{W}$, we define weight-restricted subgraphs $G[w'] = (V, E[w'])$ of G with $E[w'] := \{e \in E : w(e) \leq w'\}$. Checking whether some $G[w']$ constitutes a feasible spanner, i.e., $d^{\ell}_{G[w']}(u, v) \leq \delta(u, v)$ for all $(u, v) \in K$, is trivial. Thus, using binary search over sorted $\mathcal{W}$, we can easily find the smallest weight $\hat{w}$ such that $G[\hat{w}]$ is a feasible spanner. Clearly, every feasible spanner contains an edge of weight at least $\hat{w}$ and $\hat{w} \leq \text{OPT}$. Further, if the problem definition ensures that a spanner is undirected and must be spanning and connected (e.g., for undirected multiplicative α-spanner problems), the weight of a minimum-weight spanning tree is a natural lower bound for OPT. In such a case, we may further increase $\hat{w}$ to that weight, overall knowing that any feasible solution will require a total weight of at least $\hat{w}$. Phase 2 further sparsifies $G[\hat{w}]$. For all $(u, v) \in K$, to ensure $d^{\ell}_H(u, v) \leq \delta(u, v)$ in the final spanner H, distances in $G[\hat{w}]$ generally must be preserved exactly. This gives us a lower bound on the best achievable size guarantee of $\Omega(n^{2/3}|K|^{2/3})$, i.e., $\Omega(n^2)$ or $\Omega(m)$ for $|K| \in \Omega(n^2)$ [11]. Hence, there is no way to guarantee further sparsification of $G[\hat{w}]$. Yet, by calling Greedy on this restricted graph $G[\hat{w}]$, but with unaltered terminal pairs K, lengths ℓ, and distance demands δ, we have a simple way to generate a feasible spanner that does obtain stronger weight and size guarantees for several interesting instance classes.

Theorem 3. *AugmentedGreedy is an m-approximation for all (un)directed decoupled freeform spanner problems.*

Proof. $G[\hat{w}]$ is a feasible spanner. Thus, for all $(u, v) \in K$, the shortest uv-path in $G[\hat{w}]$ has length at most $\delta(u, v)$ and Algorithm 2 returns a feasible spanner. The weight guarantee follows from $w(H) \leq w(E[\hat{w}]) \leq |E[\hat{w}]| \cdot \hat{w} \leq m \cdot \text{OPT}$.

The running time is dominated by $|K| \log m$ and $2|K|$ many shortest path computations in phase 1 and 2, respectively. Thus, it is in $\mathcal{O}(|K|(m + n \log n) \log m)$.

Theorem 3 is the first approximation algorithm for the decoupled freeform spanner problem that does not require polynomially bounded integer lengths and, e.g., is applicable to instances with Euclidean length. Further, it works for both directed and undirected graphs. As previously mentioned, using known bounds for distance preservers, i.e., Theorem 1.1(2) in [11], we can construct instances for which our analysis of ratio m is tight. The running time matches GREEDY up to a factor of $\mathcal{O}(\log m)$. Note that sparsifiying G prior to phase 1 generally leads to the loss of any weight guarantee.

In many applications, edge weight and length are strongly anti-correlated; e.g., in network design, low-delay connections are typically more expensive. Here, GREEDY tends to struggle, as it hastily selects all of the shortest, and thus most expensive, connections. Hence, it cannot yield any weight guarantees. Conversely, AUGMENTEDGREEDY and its analysis can benefit from such scenarios: Especially if distance demands are somewhat similar to the original distances, it is in practice often reasonable to assume that one has to select at least one short but comparably expensive edge into the spanner. This yields a large $\hat{w}$, and thus there may be only few *high-weight* edges $E^{>}{:=}E \setminus E[\hat{w}]$. In such settings, we can use GREEDY's absolute size guarantees to deduce stronger weight guarantees. Recall that for decoupled multiplicative α-spanners, by Observation 2 and [27], we only have to enforce distance demands for all metric edges $M \subseteq E$.

Lemma 4. *Consider the undirected decoupled multiplicative α-spanner problem. AUGMENTEDGREEDY (for $\alpha = 2k-1$, $k \in \mathbb{N}_{\geq 1}$) yields a spanner both of size and of weight ratio $\mathcal{O}(n^{1+1/k} + \sqrt{n}|E^{>}|)$. In the geometric setting (for $\alpha > 1$), this improves to $\mathcal{O}(n + \sqrt{n}|E^{>}|)$.*

Proof. For each $\{u, v\} \in M$, the shortest uv-path in G by definition consists only of the edge itself. Crucially, all $M \cap E[\hat{w}]$ still exist in the weight-restricted graph $G[\hat{w}]$. Thus, covering these edges in phase 2 is analogous to using GREEDY on $G[\hat{w}]$ with terminal pairs $M \cap E[\hat{w}]$. By GREEDY's size guarantees, this introduces at most $\mathcal{O}(n^{1+1/k})$ edges, in general, or $\mathcal{O}(n)$ edges in the geometric setting. In contrast, for each edge in $M \cap E^{>}$, we may have to add a path consisting of up to $n-1$ low-weight edges. However, recall that we compute unique shortest paths achieved by tie-breaking. Thus, by [11, Corollary 7.8], originally formulated in the context of pairwise distance preservers, the overall number of added edges can be upper bounded by $\mathcal{O}(n + \sqrt{n}|E^{>}|)$.

Corollary 5. *If $|E^{>}| \in o(m/\sqrt{n})$, better approximation guarantees than ratio m become possible. The weight ratio and size are in $\mathcal{O}(n^{1+1/k})$ in general for $|E^{>}| \in \mathcal{O}(\sqrt{n} \cdot n^{1/k})$, and in $\mathcal{O}(n)$ in the geometric setting for $|E^{>}| \in \mathcal{O}(\sqrt{n})$.*

Note that even for $|E^{>}| \in \mathcal{O}(1)$, in fact even $|E^{>}| = 1$, GREEDY has no weight guarantee on decoupled instances. Next, we show that, despite it working on a restricted graph, AUGMENTEDGREEDY retains GREEDY's size and lightness guarantees in the established contexts.

Lemma 6. *AUGMENTEDGREEDY has the same size and lightness guarantees as GREEDY for undirected coupled multiplicative α-spanners, for all $\alpha \geq 1$.*

Proof. Let GREEDY and AUGMENTEDGREEDY run in parallel. Recall that $K = M$ suffices. Both process edges in the same order, starting with $E[\hat{w}]$. For all $e \in E[\hat{w}]$, they behave identically. GREEDY always includes an MST [2] and note that $\hat{w} \geq w(MST(G, w))$. Hence, after covering all $e \in E[\hat{w}]$, AUGMENTEDGREEDY contains an MST, too, and the weight and thus length of every $e \in E^>$ is larger than $w(MST(G, w))$. Thus, for any $e \in E^>$, a sufficiently short path through the MST already exists and AUGMENTEDGREEDY does not add any more edges. For coupled α-spanners, GREEDY and AUGMENTEDGREEDY yield the same spanner.

Corollary 7. *AUGMENTEDGREEDY yields undirected coupled α-spanners of lightness (and thus weight ratio) $\mathcal{O}(n/k)$ and size $\mathcal{O}(n^{1+1/k})$, where $\alpha = 2k - 1$, $k \in \mathbb{N}_{\geq 1}$. In the geometric setting, these improve to $\mathcal{O}(\log n)$ and $\mathcal{O}(n)$, respectively, for all $\alpha > 1$. For $\alpha \geq \log n$ and $\alpha \geq (\log n)^2$, lightness even is in $\mathcal{O}(\log n)$ and $\mathcal{O}(1)$, respectively. Violating distance demands by a factor of $(1+\varepsilon)$, $\varepsilon > 0$, we yield $(1+\varepsilon)(2k-1)$-spanners of lightness $\mathcal{O}(n^{1/k}(1 + k/(\varepsilon^{1+1/k} \log k)))$.*

These four statements follow from combining Lemma 6 with [2,6], and [15], respectively. Corollaries 7 and 5 also hold for undirected decoupled linear (α, β)-spanners, for all $\beta \geq 0$.

For undirected basic additive $+\beta$-spanners and $\beta = 2$, GREEDY guarantees size in $\mathcal{O}(n^{3/2})$ [22]. The proof can easily be generalized to accommodate AUGMENTEDGREEDY by adjusting some minor details, even in the unit-length setting and for any constant $\beta \geq 2$. The resulting proof works analogous to [22], see [5]. The key observation is that, to achieve size $\mathcal{O}(n^{3/2})$, the uv-paths added between $\{u, v\} \in K$ during construction, do not necessarily have to be shortest paths. For any constant $\beta \geq 2$, paths of length at most $\delta(u, v) = d_G^\ell(u, v) + \beta$ suffice.

Corollary 8. *AUGMENTEDGREEDY yields undirected unit-length additive $+\beta$-spanners of size and weight ratio $\mathcal{O}(n^{3/2})$, for any constant $\beta \geq 2$.*

Hence, AUGMENTEDGREEDY not only matches the best possible [30] size guarantee for $\beta = 2$, but derives new weight guarantees in the unit-length setting, even for all constant $\beta \geq 2$.

4 Randomized Rounding

We now describe our $\ln(n\bar{\mu}|K|) \in \mathcal{O}(n \log n)$-approximation algorithm for the (un)directed decoupled freeform spanner problem, with maximum distance demand $\bar{\delta}$ and $\bar{\mu} = (\bar{\delta} + 2)^{n-2}$. As in the other state-of-the-art algorithms for decoupled spanner problems [8,20] we assume integer lengths (and, thus, w.l.o.g. integer distance demands) and polynomially bounded distance demands. The running time is polynomial in n, and the approximation ratio of $\ln(n\bar{\mu}|K|)$ is in $\mathcal{O}(n \log n)$. For simplicity, we assume a directed setting. Undirected settings work analogously, by bi-directing the graph, see [5].

We use multicommodity flow in layered graphs. Dodis and Khanna [14] apply a similar technique to unit-length multiplicative α-spanners. Their algorithms

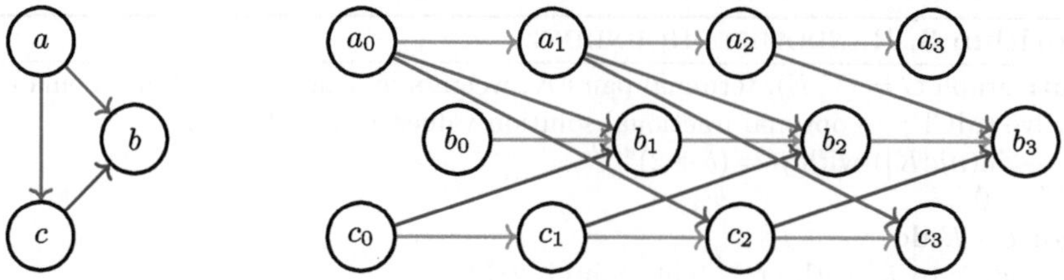

Fig. 1. The graph G (left) and its 3-extension (right), as described in Example 10.

generalization to non-unit lengths via subdividing edges is flawed, see [5]. Our generalization can handle both non-unit lengths and freeform distance demands. A key challenge lies in the non-unit lengths. In contrast to [14], we do not subdivide edges and we do not use a unit-length spanner algorithm as a subroutine. Instead, we encode lengths directly in the graph extension. Ultimately, our construction allows us to effortlessly enforce distance demands of any form for any subset of node pairs. We establish the following graph extension; for easier distinction, we use the term *arc* to refer to the directed edges in the extension.

Definition 9 ($\bar{\delta}$-extension). *The $\bar{\delta}$-extension of a directed graph $G = (V, E)$ with integer lengths and distance demands is a layered directed graph $G^{\bar{\delta}} = (V^{\bar{\delta}}, E^{\bar{\delta}})$ with:*

- *$\bar{\delta} + 1$ layers of nodes $V_0^{\bar{\delta}}, V_1^{\bar{\delta}}, \ldots, V_{\bar{\delta}}^{\bar{\delta}}$ where each $V_i^{\bar{\delta}}$ is a copy of V. For $q \in V$, q_i denotes the copy of q in $V_i^{\bar{\delta}}$.*
- *For each edge $(s, t) = e \in E$, there are arcs $(s_i, t_{i+\ell(e)})$, for all $0 \le i \le \bar{\delta} - \ell(e)$. Additionally, for all $q \in V$, there are self-arcs (q_i, q_{i+1}), for all $0 \le i < \bar{\delta}$.*

Clearly, $|V^{\bar{\delta}}| = n(\bar{\delta} + 1)$ and $|E^{\bar{\delta}}| \le (n + m)\bar{\delta}$. Consider Example 10.

Example 10. Figure 1 shows a simple example. We consider a directed decoupled freeform spanner problem instance with graph $G = (V, E)$, $V = \{a, b, c\}$, $E = \{(a, b), (a, c), (c, b)\}$, terminal pairs $K = E$, weights $w(a, b) = 5$, $w(a, c) = w(c, b) = 1$, lengths $\ell(a, b) = \ell(c, b) = 1$, $\ell(a, c) = 2$, and distance demands $\delta(a, b) = 3$, $\delta(a, c) = \delta(c, b) = 2$. Thus, $\bar{\delta} = 3$. Clearly, the optimum solution to this instance is $H = (V, E')$ with $E' = \{(a, c), (c, b)\}$ of total weight 2. ◇

We state the spanner problem as a standard multicommodity flow LP in the $\bar{\delta}$-extension of G, where we ship one commodity from u_0 to $v_{\delta(u,v)}$, for each $(u, v) \in K$. We then round the fractional solution to obtain a subgraph $H = (V, E')$ and prove that H forms a feasible spanner w.h.p. via a simple cut argument. The full algorithm is shown in Algorithm 3. Let us first discuss the formulation (MCF) below: There are *flow variables* (4) and *edge variables* (5). We use standard flow conservation constraints (3), where the indicator function $\mathbb{1}_\varphi$ is 1 if the Boolean expression φ is true, and 0 otherwise. Constraints (2)

Algorithm 3. RandomizedRounding.

Input: graph $G = (V, E)$, terminal pairs K, weights w, lengths ℓ, distance demands δ
1: solve (MCF) $\rightarrow$ optimal fractional solution values x_e^* for all $e \in E$
2: $\gamma \leftarrow \ln(n\bar{\mu}|K|)$ with $\bar{\mu} = (\bar{\delta} + 2)^{n-2}$
3: $E' \leftarrow \emptyset$
4: **for** $e \in E$ **do**
5: add e to E' with probability $\min\{1, \gamma x_e^*\}$
6: **return** $H = (V, E')$

ensure that if any non-self-arc in $E^{\bar{\delta}}$ carries flow, its corresponding edge variable must be active. The objective function (1) minimizes the weight of the spanner. In total, (MCF) contains $\mathcal{O}(|K|m\bar{\delta})$ variables, $\mathcal{O}(|K|(m + n\bar{\delta}))$ constraints, and the largest coefficient stems from the input weights. It can directly be solved using any standard (polynomial-time) LP solver.

$$\text{(MCF)} \quad \min \sum_{e \in E} w(e) x_e \tag{1}$$

$$\text{s.t.} \quad \sum_{i=0}^{\bar{\delta}-\ell(e)} f^{uv}_{(s_i, t_{i+\ell(e)})} \leq x_e \qquad \forall (u,v) \in K, \forall e = (s,t) \in E \tag{2}$$

$$\sum_{(q,\cdot) \in E^{\bar{\delta}}} f^{uv}_{(q,\cdot)} - \sum_{(\cdot,q) \in E^{\bar{\delta}}} f^{uv}_{(\cdot,q)} = \mathbb{1}_{q=u_0} - \mathbb{1}_{q=v_{\delta(u,v)}} \quad \forall (u,v) \in K, \forall q \in V^{\bar{\delta}} \tag{3}$$

$$0 \leq f^{uv}_{(q,r)} \leq 1 \qquad \forall (u,v) \in K, \forall (q,r) \in E^{\bar{\delta}} \tag{4}$$

$$0 \leq x_e \leq 1 \qquad \forall e \in E \tag{5}$$

We now derive a subgraph $H = (V, E')$ from an optimal solution to (MCF). Let x_e^* denote an optimal (fractional) solution value for variable x_e. Let γ be a value defined later. For each edge $e \in E$, we include e in E' with probability $\min\{1, \gamma x_e^*\}$. The expected total weight of E' is then at most a factor of γ away from the optimum spanner.

We show that H forms a feasible spanner with high probability by a suitable choice for γ. To this end, we use a simple cut argument in the $\bar{\delta}$-extension $H^{\bar{\delta}} = (V^{\bar{\delta}}, E'^{\bar{\delta}})$ of H. A $u_0 v_{\delta(u,v)}$-cut $(A{:}B)$ partitions the nodes $V^{\bar{\delta}}$ into disjoint subsets A and B with $u_0 \in A$ and $v_{\delta(u,v)} \in B$. A cut $(A{:}B)$ is *satisfied* if at least one arc in $E'^{\bar{\delta}}$ goes from A to B, see Fig. 2.

Recall that H is feasible if and only if $d^{\ell}_H(u,v) \leq \delta(u,v)$, for all $(u,v) \in K$.

Lemma 11. *Let $H = (V, E') \subseteq G$ with $\bar{\delta}$-extension $H^{\bar{\delta}}$ and any $(u,v) \in K$. Then*

(1) $d^{\ell}_H(u,v) \leq \delta(u,v) \Leftrightarrow$ **(2)** every $u_0 v_{\delta(u,v)}$-cut in $H^{\bar{\delta}}$ is satisfied.

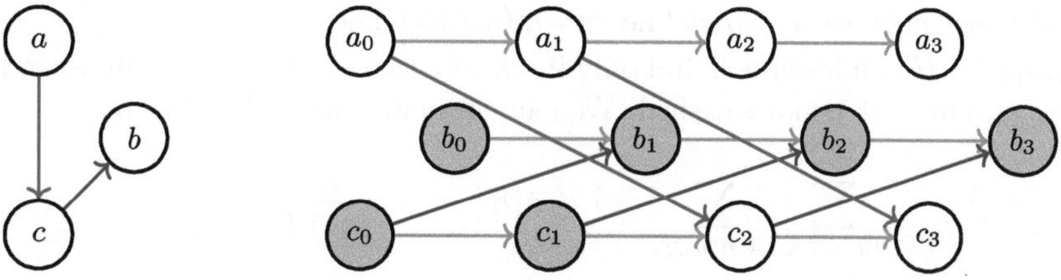

Fig. 2. The optimum solution $H \subset G$ of Example 10 (left) and an a_0b_3-cut $(A{:}B)$ (A is white and B is gray) in its 3-extension satisfied by arc (c_2, b_3) (right). The cut is ascending, induced by the node labeling $\pi(a) = 0$, $\pi(b) = \delta(a,b) + 1 = 4$, and $\pi(c) = 2$.

Proof. Assume (1) and let $P \subseteq E'$ be a shortest uv-path in $H = (V, E')$. Then $\ell(P) \leq \delta(u,v)$. We can construct a $u_0v_{\delta(u,v)}$-path $P^{\bar{\delta}} \subseteq E'^{\bar{\delta}}$ in $H^{\bar{\delta}}$ using the arcs corresponding to edges in P (and self-arcs), and (2) holds. Conversely, if (2), there is a path $Q^{\bar{\delta}}$ from u_0 to $v_{\delta(u,v)}$ in $H^{\bar{\delta}}$. $Q^{\bar{\delta}}$ induces a uv-path of length at most $\delta(u,v)$ in H (removing self-arcs).

We argue that only a certain kind of cuts may ever be non-satisfied in $H^{\bar{\delta}}$.

Definition 12 (Ascending $u_0v_{\delta(u,v)}$-cut). *Let $H = (V, E')$ with $\bar{\delta}$-extension $H^{\bar{\delta}}$ and $u, v \in V$. A $u_0v_{\delta(u,v)}$-cut $(A{:}B)$ in $H^{\bar{\delta}}$ is ascending if it is induced by a node labeling $\pi\colon V \to \{0, 1, \ldots, \delta(u,v) + 1\}$, i.e., for every $q \in V$ we have that $q_i \in B$ for all $0 \leq i < \pi(q)$, and $q_i \in A$ for all $\pi(q) \leq i \leq \delta(u,v)$.*

Necessarily, $\pi(u) = 0$ and $\pi(v) = \delta(u,v) + 1$. Observe that Fig. 2 indeed shows an ascending cut. A $u_0v_{\delta(u,v)}$-cut $(A{:}B)$ is non-ascending if and only if there is a $q \in V$ with $q_i \in A$ and $q_{i+1} \in B$, for some $0 \leq i \leq \delta(u,v)$. Such a cut is trivially satisfied by self-arc (q_i, q_{i+1}). Hence, we only have to consider ascending cuts. For $(u,v) \in K$, let $\mathcal{C}_{uv}$ denote the set of all ascending $u_0v_{\delta(u,v)}$-cuts in $H^{\bar{\delta}}$. By construction via the node labeling π, we get $|\mathcal{C}_{uv}| = (\delta(u,v) + 2)^{n-2}$. Define $\bar{\mu} := \max_{(u,v)\in K} |\mathcal{C}_{uv}| \leq (\bar{\delta} + 2)^{n-2}$ as the maximum number of ascending $u_0v_{\delta(u,v)}$-cuts for any $(u,v) \in K$. Next, we bound the probability that an ascending cut is non-satisfied in the $\bar{\delta}$-extension of our rounded subgraph.

Observation 13. *Let $X_1, \ldots, X_k$ be independent, binary, random variables with $\Pr(X_i) = \min\{1, p_i\}$ and $\sum_{i=1}^{k} p_i \geq S$. Then $\Pr[\bigwedge_{i=1}^{k}(X_i = 0)] = \prod_{i=1}^{k}(1-p_i) \leq e^{-(\sum_{i=1}^{k} p_i)} \leq e^{-S}$.*

Lemma 14. *For $(u,v) \in K$, a given ascending $u_0v_{\delta(u,v)}$-cut $(A{:}B)$ is non-satisfied in $H^{\bar{\delta}}$ with probability at most $e^{-\gamma}$.*

Proof. Let $E_c'^{\bar{\delta}} \subseteq E'^{\bar{\delta}}$ be the set of arcs that go from A to B in $H^{\bar{\delta}}$. Let $E_c' \subseteq E'$ be their corresponding edges in H. Since one unit of flow must cross from A to B in $H^{\bar{\delta}}$, we get $1 \leq \sum_{a\in E_c'^{\bar{\delta}}} f_a^{uv*} \leq \sum_{e\in E_c'} x_e^*$, where f_a^{uv*} are the optimal solution values for f_a^{uv}, respectively. Each $e \in E_c'$ is chosen with probability $\min\{1, \gamma x_e^*\}$ and $\sum_{e\in E_c} \gamma x_e^* \geq \gamma$. Thus, by Observation 13, we get $\Pr[(A{:}B)$ is not satisfied$] = \Pr[E_c \cap E' = \emptyset] \leq e^{-\gamma}$.

We are now ready to define $\gamma{:=}\ln(n\bar{\mu}|K|)$ and get $e^{-\gamma} = \frac{1}{n\bar{\mu}|K|}$. By Lemma 11, H is infeasible if and only if, for any $(u,v) \in K$, there is an ascending $u_0v_{\delta(u,v)}$-cut that is not satisfied. We can estimate this probability as

$$\sum_{(u,v)\in K}\sum_{(A:B)\in\mathcal{C}_{uv}} e^{-\gamma} \leq \bar{\mu}|K| \cdot e^{-\gamma} = \frac{\bar{\mu}|K|}{n\bar{\mu}|K|} = \frac{1}{n}. \tag{6}$$

Hence, our rounded solution $H = (V, E')$ forms a feasible spanner with probability at least $1 - \frac{1}{n}$. The running time of Algorithm 3 is in $\mathcal{O}(\mathrm{LP}(|K|m\bar{\delta}, |K|(m + n\bar{\delta}), w))$; in particular, it is polynomial in n if $\bar{\delta} \in \mathcal{O}(\mathrm{poly}(n))$. Using $\bar{\delta} \leq nL$ and $n \leq m$ yields an upper bound of $\mathcal{O}(\mathrm{LP}(|K|m^2\bar{\ell}, |K|m^2\bar{\ell}, w))$. Thus

Theorem 15. *For (un)directed decoupled freeform spanner problems with integer lengths and polynomially bounded distance demands, Algorithm 3 returns a feasible spanner w.h.p., with expected approximation ratio* $\gamma = \ln(n\bar{\mu}|K|) \leq \ln(n^3(\bar{\delta}+2)^{n-2}) \in \mathcal{O}(n\log\bar{\delta}) \subseteq \mathcal{O}(n\log n)$.

Comparison to Grigorescu et al. [20]. It is worthwhile to compare our algorithm to the currently theoretically strongest known approach [20] for this setting. Amongst other steps, the latter requires the computation of up to $n\log(m\bar{w})$ minimum-weight distance-preserving junction trees. Each such tree requires solving an LP with orders of magnitude more variables and constraints; namely, each such computation requires $\Omega(\mathrm{LP}(|K|^5m^8\bar{\ell}^8, |K|^3m^4\bar{\ell}^4, \max\{\ell, w\}))$ time. Clearly, these computations alone, even disregarding all other necessary subroutines (solving Path-LPs, etc.), drastically exceed our time guarantee. Also, neglecting all additional (expected) weight for covering the remaining terminal pairs, the weight introduced by their constrained shortest path trees is lower bounded by $12 \cdot n^{4/5}\ln n \cdot \mathrm{OPT}$. Compare this to our expected ratio of $\gamma = \ln(n\bar{\mu}|K|) \leq \ln(n^3(\bar{\delta}+2)^{n-2})$. For, e.g., instances with $\bar{\delta} \leq n$, $\sqrt{n}$, $\log n$, or 10, our algorithm expects a superior ratio for all $n \leq 248\,000$, 10^6, 10^9, and 10^{10}, respectively. Clearly, this encompasses many practically relevant instances.

Constant-Bounded Instances. Consider any instance in the scenario of Theorem 15; we say it is *constant-bounded* if the maximum distance demand $\bar{\delta}$ is bounded by some constant.

Corollary 16. *Algorithm 3 is an* $\mathcal{O}(n)$*-approximation for constant-bounded instances.*

Currently, among constant-bounded spanner problems, only undirected unit-length multiplicative 2-spanners (where thus $\delta(u,v) = 2$ for all $\{u,v\} \in K = E$) are known to allow approximation ratios in $\mathcal{O}(\log n)$ [23,24]. We extend this result to all constant-bounded decoupled freeform spanner problems with constant maximum (out)degree Δ, even on directed graphs. Recall that under these restriction, even the undirected basic multiplicative α-spanner problem is **NP**-hard [19]. Further, without restricting the (out)degrees, there cannot be a polylogarithmic approximation ratio unless $\mathbf{NP} \subseteq \mathbf{BPTIME}(n^{\mathrm{polylog}(n)})$ [12].

For $(u, v) \in K$ and $G = (V, E)$, let subgraph $G[uv] = (V[uv], E[uv])$, with $V[uv] \subseteq V$ and $E[uv] \subseteq E$, be induced by all uv-paths P in G with $\ell(P) \leq \delta(u, v)$. Then $G[uv]$ contains all nodes and edges that affect whether terminal pair (u, v) satisfies its distance demand.

Lemma 17. *Every ascending $u_0 v_{\delta(u,v)}$-cut in $\mathcal{C}_{uv}$ in $H^{\bar{\delta}}$ is satisfied if and only if every ascending $u_0 v_{\delta(u,v)}$-cut in $\mathcal{C}'_{uv}$ in $H[uv]^{\bar{\delta}}$ is satisfied.*

Proof. $\mathcal{C}_{uv} \subseteq \mathcal{C}'_{uv}$ gives necessity. For sufficiency, note that no satisfied cut in $\mathcal{C}'_{uv}$ induced by a labeling π of nodes $V[uv]$ can be dissatisfied by expanding π to label further nodes.

By Lemma 17, we only have to choose γ such that all ascending $u_0 v_{\delta(u,v)}$-cuts in $H[uv]^{\bar{\delta}}$ are satisfied w.h.p. Note that $n_{uv} := |V[uv]| \leq \Delta^{\delta(u,v)}$, for every $(u, v) \in K$. Hence, the number of relevant ascending cuts $|\mathcal{C}'_{uv}|$ in $H[uv]^{\bar{\delta}}$ is bounded by $|\mathcal{C}'_{uv}| \leq (\delta(u, v) + 2)^{n_{uv}-2}$. Let $\bar{\mu}' := \max_{(u,v)\in K} |\mathcal{C}'_{uv}|$ the maximum number of relevant cuts for any $(u, v) \in K$. Then, $\gamma = \ln(n\bar{\mu}'|K|) \in \mathcal{O}(\log n + \Delta^{\bar{\delta}} \log \bar{\delta})$ suffices to guarantee feasible solutions w.h.p. Hence, the expected approximation ratio improves with decreasing $\Delta^{\bar{\delta}}$ and, e.g., beats the best known ratio for unit-length multiplicative 3-spanners of $\widetilde{\mathcal{O}}(n^{1/2})$ [13] if $\Delta \in o(n^{1/6})$. Crucially, for constant $\bar{\delta}$ and constant Δ, choosing $\gamma = \ln(n\bar{\mu}'|K|) \in \mathcal{O}(\log(n|K|)) \subseteq \mathcal{O}(\log n)$ suffices.

Corollary 18. *Algorithm 3 is an $\mathcal{O}(\log n)$-approximation for constant-bounded instances with constant maximum (out)degree.*

Choosing $\gamma = \ln(2\bar{\mu}'|K|)$ still yields feasible solutions with probability at least $1 - \frac{1}{2} = \frac{1}{2}$ (see (6)), but no longer w.h.p. Then, for constant-bounded instances with constant maximum (out)degree, if the number of terminal pairs $|K|$ is sublinear in n or constant, the so-modified Algorithm 3 achieves expected weight ratios in $o(\log n)$ or $\mathcal{O}(1)$, respectively.

Acknowledgments. M. Chimani and H. Jasper were supported by the German Research Foundation (DFG), project number 517835933.

References

1. Ahmed, R., et al.: Graph spanners: a tutorial review. Comput. Sci. Rev. **37** (2020)
2. Althöfer, I., Das, G., Dobkin, D., Joseph, D., Soares, J.: On sparse spanners of weighted graphs. Discrete Comput. Geom. **9**(1), 81–100 (1993)
3. Archambault, D., Liotta, G., Nöllenburg, M., Piselli, T., Tappini, A., Wallinger, M.: Bundling-aware graph drawing. In: GD 2024, vol. 320, pp. 15:1–15:19 (2024)
4. Bökler, F., Chimani, M., Jasper, H., Wagner, M.H.: Exact minimum weight spanners via column generation. In: ESA 2024. LIPIcs, vol. 308, pp. 30:1–30:17 (2024)
5. Bökler, F., Chimani, M., Jasper, H.: Simple approximations for general spanner problems (2025). https://arxiv.org/abs/2506.23638

6. Chandra, B., Das, G., Narasimhan, G., Soares, J.: New sparseness results on graph spanners. In: SoCG 1992, pp. 192–201. ACM (1992)
7. Chekuri, C., Hajiaghayi, M.T., Kortsarz, G., Salavatipour, M.R.: Approximation algorithms for nonuniform buy-at-bulk network design. SIAM J. Comput. **39**(5), 1772–1798 (2010)
8. Chimani, M., Spoerhase, J.: Network design problems with bounded distances via shallow-light Steiner trees. In: STACS 2015. LIPIcs, vol. 30, pp. 238–248 (2015)
9. Chimani, M., Stutzenstein, F.: Spanner approximations in practice. In: ESA 2022. LIPIcs, vol. 244, pp. 37:1–37:15 (2022)
10. Chlamtáč, E., Dinitz, M., Kortsarz, G., Laekhanukit, B.: Approximating spanners and directed Steiner forest: upper and lower bounds. ACM Trans. Algorithms **16**(3) (2020)
11. Coppersmith, D., Elkin, M.: Sparse sourcewise and pairwise distance preservers. SIAM J. Discret. Math. **20**(2), 463–501 (2006)
12. Dinitz, M., Kortsarz, G., Raz, R.: Label cover instances with large girth and the hardness of approximating basic k-spanner. ACM Trans. Algorithms **12**(2), 1–16 (2015)
13. Dinitz, M., Krauthgamer, R.: Directed spanners via flow-based linear programs. In: STOC 2011, pp. 323–332. ACM (2011)
14. Dodis, Y., Khanna, S.: Design networks with bounded pairwise distance. In: STOC 1999, pp. 750–759. ACM (1999)
15. Elkin, M., Neiman, O., Solomon, S.: Light spanners. SIAM J. Discret. Math. **29**(3), 1312–1321 (2015)
16. Elkin, M., Peleg, D.: The hardness of approximating spanner problems. Theory Comput. Syst. **41**(4), 691–729 (2007)
17. Garey, M.R., Johnson, D.S.: Computers and Intractability: A Guide to the Theory of NP-Completeness. A Series of Books in the Mathematical Sciences. W. H. Freeman and Company, New York (1979)
18. Garg, N., Konjevod, G., Ravi, R.: A polylogarithmic approximation algorithm for the group Steiner tree problem. J. Algorithms **37**(1), 66–84 (2000)
19. Gómez, R., Miyazawa, F.K., Wakabayashi, Y.: Improved NP-hardness results for the minimum t-spanner problem on bounded-degree graphs. Theoret. Comput. Sci. **947**, 113691 (2023)
20. Grigorescu, E., Kumar, N., Lin, Y.S.: Approximation algorithms for directed weighted spanners. In: APPROX/RANDOM 2023. LIPIcs, vol. 275, pp. 8:1–8:23 (2023)
21. Heinrich, I., Herrala, O., Schiewe, P., Terho, T.: Using light spanning graphs for passenger assignment in public transport. In: ATMOS 2023. OASIcs, vol. 115, pp. 2:1–2:16 (2023)
22. Knudsen, M.B.T.: Additive spanners: a simple construction. In: Ravi, R., Gørtz, I.L. (eds.) SWAT 2014. LNCS, vol. 8503, pp. 277–281. Springer, Cham (2014). https://doi.org/10.1007/978-3-319-08404-6_24
23. Kortsarz, G.: On the hardness of approximating spanners. Algorithmica **30**, 432–450 (2001)
24. Kortsarz, G., Peleg, D.: Generating sparse 2-spanners. J. Algorithms **17**(2), 222–236 (1994)
25. Lorenz, D.H., Raz, D.: A simple efficient approximation scheme for the restricted shortest path problem. OPERRL **28**(5), 213–219 (2001)
26. Peleg, D.: Proximity-preserving labeling schemes and their applications. In: Widmayer, P., Neyer, G., Eidenbenz, S. (eds.) WG 1999. LNCS, vol. 1665, pp. 30–41. Springer, Heidelberg (1999). https://doi.org/10.1007/3-540-46784-X_5

27. Peleg, D., Schäffer, A.: Graph spanners. J. Graph Theory **13**(1), 99–116 (1989)
28. Sigurd, M., Zachariasen, M.: Construction of minimum-weight spanners. In: Albers, S., Radzik, T. (eds.) ESA 2004. LNCS, vol. 3221, pp. 797–808. Springer, Heidelberg (2004). https://doi.org/10.1007/978-3-540-30140-0_70
29. Wojtczak, D.: On strong NP-completeness of rational problems. In: Fomin, F.V., Podolskii, V.V. (eds.) CSR 2018. LNCS, vol. 10846, pp. 308–320. Springer, Cham (2018). https://doi.org/10.1007/978-3-319-90530-3_26
30. Woodruff, D.P.: Lower bounds for additive spanners, emulators, and more. In: IEEE FOCS 2006, pp. 389–398. IEEE (2006)

Hardness of Dynamic Core and Truss Decompositions

Yan S. Couto and Cristina G. Fernandes(✉)

University of São Paulo, São Paulo, SP, Brazil
{yancouto,cris}@ime.usp.br

Abstract. The k-core of a graph is its maximal subgraph with minimum degree at least k, and the core value of a vertex u is the largest k for which u is contained in the k-core of the graph. Among cohesive subgraphs, k-core and its variants have received a lot of attention recently, particularly on dynamic graphs, as reported by Hanauer, Henzinger, and Schulz in their recent survey on dynamic graph algorithms. We answer questions on k-core stated in the survey, proving that there is no efficient dynamic algorithm for k-core or to find $(2-\epsilon)$-approximations for the core values, unless we can improve decade-long state-of-the-art algorithms in many areas including matrix multiplication and satisfiability, based on the established OMv conjecture and SETH. Our results show that there is no dynamic algorithm for k-core asymptotically faster than the trivial ones, and we derive them via proving hardness results for a dynamic version of the circuit value problem. This explains why most recent research papers in this area focus not on a generic efficient dynamic algorithm, but on finding a bounded algorithm, which is fast when few core values change per update. However, we also prove that such bounded algorithms do not exist, based on the OMv conjecture. We also obtain lower bounds for a directed version of the problem, and for the edge variant of the problem, known as k-truss. On the positive side, we present a polylogarithmic dynamic algorithm for 2-core.

Keywords: community search · dynamic graphs · core decomposition

1 Introduction

Graphs are widely used to model data in different applications, and the search for communities, through the extraction of cohesive subgraphs, has a variety of applications in the analysis of social, biological, and other networks [15]. Most classic cohesive subgraphs, such as maximum cliques, quasi-cliques, and k-clans are NP-hard to find or sometimes even to approximate [16]. A looser definition of a cohesive subgraph is the k-core, which is the maximum subgraph of a graph of minimum degree at least k [26]. An efficient $\mathcal{O}(m)$ algorithm exists [3] to compute the cores of a graph on m edges. Due to this, cores have been widely used in many community search applications, and they have been studied extensively [21,31].

J. Matuschke and J. Verschae (Eds.): WAOA 2025, LNCS 16077, pp. 64–80, 2026.
https://doi.org/10.1007/978-3-032-06706-7_5

In the *core decomposition* problem, the task is to compute, for each vertex v, the maximum k for which v is in the k-core, called the *core value* of v. In its approximate version, the task is relaxed, and the goal is to compute numbers that approximate the core values within an interval. There is also a directed version, and a generalized version of k-cores, called k-truss, whose goal is to find subgraphs in which all edges are in at least $k - 2$ triangles.

In many real-world applications, data is constantly changing, and thus there has been increased interest in dynamic graph algorithms, where vertices/edges can be inserted/deleted over time, and one wishes to answer queries on the graph that is being modified. It is undesirable, at each query, to apply a static algorithm from scratch on the current graph. Hanauer et al. [12] recently presented a broad survey on dynamic graph problems that addresses, among others, the k-core, summarizing what was known and stating open questions. There exists a dynamic algorithm for the $(4+\epsilon)$-approximate core decomposition problem that takes polylogarithmic time per graph update and query [27]. Mostly, research has turned to the *core maintenance* problem, in which the core value for all vertices must be maintained explicitly after each update (edge insertion or deletion) in the graph. For this problem, research focused on polynomially bounded algorithms [18,25,34], whose running time per update is bounded by a polynomial on the size of the neighborhood of the vertices whose core value changed due to the update. This is under the assumption that, in practice, on real dynamic graphs, the core values of vertices change very rarely, though there are examples of inputs in which all core values change with every update [27].

Parallel to the search for efficient algorithms, it is desirable to find good lower bounds to either prove that an algorithm is optimal or that it is pointless to look for better algorithms to some problems. The core maintenance problem is shown to be unbounded considering only insertions, when under a specific model of computation [33], as discussed below. The survey by Hanauer et al. [12] explicitly notes that there are no known lower bounds for the k-core and related problems matching their upper bounds.

For other dynamic graph problems, lower bounds are usually achieved by using well-established conjectures. The OMv conjecture [13] states that there is no truly subcubic algorithm for "online" Boolean matrix multiplication (defined in Sect. 4), even when one has arbitrary polynomial preprocessing time. The OMv conjecture is as strong as many other conjectures around matrix multiplication, triangle detection in graphs, and problems such as 3SUM [15]. It is widely used in dynamic graph problems to give conditional lower bounds, such as for reachability, shortest path, and triangle detection problems [12,13]. The well-known SETH [17] states that there is no truly sub-exponential algorithm for k-SAT. This hypothesis has been used to prove complexity results for both NP-hard [6] and polynomial [1,23] problems, including many dynamic graph problems [1]. Falsifying the OMv conjecture or the SETH would improve decade-long state-of-the-art algorithms in several areas.

Our Contributions

We show conditional lower bounds for the dynamic core decomposition problem, as well as for its $(2-\epsilon)$-approximate version, for its directed version, and for the k-truss generalization, with the following results:

- These problems cannot be solved in polylog time per update or query unless dynamic versions of any problem in **P** can be solved in polylog time per update or query.
- Under the OMv conjecture, these problems require $\Omega(m^{\frac{1}{2}-\epsilon})$ time per update or $\Omega(m^{1-\epsilon})$ time per query.
- Under the SETH, these problems require $\Omega(m^{1-\epsilon})$ time per update or query.
- Under the OMv conjecture, no bounded algorithm exists for dynamic core or truss maintenance.

Additionally, we show a polylog algorithm for deciding if the core value of any vertex in a dynamic graph is at least two, so our reductions are tight in a sense, as they use the 3-core.

The rest of the paper is organized as follows. Section 2 provides some preliminaries. Section 3 describes reductions from a dynamic version of the classic circuit value problem to our core and truss decomposition problems. Sections 4 to 6 prove the lower bounds from the OMv conjecture, the SETH, and cell probe models, respectively, while Sect. 7 presents the unboundedness result for core and truss maintenance. Finally, Sect. 8 summarizes the work and highlights open problems in the area. Due to space constraints, the polylog algorithm for the 2-core in dynamic graphs [30] and some of the proofs are left to the full version of the paper.

Related Work

Dynamic Core Decomposition. A polylogarithmic algorithm for the $(4+\epsilon)$-approximate core decomposition on dynamic graphs [27] uses similar techniques previously applied to dynamic densest subgraphs [14]. Another algorithm for the $(4+\epsilon)$-approximate core decomposition was presented by Liu et al. [20], with focus on parallelism and batching queries.

Core Maintenance. A bounded deletion-only algorithm for core maintenance [34] exists, but no bounded algorithm supporting insertions exists. There are, however, progressively better heuristics aiming boundedness: Li et al. [18] showed an algorithm based on the fact that each insertion or deletion changes by one the core value of at most a connected set of vertices. This was later optimized by Sarıyüce et al. [25] with an elaborate graph traversal, and further optimized by Zhang et al. [34] by explicitly storing a deletion order for the vertices using an ordered-list data structure. This idea has since been reused [11] and applied broadly in the area, on parallel algorithms [10], weighted graphs [32], hierarchical core decompositions [7], partial maintenance problems [35], truss maintenance problems [33], etc.

Lower Bounds and Unboundedness. Zhang and Yu [33] proved the unboundedness of core and truss maintenance problems under the locally persistent computation model, a very limited model in which an algorithm can store data only on vertices and move only from a node to an adjacent one. This result was extended to the directed case of truss maintenance by Tian et al. [28]. Even though lower bounds for dynamic graph problems have been studied extensively [1,12,13], to our knowledge, there is no such result for any dynamic core decomposition variant.

2 Preliminaries

We use $G = (V(G), E(G))$ to denote a graph, and uv to denote an edge $\{u, v\} \in E(G)$. The number of edges incident to a vertex is its *degree*. For directed graphs (digraphs), each vertex has both an *in-degree* and an *out-degree*, for incoming and outgoing arcs, respectively. When the (di)graph is clear from the context, we use n and m for its number of vertices and edges (arcs) respectively. For complexity functions, we assume $m = \Omega(n)$.

Core Decomposition Problems. A k-core of a graph G is a maximal subgraph of G with minimum degree at least k. The k-core is unique, otherwise the union of two k-cores would be a larger k-core. The *core value* of a vertex u, denoted as K_u, is the largest k for which u is contained in the k-core. The *core decomposition problem* is the task of determining K_u for each vertex u, and it can be solved in $\mathcal{O}(m)$ time [3] by repeatedly removing a lowest degree vertex.

Dynamic Graph Problems. In a *dynamic graph problem*, we are given a graph subject to edge insertions/deletions, and we have to answer queries about the current state of the graph. We are interested in *partially* and *fully dynamic* variants of the problems. Partially dynamic variants might be *incremental* (insertions-only) or *decremental* (deletions-only), while fully dynamic allow both insertions and deletions. The following is a dynamic version of the core decomposition problem:

Problem 1 (CoreValue). Given a graph, process a series of edge insertions/deletions, and queries: Given a vertex u, what is the value of K_u?

The running time of a dynamic graph algorithm is $\langle p(n,m), u(n,m), q(n,m)\rangle$ if $p(n,m)$ is the time the algorithm takes preprocessing the graph before receiving any updates or queries, $u(n,m)$ is the time per update, and $q(n,m)$ is the time per query.

We say the running time of a dynamic graph algorithm is amortized if the update and the query running times are amortized over all updates and queries. Specifically, the total running time over x updates and y queries is $\mathcal{O}((x + m_0)u(n, m_0) + (y + m_0)q(n, m_0))$, where m_0 is the maximum number of edges at any time.[1]

[1] Our definition of amortization is more lenient than some in the literature, which yields better lower bounds. See [13, Sect. 1.2] for more details.

Approximation Variants. The goal of CoreValue is to answer queries on the core value of any vertex. If an efficient algorithm for that is hard to find, it is natural to consider algorithms that give approximations to the core values.

Given $0 < \alpha_1 \leq 1$ and $\alpha_2 \geq 1$, a value s is an (α_1, α_2)-*approximation* for some non-negative value c if $\alpha_1 c \leq s \leq \alpha_2 c$. We also say s is an $\frac{\alpha_2}{\alpha_1}$-approximation for c.

Lemma 1 (Gap reduction). *If an unknown value c satisfies $0 \leq c \leq X$ or $c \geq Y$, then we can use any α-approximation for c, with $\alpha < \frac{Y}{X}$, to determine whether $c \leq X$ or $c \geq Y$.*

3 Reductions from Circuit Value to Core Decomposition

One of the most basic and complete problems in **P** is the circuit value problem, which is as hard as its bounded monotone version [9]. A *bounded monotone Boolean circuit* is a directed acyclic graph (DAG) with a label on each vertex taken from the set $\{0, 1, \text{AND}, \text{OR}\}$ and a specified output vertex. We refer to vertices and arcs of the DAG as *gates* and *wires*, respectively. The output gate has out-degree 0, 0- and 1-gates have in-degree 0 and out-degree at most 1, and all other gates have in- and out-degree at most 2. The propagation of the values of the 0- and 1-gates through all the AND- and OR-gates results in a 0 or 1 value at each gate, and the value of the output gate is the *circuit value*. AND- and OR-gates with in-degree less than 2 work as if their remaining inputs were 0, and gates with out-degree 2 output the same value on both wires. The *size* of the circuit is its number of gates. Here is a dynamic version of the bounded monotone circuit value problem:

Problem 2 (Monotone circuit value (MCVP)). Given a bounded monotone Boolean circuit and a fixed output gate, process a series of wire insertions/deletions, and queries: Is the circuit value 1?

3.1 Exact Core Decomposition

We will now show how to solve MCVP using an easier variant of CoreValue that only allows queries on the relation between a fixed k and the core value of a fixed vertex.

Problem 3 (k-core). Given a graph with a fixed vertex s^*, process a series of edge insertions/deletions, and queries: Is $K_{s^*} \geq k$?

Proposition 1. *For any $k \geq 3$, if fully (partially) dynamic k-core on a graph on n vertices and m edges can be solved in $\langle p(n,m), u(n,m), q(n,m) \rangle$ time, then fully (partially) dynamic MCVP on a circuit of size N can be solved in $\langle \mathcal{O}(p(N,N)), \mathcal{O}(u(N,N)), \mathcal{O}(q(N,N)) \rangle$ time.*

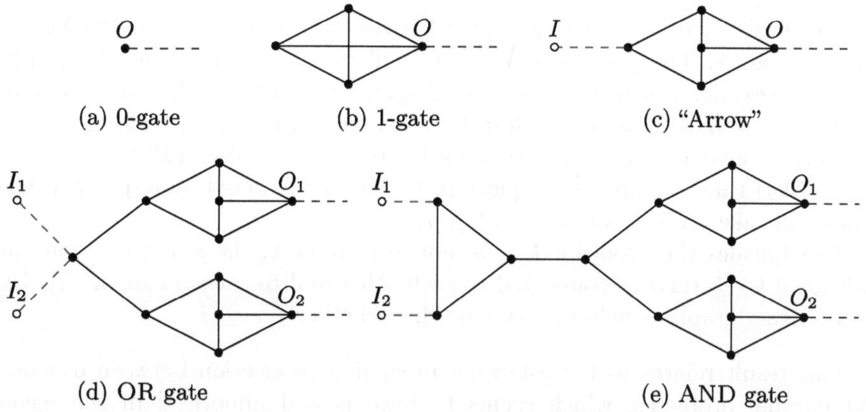

Fig. 1. Building blocks of the reduction from MCVP to 3-CORE. Input edges are red and output edges are blue. Input vertices either do not exist or are output vertices of other gates. (Color figure online)

Proof. Let us consider the case $k = 3$. Given a circuit C with output gate g^*, we will describe a graph G_C with a vertex s^* such that $K_{s*} \geq 3$ if and only if the circuit value is 1.

Each 0-gate in C corresponds to a vertex in G_C, and each 1-gate in C corresponds to a K_4 in G_C with a specified output vertex O, as in Figs. 1a and 1b. Each OR- and AND-gate in C corresponds to a graph as depicted in Figs. 1d and 1e, with their specified input and output vertices. Additionally, for each wire from a gate g to a gate h in C, identify an output vertex of the graph for g and an input vertex of the graph for h, and let this wire correspond to the edge between these two vertices. This completes the description of the graph G_C. Let s^* be an output vertex of the graph for g^*. The insertion or deletion of wires in C will be simulated by the insertion or deletion of the corresponding edge of G_C, preserving the relation between C and G_C.

Let us argue that, at any time, $K_{s^*} \geq 3$ if and only if the circuit value is 1. First note that all vertices in the graph for a 1-gate are in the 3-core of G_C, and no vertex corresponding to a 0-gate is in the 3-core, because it has maximum degree 1. Now note that the arrow graph shown in Fig. 1c, which is used in both the AND- and OR-gates, has the property that its output vertex is in the 3-core of G_C if and only if its input vertex is in the 3-core of G_C. This implies that the output vertices of the graph for an OR-gate are in the 3-core of G_C if and only if one of its two input vertices is in the 3-core of G_C. Similarly, the output vertices of the graph for an AND-gate are in the 3-core of G_C if and only if both its input vertices are in the 3-core of G_C. The way the graphs for the gates of C are connected in G_C implies that, in the same way the 0- and 1-gate values propagate towards the output gate, the fact that all vertices from the 1-gate graph are in the 3-core of G_C propagates the 3-core, reaching vertex s^* if and only if the circuit value is 1.

If C has N gates, then C has $\mathcal{O}(N)$ wires, and clearly G_C has $\mathcal{O}(N)$ vertices and edges. The preprocessing of C will be the construction of G_C plus its preprocessing, which will take $p(\mathcal{O}(N), \mathcal{O}(N)) = \mathcal{O}(p(N, N))$ time, assuming $p(n, m) = \Omega(m)$. Each wire insertion/deletion in C corresponds to a single edge insertion/deletion in G_C, thus each update will take $u(\mathcal{O}(N), \mathcal{O}(N)) = \mathcal{O}(u(N, N))$ time. Finally, each query in C will be answered by a query in G_C, which will take $q(\mathcal{O}(N), \mathcal{O}(N)) = \mathcal{O}(q(N, N))$ time.

This finishes the proof for $k = 3$. For any larger k, the graph G_C contains additional $k - 3$ vertices connected to each other and to each vertex in G_C. Use the resulting graph as before, now querying whether $K_{s^*} \geq k$.

This result relates to the following intriguing connection between dynamic and parallel problems, which seems to have passed unnoticed in the recent years. In 1994, Miltersen et al. [22] defined the following complexity class for dynamic problems. A decision problem π is in **incr-POLYLOGTIME** (we use **IPL**, for short) if, after some polynomial preprocessing time over instance I (a binary string), we can maintain $\pi(I)$ given a sequence of 1-bit updates to I in polylog($|I|$) time per update. A *polylog incremental reduction* from a decision problem π_1 to a decision problem π_2 is a polynomial algorithm that maps each instance I_1 of π_1 to an instance I_2 of π_2 with $\pi_1(I_1) = \pi_2(I_2)$, and a polylogarithmic algorithm that maps a 1-bit update to I_1 to 1-bit updates to I_2, such that $\pi_1(I_1) = \pi_2(I_2)$ after the updates are applied to both instances. A decision problem π is **P**-*complete under* **IPL** *reductions* if π is in **P** and there is a polylog incremental reduction from any problem in **P** to π. This implies that if π is in **IPL** then so is *every* problem in **P**, which is highly unlikely. Most **P**-complete problems (under **NC** reductions), which cannot be parallelized efficiently, are also **P**-complete under **IPL** reductions [22].

The polylog incremental reductions described refer to classic decision problems, but it addresses dynamic versions of these problems, defined only implicitly. The approach in Proposition 1 is different, focused on specific dynamic versions of decision problems, but it can be seen as a polylog incremental reduction from (the static versions of) MCVP to k-CORE. It implies that COREVALUE and k-CORE for any fixed $k \geq 3$ are **P**-*hard under* **IPL** *reductions*, meaning that solving them in $\langle \mathcal{O}(\text{poly}), \mathcal{O}(\text{polylog}), \mathcal{O}(\text{polylog}) \rangle$ time would imply that all problems in **P** are in **IPL**, that is, dynamic versions of *all* problems in **P** would be solvable in $\langle \mathcal{O}(\text{poly}), \mathcal{O}(\text{polylog}), \mathcal{O}(\text{polylog}) \rangle$ time. This is similar to the **P**/**NP** and the **NC**/**P** relation. One might wonder if the same holds for 2-CORE. In the full version, we show that this is unlikely, as there is a $\langle \mathcal{O}(\text{poly}), \mathcal{O}(\text{polylog}), \mathcal{O}(\text{polylog}) \rangle$ time algorithm for 2-CORE.

In the next sections, we use MCVP to prove conditional and unconditional lower bounds for COREVALUE. In the remaining of this section, we reduce MCVP to other k-core variants, which allows us to deduce the same lower bounds for other k-core variants, and it also shows they are **P**-hard under **IPL** reductions as well.

3.2 Approximate Core Decomposition

Proposition 1 and Lemma 1 imply that finding any $(\frac{3}{2} - \epsilon)$-approximation for the core values is also **P**-hard under **IPL** reductions. In fact, it is possible to show an even stronger lower bound for the achievable approximation ratio. Let us first define a simplified approximation version of CoreValue.

Problem 4 (α-ApproxCoreValue). Given a graph with a fixed vertex s^*, process a series of edge insertions/deletions, and queries: Compute an α-approximation for K_{s^*}.

Proposition 2. *For any $\epsilon > 0$, if fully (partially) dynamic $(2 - \epsilon)$-ApproxCoreValue on a graph with n vertices and m edges can be solved in $\langle p(n,m), u(n,m), q(n,m)\rangle$ time, then fully (partially, resp.) dynamic MCVP on a circuit of size N can be solved in $\langle \mathcal{O}(p(N,N)), \mathcal{O}(u(N,N)), \mathcal{O}(q(N,N))\rangle$ time.*

Proof. Given a circuit C with output gate g^*, for any fixed $k \geq 2$, we will describe a graph G_C with a vertex s^* such that $K_{s^*} \geq 2k$ if the circuit value of C is 1 and $K_{s^*} \leq k+1$ otherwise. Then, by Lemma 1, any α-approximation for K_{s*} for $\alpha < \frac{2k}{k+1}$ can be used to determine the circuit value of C.

Each 0-gate in C corresponds to k vertices in G_C with k output edges each, and each 1-gate corresponds to a K_{2k+1} with k^2 output edges, as in Figs. 2a and 2c. Each AND-gate corresponds to a graph like Fig. 2b, with inputs I_1 and I_2, and outputs O_1 and O_2, and it may be generalized as a sequence of $2k$ groups of k vertices with $2k^2$ output edges, one from each of its $2k^2$ vertices. Each OR-gate corresponds to a graph like Fig. 2d, and may be generalized as a sequence of $2k$ groups of k vertices with $2k^2$ output edges, plus a K_{2k+1} connected to each vertex in the last group with k edges.

For each wire from a gate g to a gate h in C, the k^2 output edges from O_1 or O_2 of the graph for g are identified with the k^2 input edges from either I_1 or I_2 of the graph for h, that is, O_1 or O_2 from g will play the role of I_1 or I_2 for h. This completes the description of G_C. Each vertex in the graph for a gate incident to some of its output edges is called an output vertex for the gate. Let s^* be an output vertex in the graph for g^*. The insertion or deletion of wires in C will be simulated by the insertion or deletion of $k^2 = \mathcal{O}(1)$ edges in G_C (from the input to the output of the wire), preserving the relation between C and G_C.

We argue that, at any time, if the circuit value of C is 1, then $K_{s^*} \geq 2k$, otherwise $K_{s^*} \leq k+1$. First note that all vertices in a 1-gate graph are in the $2k$-core of G_C, and no vertex in a 0-gate graph is in the $(k+1)$-core of G_C, because they have maximum degree k. The output vertices of the graph for an AND-gate are in the $2k$-core if the vertices in both I_1 and I_2 are in the $2k$-core, and are not in the $(k+2)$-core if no vertex in I_1 or no vertex in I_2 is in the $(k+2)$-core. An analogous argument holds for an OR-gate (the K_{2k+1} on the right assures that only one between I_1 and I_2 is needed to push all vertices of the graph into the $2k$-core). The fact that all vertices from the 1-gate graph are in the $2k$-core propagates through G_C reaching s^* if the circuit value is 1, otherwise it does

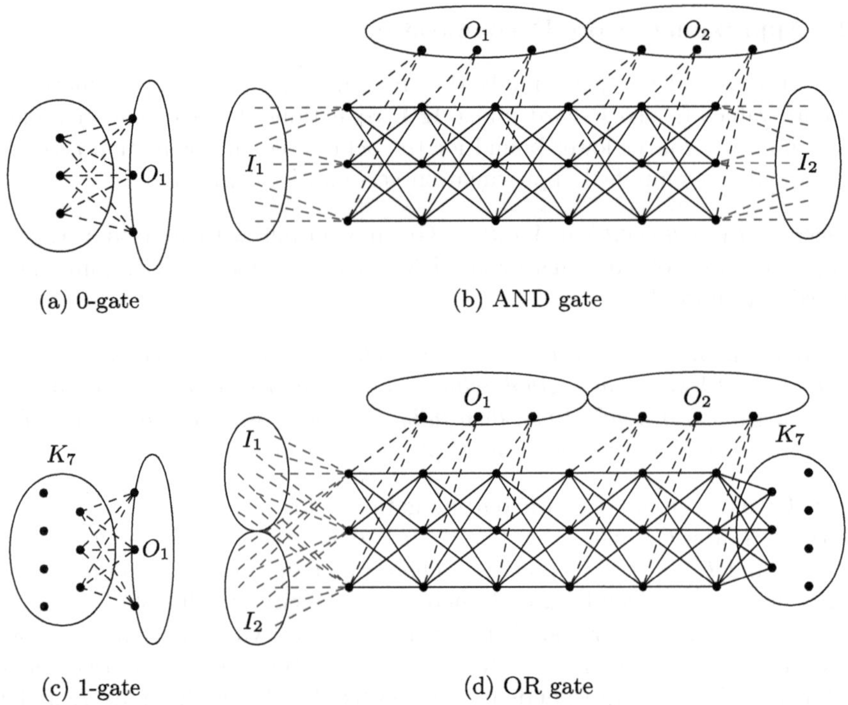

Fig. 2. Building blocks of the reduction from MCVP to $(2-\epsilon)$-*ApproxCoreValue* for $k = 3$.

not propagate and s^* is not in the $(k+2)$-core. This proves the correctness of the reduction.

If C has size N, then G_C will have $\mathcal{O}(Nk^3) = \mathcal{O}(N)$ vertices and edges. Each update in C corresponds to $\mathcal{O}(k^2) = \mathcal{O}(1)$ updates in G_C and each query in C corresponds to a single query in G_C. So MCVP can be solved in $\langle \mathcal{O}(p(N,N)), \mathcal{O}(u(N,N)), \mathcal{O}(q(N,N))\rangle$ time.

This result shows that $(2-\epsilon)$-*ApproxCoreValue* is also **P**-hard under **IPL** reductions, and thus unlikely to have a polylogarithmic algorithm. The bounds on approximation ratios for polylogarithmic algorithms are not yet tight. Indeed, we have shown a polylogarithmic algorithm with ratio $2-\epsilon$ is unlikely to exist, Sun et al. [27] gave a polylogarithmic algorithm with ratio $4+\epsilon$, and it is still not known whether a polylogarithmic algorithm with ratio between 2 and 4 exists. Sun et al. [27] also gave a polylogarithmic algorithm for the incremental $(2+\epsilon)$, so the gap for incremental algorithms is much smaller.

These bounds are somewhat similar to the ones by Anderson and Mayr [2] for **NC** approximation algorithms for core value on static graphs, who show an **NC** algorithm with ratio $2-\epsilon$ is unlikely to exist and an **NC** algorithm with ratio $2+\epsilon$ exists. This highlights the similarity between hardness to parallelize (**P**-hardness) and hardness to dynamize (**P**-hard under **IPL** reductionsness), a topic explored by Miltersen et al. [22]. We believe this connection, that seems to have been forgotten, should be investigated more deeply.

3.3 Truss Decomposition

The notion of k-cores and core value can be generalized to edges [4,24] instead of vertices. The k-truss of a graph is its maximal subgraph in which each edge is in at least $k-2$ triangles. This definition captures a stricter notion of connectedness, and identifies denser cohesive subgraphs than k-cores. For example, the k-truss of a graph is contained in its $(k-1)$-core, but the opposite does not necessarily hold. It is possible to define truss value and a TRUSSVALUE problem analogous to COREVALUE, and an $\mathcal{O}(m^{1.5})$ algorithm [29] exists for truss decomposition. In the full version of the paper, we prove lower bounds on a weaker version of this problem as follows.

Problem 5 (k-TRUSS). Given a graph G with a fixed edge e^*, process a series of edge insertions/deletions, and queries: Is e^* in the k-truss of G?

Proposition 3. *For any $k \geq 4$, if fully (partially) dynamic k-TRUSS on a graph with n vertices and m edges can be solved in $\langle p(n,m), u(n,m), q(n,m)\rangle$ time, then fully (partially) dynamic MCVP on a circuit of size N can be solved in $\langle \mathcal{O}(p(N,N)), \mathcal{O}(u(N,N)), \mathcal{O}(q(N,N))\rangle$ time.*

3.4 Directed Core Decomposition

Giatsidis et al. [8] extended the notion of k-cores to directed graphs. The (k,l)-core of a directed graphs is its maximal subgraph in which each vertex has in-degree at least k and out-degree at least l, and the (k,l)-CORE problem is defined analogously to k-CORE. Proposition 1 trivially holds for $(3,3)$-CORE as well, but in the full version we show this result for $(2,0)$-CORE.

4 OMV-Based Lower Bounds

In the OMV problem for "online" $N \times N$ Boolean matrix multiplication, one is given the first matrix, and then the columns of the second matrix one at a time. The first matrix may be preprocessed, and the task is to output the multiplication of each column vector by the matrix before receiving the next column. In Boolean matrix multiplication, AND plays the role of multiplication and OR plays the role of addition. The OMv conjecture essentially states that there is no algorithm faster than the trivial $\Theta(N^3)$ one by a polynomial factor.

Conjecture 1 (OMv [13]). For any $\epsilon > 0$, there is no algorithm for OMV with polynomial preprocessing time and $\mathcal{O}(N^{3-\epsilon})$ computation time, with an error probability of at most $\frac{1}{3}$.

We will work with a simpler variant of the same problem which makes reductions easier, but that is as hard as the original problem [13].

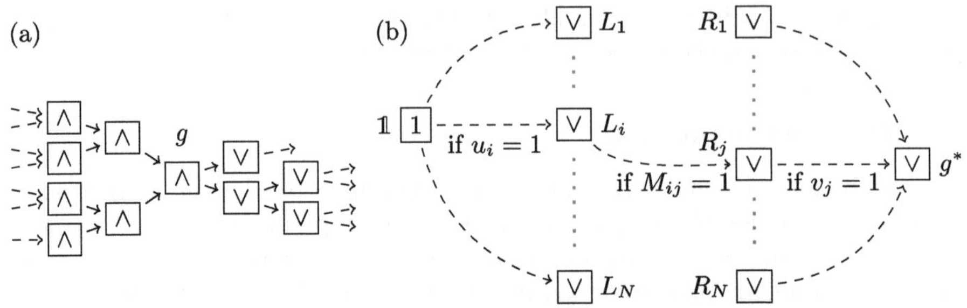

Fig. 3. (a) Simulating an AND-gate with in-degree 7 and out-degree 5. (b) Sketch of $C_M(u, v)$, omitting binary trees that simulate unbounded in- and out-degree.

Problem 6 (OuMv). Given an $N \times N$ Boolean matrix M and an online sequence of Q Boolean vectors pairs $(u^1, v^1), \ldots, (u^Q, v^Q)$, compute each $(u^k)^\top M v^k$ before receiving (u^{k+1}, v^{k+1}). Let $p(N)$ be the preprocessing time before receiving the first vector pair and $c(N, Q)$ be the computation time.

We say OuMv has *size* N and Q queries. Notice that $u^\top M v = 1$ if and only if there exist $i, j \in [N]$ such that $u_i = M_{ij} = v_j = 1$, and (i, j) is called a *witness* in this case.

Proposition 4 (Theorem 2.7 in [13]). *For any $\epsilon > 0$, there is no algorithm for OuMv of size N and Q queries with $p(N) = \text{poly}(N)$ preprocessing time and $c(N, Q) = \mathcal{O}(N^{2-\epsilon}Q + N^2Q^{1-\epsilon})$ computation time unless the OMv conjecture is false.*

In this section, we use the OMv conjecture to prove lower bounds for MCVP and thus CoreValue. We start by arguing that it is possible to simulate gates with unbounded in- or out-degree, which will make our reductions simpler to understand. Note the AND and OR functions are commutative and associative, so their inputs may be combined in any order. Thus, we can replace any gate g of fixed maximum in-degree d with a leaves-to-root balanced[2] binary tree made of gates of the same type as g and with $\lceil \frac{d}{2} \rceil$ leaves, so at most d gates. Then we connect input wires to the leaf gates and the value of the root gate will always have the same value as g. Similarly, for unbounded out-degree, we use a root-to-leaves binary tree of OR-gates, and connect output wires to the leaves, which will all have the same value as g. See Fig. 3a.

Proposition 5. *If fully (partially) dynamic MCVP on a circuit of size n can be solved in $\langle p(n), u(n), q(n) \rangle$ amortized (worst-case, resp.) time, then OuMv of size N and Q queries can be solved in $\mathcal{O}(p(N^2))$ preprocessing time and $\mathcal{O}((NQ + N^2)u(N^2) + (Q + N^2)q(N^2))$ computation time.*

[2] In fact, the tree does not need to be balanced for this to work, but it does keep the depth of the circuit logarithmic.

Proof. Given an $N \times N$ Boolean matrix M, we will describe a circuit C_M with output gate g^* and, when given a vector pair (u, v), we will show how to modify C_M to build $C_M(u, v)$ such that the circuit value of $C_M(u, v)$ is 1 if and only if $u^\top M v = 1$. See Fig. 3b.

Build C_M with a 1-gate $\mathbb{1}$ and OR-gates $L_1, \ldots, L_N, R_1, \ldots, R_N$, and g^*. Add wire $L_i R_j$ for each $i, j \in [N]$ such that $M_{ij} = 1$. To build $C_M(u, v)$ from C_M, u, and v, add wires $\mathbb{1}L_i$ for each $i \in [N]$ with $u_i = 1$ and wires $R_j g^*$ for each $j \in [N]$ with $v_j = 1$. This concludes the description of C_M and $C_M(u, v)$. Note that C_M and $C_M(u, v)$ can be simulated with $\mathcal{O}(N^2)$ gates of bounded in- and out-degree, and $C_M(u, v)$ adds at most $2N$ wires to C_M.

The circuit value of $C_M(u, v)$ is 1 if and only if a directed path exists from the single 1-gate $\mathbb{1}$ to g^*, as other gates are OR-gates. This occurs if and only if there are $i, j \in [N]$ for which wires $\mathbb{1}L_i$, $L_i R_j$, and $R_j g^*$ exist, i.e., $u_i = M_{ij} = v_j = 1$, so (i, j) is a witness to $u^\top M v = 1$.

For MCVP, build C_M and preprocess it in $\mathcal{O}(p(N^2))$ time. Then, after receiving each vector pair (u, v), build $C_M(u, v)$ by adding at most $2N$ wires, perform a single query to decide whether $u^\top M v = 1$, and finally delete at most $2N$ wires to recover C_M. Do this for each of the Q vector pairs, for a total of $\mathcal{O}(NQ)$ updates and Q queries, which by our definition of amortization takes $\mathcal{O}((NQ + N^2)u(N^2) + (Q + N^2)q(N^2))$ computation time.

For the incremental case of MCVP, proceed the same way but, when building $C_M(u, v)$ for a vector pair (u, v), record all changes done to the data structures of the algorithm, which sum up to $\mathcal{O}(Nu(N^2) + q(N^2))$ and then undo the changes (rollback) to restore C_M.[3] For the decremental case, before preprocessing, add wires $\mathbb{1}L_i$ and $R_j g^*$ for each $i, j \in [N]$ to C_M. Then, to build $C_M(u, v)$, instead delete all wires $\mathbb{1}L_i$ for each $i \in [N]$ such that $u_i = 0$, and $R_j g^*$ for each $j \in [N]$ such that $v_j = 0$. Finally, use the same rollback technique to restore the initial circuit. Both partially dynamic cases take $\mathcal{O}(Nu(N^2) + q(N^2))$ per vector pair, and thus take $\mathcal{O}(NQu(N^2) + Qq(N^2))$ computation time.

By using this reduction, we are able to obtain lower bounds on MCVP and on 3-CORE using the OMv conjecture.

Theorem 1. *For any $\epsilon > 0$, there exists no algorithm for the fully (partially) dynamic MCVP on a circuit of size n with $\langle \text{poly}(n), \mathcal{O}(n^{\frac{1}{2}-\epsilon}), \mathcal{O}(n^{1-\epsilon}) \rangle$ amortized (worst-case, resp.) time, unless the OMv conjecture is false.*

Proof. Suppose there is an algorithm $\mathcal{A}$ for MCVP with complexity $\langle \text{poly}(n), \mathcal{O}(n^{\frac{1}{2}-\epsilon}), \mathcal{O}(n^{1-\epsilon}) \rangle$. Choose $Q = N^2$ and, by Proposition 5, we can solve OuMv with size N and Q queries with $\mathcal{O}(\text{poly}(N^2)) = \text{poly}(N)$ preprocessing time and $\mathcal{O}((NQ + N^2)u(N^2) + (Q + N^2)q(N^2)) = \mathcal{O}(NQN^{1-2\epsilon} + QN^{2-2\epsilon}) = \mathcal{O}(N^{2-\epsilon}Q)$, computation time which, by Proposition 4, falsifies the OMv conjecture.

By directly applying Propositions 2 and 3, we derive the following.

[3] This rewinding forbids an amortized analysis over the whole sequence of insertions, so we require a worst-case time bound for partially dynamic cases, otherwise building $C_M(u, v)$ could take $\Theta(N^2 u(N^2))$ time.

Corollary 1. *For any ϵ, $\delta > 0$, there exists no algorithm for fully (partially) dynamic* TrussValue *or* $(2-\epsilon)$-ApproxCoreValue *on a graph on n vertices and m edges with $\langle \text{poly}(n,m), \mathcal{O}(m^{\frac{1}{2}-\delta}), \mathcal{O}(m^{1-\delta})\rangle$ amortized (worst-case, resp.) time, unless the OMv conjecture is false.*

5 SETH-Based Lower Bounds

In the k-SAT problem, we are given a k-CNF formula on N variables and M clauses, that is, the formula is a conjunction where all clauses have k literals. The task is to determine whether the formula is satisfiable by some assignment for its variables. The strong exponential time hypothesis (SETH) essentially states that there is no algorithm for SAT faster than the trivial $\Theta^*(2^N)$ one[4] by an exponential factor:

Conjecture 2 (SETH [17]*).* For every $\epsilon > 0$, there is a k such that there is no algorithm for k-SAT with $\mathcal{O}^*(2^{(1-\epsilon)N})$ running time.

By using the SETH, we can show stronger conditional lower bounds, indicating that there is essentially no algorithm for MCVP or CoreValue better than the trivial one, in which we just apply the static algorithm to the current circuit/graph to answer every query.

Theorem 2. *For any $\epsilon > 0$, there exists no algorithm for the fully (partially) dynamic MCVP on a circuit of size n with $\langle \text{poly}(n), \mathcal{O}(n^{1-\epsilon}), \mathcal{O}(n^{1-\epsilon})\rangle$ amortized (worst-case, resp.) time, unless the SETH is false.*

By directly applying Propositions 2 and 3, we derive the following.

Corollary 2. *For any ϵ, $\delta > 0$, there exists no algorithm for fully (partially) dynamic* TrussValue *or* $(2-\epsilon)$-ApproxCoreValue *on a graph on n vertices and m edges with $\langle \text{poly}(n,m), \mathcal{O}(m^{1-\delta}), \mathcal{O}(m^{1-\delta})\rangle$ amortized (worst-case, resp.) time, unless the SETH is false.*

6 Unconditional Lower Bounds

The other lower bounds in this paper are either based on famous conjectures such as OMv, SETH, or the fact that **P**-hard under **IPL** reductions problems are unlikely to have an efficient algorithm. However, we can prove, based only on the memory accesses, there exists no algorithm for fully dynamic $(2-\epsilon)$-*ApproxCoreValue* or TrussValue on a graph on n vertices and m edges with $\langle \text{poly}(n,m), o(\frac{\lg m}{\lg\lg m}), o(\frac{\lg m}{\lg\lg m})\rangle$ time. We use a well-known technique based on the *dynamic XOR prefix* problem [22]. The proof is left to the full version.

[4] Recall that $\mathcal{O}^*$ omits polynomial factors, that is, $\mathcal{O}^*(f(n)) = \mathcal{O}(f(n)\,\text{poly}(n))$. The same holds for Θ^*.

7 Core Maintenance Unboundedness

Many papers on dynamic k-core in graphs [11,25,35] approach the problem from a different perspective. They consider a *maintenance* version of the problem.

Problem 7 (Full Core Maintenance (FCM) [35]). Given a graph, process a series of edge insertions/deletions, keeping the core values of all vertices explicitly and up to date.

For any edge insertion or deletion on a graph G, its *affected set* $V^* \subseteq V(G)$ is the set of vertices of the inserted or deleted edge together with any vertices which had their core value changed due to the operation. Let $||V^*||_d$ be the number of vertices within distance d in G from any vertex in V^*. An algorithm $\mathcal{A}$ for FCM is *bounded* if there exist a constant $d \in \mathbb{N}$ and a non-decreasing function f such that $\mathcal{A}$ takes at most $f(||V^*||_d)$ time on any operation with affected set V^*. Algorithm $\mathcal{A}$ is *polynomially bounded* if f is a polynomial function. A problem is called *bounded* if a bounded algorithm exists for it, and *unbounded* otherwise. The idea is that operations should take time proportional to how much they affect the core values. The disadvantage is that we always explicitly store the core values of all vertices, so operations that change the core value of many vertices may be slow even in a bounded algorithm. This is similar to what we call "output-sensitive algorithms".

Zhang and Yu [33] proved that, under the locally persistent (LP) model of computation, incremental FCM is unbounded, but decremental FCM is bounded. The LP model, however, is not a very strong model of computation, as it disallows global data or pointers from a vertex to other non-adjacent vertices. Algorithms like the one in [34] are *not* in LP, so this result does not show that they could not be turned into a bounded algorithm. We will instead adopt the RAM model [5], which is much more generic and stronger, and we will show the unboundedness of incremental FCM under the OMv conjecture.

Theorem 3. *There exists no bounded algorithm in the RAM model for incremental* FCM*, unless the OMv conjecture is false.*

We can also extend the unboundedness result for the truss maintenance problem, as below, and also for directed and $(2-\epsilon)$-approximate core maintenance.

Problem 8 (Full truss maintenance (FTM) [33]). Given a graph, process a series of edge insertions/deletions, keeping the truss values of all edges explicitly and up to date.

Theorem 4. *There exists no bounded algorithm in the RAM model for incremental* FTM*, unless the OMv conjecture is false.*

Theorems 3 and 4 show that the core and truss maintenance problems are unlikely to be bounded, even exponentially. The current line of research has been attempting to obtain a bounded algorithm for core maintenance, however, this is fated to fail. An instance that proves the unboundedness of the state-of-the-art core maintenance algorithm is left to the full version. The algorithms in [11,25,34,35] are useful, and work fast in practice, but they inevitably take $\tilde{\Omega}(n)$ time in the worst case, even when no vertex changes core value, and our results show this will likely always be the case.

8 Conclusions

We investigated several questions on k-core and related problems, some coming from the survey of Hanauer et al. [12], and derived strong hardness results for most of these problems. We proved lower bounds for dynamic core and truss decomposition problems. We proved that, for any $\epsilon > 0$, even with arbitrary polynomial preprocessing, $(2-\delta)$-*ApproxCoreValue* and TrussValue need $\Omega(m^{1-\epsilon})$ time for updates or queries under SETH, and $\Omega(n^{\frac{1}{2}-\epsilon})$ time for updates or $\Omega(n^{1-\epsilon})$ time for queries under the OMv conjecture. We also proved that FCM and FTM are unbounded under the OMv conjecture. We are currently investigating yet another variant of k-core, on a bipartite setting [19], for which we believe similar hardness results hold. On the positive side, we designed a novel polylogarithmic algorithm for determining if a given vertex is in the 2-core.

Several open questions remain on dynamic core decomposition and maintenance problems, a few being:

- Are there better approximations? The best polylog approximation has ratio $4+\epsilon$ [27], and our lower bound only prohibits a ratio of $2-\epsilon$. A possible line of investigation is to try to design a 2- or 4-approximation that runs in polylogarithmic time, or prove no such algorithm exists.
- Is it possible to find an instance that is unbounded for *any* core maintenance algorithm?

Acknowledgments. This study was financed, in part, by the São Paulo Research Foundation (FAPESP), Brazil, grants number 2024/18049-0 and 2023/03167-5, by Conselho Nacional de Desenvolvimento Científico e Tecnológico (CNPq), Brazil, grants 310979/2020-0 and 404315/2023-2, and by Coordenação de Aperfeiçoamento de Nível Superior (CAPES), Brazil, Finance Code 001.

Disclosure of Interests. The authors have no competing interests to declare that are relevant to the content of this article.

References

1. Abboud, A., Williams, V.V.: Popular conjectures imply strong lower bounds for dynamic problems. In: IEEE 55th Annual Symposium on Foundations of Computer Science (FOCS), pp. 434–443 (2014). https://doi.org/10.1109/FOCS.2014.53
2. Anderson, R., Mayr, E.W.: A P-complete problem and approximations to it. Technical report, Stanford University (1984)

3. Batagelj, V., Zaveršnik, M.: An $\mathcal{O}(m)$ algorithm for cores decomposition of networks (2003). http://arxiv.org/abs/cs/0310049
4. Cohen, J.: Trusses: cohesive subgraphs for social network analysis. Natl. Secur. Agency Tech. Rep. **16**(3.1), 1–29 (2008)
5. Cormen, T.H., Leiserson, C.E., Rivest, R.L., Stein, C.: RAM model. In: Introduction to Algorithms, 4th edn, pp. 25–26 (2022)
6. Cygan, M., et al.: On problems as hard as CNF-SAT. ACM Trans. Algorithms **12**(3), 1–24 (2016). https://doi.org/10.1145/2925416
7. Gabert, K., Pinar, A., Çatalyürek, U.V.: Batch dynamic algorithm to find k-core hierarchies. In: 5th ACM SIGMOD Joint International Workshop on Graph Data Management Experiences & Systems (GRADES) and Network Data Analytics (NDA), pp. 1–10 (2022). https://doi.org/10.1145/3534540.3534694
8. Giatsidis, C., Thilikos, D.M., Vazirgiannis, M.: D-cores: measuring collaboration of directed graphs based on degeneracy. In: IEEE 11th International Conference on Data Mining (ICDM), pp. 201–210 (2011). https://doi.org/10.1109/ICDM.2011.46
9. Greenlaw, R., Hoover, H.J., Ruzzo, W.L.: Limits to Parallel Computation: P-Completeness Theory. Oxford University Press (1995)
10. Guo, B., Sekerinski, E.: Parallel order-based core maintenance in dynamic graphs. In: 52nd International Conference on Parallel Processing (ICPP), pp. 122–131 (2023). https://doi.org/10.1145/3605573.3605597
11. Guo, B., Sekerinski, E.: Simplified algorithms for order-based core maintenance. J. Supercomput. **80**(13), 19592–19623 (2024). https://doi.org/10.1007/s11227-024-06190-x
12. Hanauer, K., Henzinger, M.R., Schulz, C.: Recent advances in fully dynamic graph algorithms – a quick reference guide. ACM J. Exp. Algorithmics **27**, 1–45 (2022). https://doi.org/10.1145/3555806
13. Henzinger, M.R., Krinninger, S., Nanongkai, D., Saranurak, T.: Unifying and strengthening hardness for dynamic problems via the online matrix-vector multiplication conjecture. In: 47th Annual ACM Symposium on Theory of Computing (STOC), pp. 21–30 (2015). https://doi.org/10.1145/2746539.2746609
14. Hu, S., Wu, X., Chan, T.H.H.: Maintaining densest subsets efficiently in evolving hypergraphs. In: 2017 ACM on Conference on Information and Knowledge Management (CIKM), pp. 929–938 (2017). https://doi.org/10.1145/3132847.3132907
15. Huang, X., Lakshmanan, L.V.S., Xu, J.: Community Search over Big Graphs. Synthesis Lectures on Data Management (2019). https://doi.org/10.1007/978-3-031-01874-9
16. Håstad, J.: Clique is hard to approximate within $n^{1-\epsilon}$. Acta Math. **182**(1), 105–142 (1999). https://doi.org/10.1007/BF02392825
17. Impagliazzo, R., Paturi, R.: On the complexity of k-SAT. J. Comput. Syst. Sci. **62**(2), 367–375 (2001). https://doi.org/10.1006/jcss.2000.1727
18. Li, R.H., Yu, J.X., Mao, R.: Efficient core maintenance in large dynamic graphs. IEEE Trans. Knowl. Data Eng. **26**(10), 2453–2465 (2014). https://doi.org/10.1109/TKDE.2013.158
19. Liu, B., Yuan, L., Lin, X., Qin, L., Zhang, W., Zhou, J.: Efficient (α, β)-core computation in bipartite graphs. VLDB J. **29**(5), 1075–1099 (2020). https://doi.org/10.1007/s00778-020-00606-9
20. Liu, Q.C., Shi, J., Yu, S., Dhulipala, L., Shun, J.: Parallel batch-dynamic algorithms for k-core decomposition and related graph problems. In: 34th ACM Symposium on Parallelism in Algorithms and Architectures (SPAA), pp. 191–204 (2022). https://doi.org/10.1145/3490148.3538569

21. Malliaros, F.D., Giatsidis, C., Papadopoulos, A.N., Vazirgiannis, M.: The core decomposition of networks: theory, algorithms and applications. VLDB J. **29**(1), 61–92 (2019). https://doi.org/10.1007/s00778-019-00587-4
22. Miltersen, P.B., Subramanian, S., Vitter, J.S., Tamassia, R.: Complexity models for incremental computation. Theoret. Comput. Sci. **130**(1), 203–236 (1994). https://doi.org/10.1016/0304-3975(94)90159-7
23. Pătraşcu, M., Williams, R.: On the possibility of faster SAT algorithms. In: 21st Annual ACM-SIAM Symposium on Discrete Algorithms (SODA), pp. 1065–1075 (2010). https://doi.org/10.1137/1.9781611973075.86
24. Saito, K., Yamada, T.: Extracting communities from complex networks by the k-dense method. In: Sixth IEEE International Conference on Data Mining - Workshops (ICDMW 2006), pp. 300–304 (2006). https://doi.org/10.1109/ICDMW.2006.76
25. Sarıyüce, A.E., Gedik, B., Jacques-Silva, G., Wu, K.-L., Çatalyürek, Ü.V.: Incremental k-core decomposition: algorithms and evaluation. VLDB J. **25**(3), 425–447 (2016). https://doi.org/10.1007/s00778-016-0423-8
26. Seidman, S.B.: Network structure and minimum degree. Soc. Netw. **5**(3), 269–287 (1983). https://doi.org/10.1016/0378-8733(83)90028-X
27. Sun, B., Chan, T.H.H., Sozio, M.: Fully dynamic approximate k-core decomposition in hypergraphs. ACM Trans. Knowl. Discov. Data **14**(4), 39:1–39:21 (2020). https://doi.org/10.1145/3385416
28. Tian, A., Zhou, A., Wang, Y., Chen, L.: Maximal D-truss search in dynamic directed graphs. Proc. VLDB Endow. **16**(9), 2199–2211 (2023). https://doi.org/10.14778/3598581.3598592
29. Wang, J., Cheng, J.: Truss decomposition in massive networks. Proc. VLDB Endow. **5**(9), 812–823 (2012). https://doi.org/10.14778/2311906.2311909
30. Yan Soares Couto: Dynamic 2-core (2024). https://github.com/yancouto/phd/tree/main/dynamic_2core
31. Zhang, F., Liu, B., Fang, Q.: Core decomposition, maintenance and applications. In: Du, D.-Z., Wang, J. (eds.) Complexity and Approximation. LNCS, vol. 12000, pp. 205–218. Springer, Cham (2020). https://doi.org/10.1007/978-3-030-41672-0_12
32. Zhang, F., Liu, B., Liu, Z., Fang, Q.: Order based algorithms for the core maintenance problem on edge-weighted graphs. Theoret. Comput. Sci. **941**, 140–155 (2023). https://doi.org/10.1016/j.tcs.2022.11.008
33. Zhang, Y., Yu, J.X.: Unboundedness and efficiency of truss maintenance in evolving graphs. In: 2019 International Conference on Management of Data (SIGMOD), pp. 1024–1041 (2019). https://doi.org/10.1145/3299869.3300082
34. Zhang, Y., Yu, J.X., Zhang, Y., Qin, L.: A fast order-based approach for core maintenance. In: IEEE 33rd International Conference on Data Engineering (ICDE), pp. 337–348 (2017). https://doi.org/10.1109/ICDE.2017.93
35. Zhang, Y., Yu, J.X., Zhang, Y., Qin, L.: Maintaining top-t cores in dynamic graphs. IEEE Trans. Knowl. Data Eng. **36**(9), 4766–4780 (2024). https://doi.org/10.1109/TKDE.2023.3332638

Controlling Tail Risk in Two-Slope Ski Rental

Qiming Cui and Michael Dinitz(✉)

Johns Hopkins University, Baltimore, MD, USA
{qcui6,mdinitz}@jhu.edu

Abstract. We study the optimal solution to a general two slope ski rental problem with a tail risk, i.e., the chance of the competitive ratio exceeding a specific value γ is bounded by a constant δ. This extends the recent study of tail bounds for ski rental initiated by [Dinitz et al. SODA 2024] to the two-slope version defined by [Lotker et al. IPL 2008]. In this version, even after "buying" we must still pay a rental cost at each time step, but the rental cost is lower after buying. This models many real-world "rent-or-buy" scenarios where a one-time investment decreases (but does not eliminate) the per-time cost. Despite this being a simple extension of the classical problem, we find that adding tail risk bounds creates a fundamentally different solution structure. For example, in our setting there is a possibility that we never buy in an optimal solution (which can also occur without tail bounds), but more strangely (and unlike the case without tail bounds or the classical case with tail bounds) we also show that the optimal solution might need to have nontrivial probabilities of buying even at finite points beyond the time corresponding to the buying cost. Moreover, in many regimes there does not exist a unique optimal solution. As our first contribution, we develop a series of structure theorems to characterize some features of optimal solutions.

The complex structure of optimal solutions makes it more difficult to develop an algorithm to compute such a solution. As our second contribution, we utilize our structure theorems to design two algorithms: one based on a greedy algorithm combined with binary search that is fast but yields arbitrarily close to optimal solutions, and a slower algorithm based on linear programming which computes exact optimal solutions.

Keywords: Ski Rental · Online Algorithms · Tail Bounds

1 Introduction

The ski rental problem is one of the most fundamental problems in online algorithms, modeling the fundamental "rent-or-buy" decision making task that arises constantly in real decision-making settings. In this problem there is an unknown number of ski days, and before each day we can choose to either rent skis for the day for \$1 or buy skis for \$b. If we have perfect knowledge then this problem is

J. Matuschke and J. Verschae (Eds.): WAOA 2025, LNCS 16077, pp. 81–96, 2026.
https://doi.org/10.1007/978-3-032-06706-7_6

trivial: we rent each day if the number of ski days is at most b, and otherwise we buy skis at the beginning of day 1. The difficulty arises from *not* knowing the number of ski days in advance, and instead being forced to make our rent/buy decision *online*. Ski rental has been the object of intensive study since its introduction [7], and has long been known to admit a deterministic 2-competitive algorithm (an algorithm that never pays more than twice the optimal cost in hindsight) [7] and a randomized algorithm that is $\frac{e}{e-1} \approx 1.58$-competitive in expectation [6]. Note that a randomized algorithm is simply a probability distribution over days; we will call this the *purchase distribution* of the algorithm.

Recently Dinitz, Im, Lavastida, Moseley, and Vassilvitskii [3] initiated the study of *tail bounds* for the classical randomized algorithm. They showed that the classical algorithm does not exhibit concentration, and has a reasonably large probability of actually performing worse than the classical deterministic algorithm. To rectify this, they considered algorithms that minimize the expected competitive ratio *subject* to satisfying explicit tail bounds which they called (γ, δ)-constraints: they required that their algorithm have probability at most δ of having competitive ratio worse than γ. They were able to construct an (efficient) algorithm to compute the optimal purchase distribution for any collection of such constraints, but perhaps more surprisingly, they showed that optimal purchase distributions satisfying even a single (γ, δ)-constraint exhibited dramatically different structure from the classical randomized algorithm (without tail bounds). For example, while the classical algorithm is a simple exponentially increasing distribution in $[1, b]$, with a single $(2, \delta)$-constraint the optimal purchase distribution becomes non-monotonic, can have arbitrarily many discontinuities, and even where there are not discontinuities the purchase distribution can grow arbitrarily quickly (i.e., have arbitrarily large derivative).

Ski rental is a fundamental problem because it abstracts a class of problems in which we can either pay a (small) repeating cost (the rental fee) or can pay a single one-time cost (the buying fee). However, many problems exhibit a different but extremely related structure: one can again pay a repeating cost, or one can pay a single one-time cost which, rather than eliminating the rental fee, instead *reduces it*. This models, for example, any manufacturing setting where one can buy a machine, piece of infrastructure, or invest in a piece of technology that makes some process more efficient: we still need to pay every day to run the better machine, but since it is more efficient we pay less per output. It is again trivial to determine whether to buy such an improved machine if we know the number of days that the factory will run, and the difficulty comes from *not knowing* this time horizon.

In order to model this, Lotker, Patt-Shamir, and Rawitz [9] introduced the *two-option generalized ski rental problem*. As usual, we first switch to a continuous setting where $b = 1$ and time is continuous starting at 0. Then the classical ski rental setting can be thought of as having a choice of two states, each of which incurs some cost per time to be in: state 1, which costs \$1 per time to be in, and state 2, which costs \$0 per time to be in. Moreover, it costs \$1 to switch from state 1 to state 2. In other words, the two states can be modeled as *lines*: if we spend t time in state 1 then we pay $1 \cdot t + 0$, while if we spend t time in state 2 we pay $0 \cdot t + 1$. Lotker et al. [9] generalized this to allow the second state to *also* have some cost $0 \leq a \leq 1$ per time and switch a switching cost of $1 - a$, corresponding to the line $a \cdot t + (1 - a)$. Note that any algorithm for this problem is similarly a purchase distribution that just specifies a distribution over when to switch from state 1 to state 2. Their main result was an optimal randomized algorithm that is $\frac{e}{e-1+a}$-competitive in expectation for this problem. Moreover, the optimal purchase distribution corresponding to this algorithm is remarkably similar to the randomized algorithm for the classical ski rental problem: it is essentially a scaled down version of the same exponential distribution in $[0, 1]$, but with some extra probability mass on the "never switch" option (corresponding to placing probability mass at time ∞).

By combining these two lines of work we get a number of interesting and important questions. For the two-state ski rental problem in the presence of tail bounds, can we still design an algorithm to find the optimal purchase distribution, as in Dinitz et al. [3]? Since the structure of the two-state solution [9] and the classical ("pure buy") solution [6] are extremely similar (just adding some probability mass at time ∞), when we add tail constraints do we similarly get the same basic structure as in [3]?

1.1 Our Contributions

In this paper we show that the structure of optimal purchase distributions in the presence of even a single tail bound is even more complicated than in Dinitz et al. [3], but that we can still design efficient algorithms to compute this distribution (even for a collection of tail bounds). For example, we prove that optimal purchase distributions can have the following structures, in contrast to the pure buy setting (we follow [9] and refer to the classical setting as the "pure buy" setting).

- There may not be a unique purchase distribution, despite the fact that there is always a unique optimal purchase distribution for either the pure buy case with a single tail bound [3] or the two-state setting without tail bounds [9].
- For some parameter settings, *every* optimal solution might include probability mass at finite points larger than 1 (or in the discrete case, larger than b). Again, this is in contrast to the pure buy case even with a collection of tail bounds [3] (where the optimal purchase distribution only has support in $[0, 1]$) and the two-state case without tail bounds [9] (where the optimal solution only has support in $[0, 1] \cup \{\infty\}$).

Note that these features make it more difficult to compute an optimal purchase distribution. In particular, since we not have an *a priori* bound on the support, it is not clear that one can even *represent* an optimal solution efficiently, much less compute it. Nevertheless, we design algorithms to compute optimal purchase distributions (in the discrete setting).

Definitions. In order to more formally present our results and techniques, we first need to introduce some notation. In this paper, we use time t and point t interchangeably. A *purchase distribution* f is a probability distribution over $\mathbb{R}_{\geq 0} \cup \{\infty\}$, where f_x represents the probability that we buy (move from state 1 to state 2) at time x (we will use f to mean both the distribution and its probability density function interchangeably). Note that since our actions are probabilistic, our competitive ratio is now a random variable. We will let $\alpha_f(x)$ denote the expected competitive ratio when the true stopping time is x and we choose our buying time from f. So our goal is to compute the purchase distribution which minimizes $\alpha_f = \sup_x \alpha_f(x)$ subject to also obeying the given tail bound(s).

Given a (γ, δ)-tail bound, the *bad interval* of x (denoted by $I_\gamma(x)$) is defined to be the set of all points t such that if we buy at time t and the adversary chooses time x to stop, then the competitive ratio is larger than γ. This is essentially the same definition of bad intervals as in [3], but it turns out that their structure is quite different: most notably, in the pure buy setting it is not hard to argue that the bad interval for x is contained in $[0, x]$, but in the two-state setting this is no longer true: the bad interval for x can contain points in (x, ∞). We will refer to an interval I of $\mathbb{R}_{\geq 0}$ as a *zero interval* for f if there is no probability mass in I, i.e., if $\int_{x \in I} f_x \, dx = 0$. When the right boundary is ∞, we call it a zero *suffix*.

Note that any purchase distribution with expected competitive ratio β that satisfies a (γ, δ)-tail constraint must satisfy two properties at every point x. First, it must be within the competitive ratio constraint: $\alpha_f(x) \leq \beta$. Second, in order to obey the tail constraint there must be at most δ probability mass in the bad interval for x: $\int_{t \in I_\gamma(x)} f_t \, dt \leq \delta$. We refer to the first of these as the competitive ratio constraint (or CR-constraint) and the second as the mass constraint. We note that the continuous and discrete settings are essentially the same for our purposes, so we move between them depending on what is convenient. The exception is for our algorithms, which only work in the discrete setting.

Structure and Greedy. In order to understand the structure of the optimal purchase distribution in the presence of tail bounds in the pure buy setting, Dinitz et al. [3] proved a structural theorem about the optimal distribution: at every point x, either the CR-constraint with $\beta = OPT$ or the mass constraint must be tight (or both). And since (as discussed) the bad interval for every point x is contained in $[0, x]$, this led them to an obvious greedy algorithm: moving from smaller to larger values of x, increase f_x until one of the two constraints becomes tight. While this is not actually an implementable algorithm as stated (since we do not know OPT), it enabled them to later design an actual algorithm as well as allowing them to reason about the optimal solution by reasoning about the behavior of the greedy algorithm.

Unfortunately, it turns out that this theorem is *not* true in the general two-state setting: unlike the pure buy setting, there can indeed be optimal purchase distributions f which have points x where $f_x > 0$ but neither of the constraints is tight at x. Moreover, as discussed, in the two-state setting the bad interval for x can actually contain points *after* x, forcing us to add a third requirement when increasing f_x to ensure that we don't violate the mass constraint for some $x' < x$. Nevertheless, despite these difficulties, we prove the following theorem.

Theorem 1. *Given the true value of OPT, the greedy algorithm gives an optimal solution.*

This theorem allows us to explore the structure of at least some optimal solutions, in particular those that are closely related to the greedy solution (the purchase distribution output by the greedy algorithm). Moreover, this allows us to prove that the structure of optimal solutions exhibits a phase transition at a specific value of δ. For large values of δ, the structure is actually quite similar to prior work:

Theorem 2 (Structure Theorem, large δ). *If* $\frac{OPT-1}{\frac{1}{a}-1} \leq \delta \leq 1$, *then there is a unique optimal solution* f, *and it has the following structure: every point in* $[0, 1]$ *has either the CR constraint or the mass constraint tight,* $f_x = 0$ *for all* $x > 1$, *and* $f_\infty > 0$.

In particular, in this range of δ, the difference between the two-state with tail constraints setting and pure buy with tail constraints setting [3] is just that extra mass is added at ∞. This is completely analogous to the difference between the two-state setting without tail constraints [9] and the pure buy setting without tail constraints [6], and shows that the structure of the optimal distribution is almost precisely as "complex" as in the pure buy setting.

When δ is below $\frac{OPT-1}{\frac{1}{a}-1}$, the two-state setting with tail bounds behaves fundamentally differently from the pure buy setting with tail bounds [3]. However, we show that in this small δ regime the greedy solution exhibits particular structure, which will enable us to actually design an algorithm in later sections. Most notably, we show that while the greedy algorithm places some probability mass at finite points beyond 1, there is actually a point beyond which the greedy algorithm will not place any mass (other than at ∞). We first give the structure theorem for general optimal purchase distributions in this small δ regime.

Theorem 3 (Structure Theorem, small δ) *If* $0 \leq \delta < \frac{OPT-1}{\frac{1}{a}-1}$, *then:*

- *There is not a unique optimal purchase distribution.*
- *There exists a finite point* $L_b = \frac{\gamma-1}{1-a\gamma} \geq 1$, *such that there must be exactly* δ *mass in the suffix* $(L_b, \infty]$ *in any optimal purchase distribution.*

We can now give a structure theorem specifically for the greedy solution. Thanks to the specificity of the greedy algorithm, we end up with a thorough understanding of the resulting solution.

Theorem 4 (Structure Theorem for the greedy solution, small δ) *If* $0 \leq \delta < \frac{OPT-1}{\frac{1}{a}-1}$, *then, there exists a finite point* $p \leq L_b$, *such that the greedy solution has the following structure:*

- *In* $[0, 1]$*, either the* CR *or mass constraint is tight at every point.*
- *In* $(1, p]$*, the* CR *constraint is tight at every point and the solution behaves as an exponential function in* $(1, p]$*. Specifically, for small time step* τ*,* $f_{1+k\tau} = (1+\tau)^{k-1} f_{1+\tau}$ *where* $1 \le k \le \lfloor \frac{p-1}{\tau} \rfloor$*.*
- *In* $(p, L_b]$*, the mass constraint is tight at every point.*
- *After* L_b*, the solution has either* CR *or mass constraint tight at each point until there is a total of* δ *mass. Moreover, it has finite support.*

One can see that the full structure of the greedy solution is quite different from the pure buy setting [3] where the support is always $[0, 1]$ even with tail bounds. In fact, the structures are even more different than they might appear. In the pure buy setting there are also regimes where the mass constraint must be tight (as in the (p, L_b) interval for us), which was shown by [3] to result in a series of exponential functions with increasing base (when combined with the fact that the bad intervals for such points are decreasing in length). In our setting, since the bad intervals for points in (p, L_b) are actually *increasing* in length, this results in exponential functions with *decreasing* base. And even more strangely, due to the discrete time setting we also end up with frequent interruptions by zero intervals (corresponding to points where the left endpoint of the bad interval does not actually move thanks to the discretization). To see an example of this, see the full version [2], where time has been discretized finely enough to result in what appear to be overlapping zeros and exponentials (although they are in reality alternating, not overlapping).

Moreover, while the greedy solution always has finite support, it is not clear whether this support can be written as a closed analytical form of the input values. So, we show that there is a closely related distribution (which is a simple modification of the greedy solution) which always has support in $[0, L_b] \cup \{\infty\}$, giving a"fixed" finite support solution. Together with the fact that there is a unique optimal solution with support $[0, 1] \cup \{\infty\}$ in the large δ regime, we have the following structural theorem for the fixed finite support solution for all δ.

Theorem 5 **(Structure Theorem for the fixed finite support solution, all** δ**).** *For every* δ*, there is an optimal purchase distribution* f *which has support in* $[0, L_b] \cup \{\infty\}$*. In other words,* $f_x = 0$ *for all* $x \in (L_b, \infty)$*.*

Algorithms. After exploring the structure of optimal solutions, we give two different algorithms to solve the problem: a binary search greedy algorithm and an LP-based algorithm. The binary search greedy algorithm takes in an accuracy parameter $\epsilon > 0$ and a guess T of OPT, and applies the aforementioned greedy algorithm using T as OPT to construct the $\{f_t\}$ values. However, since T is not necessarily equal to OPT, these values might not actually form a valid distribution. We prove that $\sum_t f_t > 1$ if and only if $T > OPT$. This allows us to do binary search for OPT, enabling us to return a feasible solution (a purchase distribution that obeys the (γ, δ)-tail constraint) with expected competitive ratio at most $OPT + \epsilon$.

Theorem 6 (Binary Search Greedy). *There is a binary search greedy algorithm that gives feasible solution with expected competitive ratio at most* $OPT + \epsilon$ *in* $O(\log(1/\epsilon))$ *iterations.*

While this algorithm is fast and effective, it does not actually return an optimal solution (just a solution that is arbitrarily close to optimal). To remedy this, we take advantage of the fact that there is a related optimal solution which has *fixed* finite support, and in particular has support in $[0, L_b] \cup \{\infty\}$ (Theorem 5). Thanks to this theorem, we can write a linear program in which the variables are the purchase distribution and the constraints are the CR and mass constraint (even without knowing OPT). By solving this LP we get a polynomial-time algorithm to find an exactly optimal purchase distribution, at the cost of larger running time to solve the LP.

Theorem 7 (LP Algorithm). *There is a linear program with* $\Theta(L_b/\tau)$ *variables and* $\Theta(L_b/\tau)$ *constraints where any optimal solution corresponds to an optimal purchase distribution.*

Experiments. Since we have efficient algorithms to compute optimal and near-optimal purchase distributions, we have the ability to experimentally explore some of the parameter space to illustrate the surprising structures of the optimal purchase distributions (as discussed in our structure theorems). We exhibit the optimal purchase distributions and the figures that illustrates CR and mass constraints in both large and small δ regimes. Specifically, in the small δ regime, we observe that the optimal purchase distributions generated by the above two algorithms not only have distinct structures after L_b as stated in the structure theorems, but also reveal differing structures before L_b: the distribution given by the LP algorithm can be non-greedy in $[1, L_b]$, thus the optimal purchase distribution is not unique even before L_b. Also, as discussed, we present a figure showing different behaviors in $(p, L_b]$, indicating that to make mass tight, the optimal purchase distribution consists of alternating exponentials and zero intervals. These results and discussions can be found in the full version [2].

1.2 Related Work

The classical ski rental problem, also known as the rent or buy problem, is the most basic problem considered in online algorithms. The literature on it and its variants is deep and vast, so we do not attempt to survey all such work, but rather only that which is most directly related to our setting. Karlin et al. [7] first analyzed the 2-competitive deterministic strategy and then gave the $e/(e-1)$-competitive randomized algorithm in [6] in the classical pure buy setting. Separately, Karlin et al. [5] applied the randomized algorithm to TCP acknowledgement and some other problems. We note that there are multiple variants of the classical ski rental problem. Lotker, Patt-Shamir, and Rawitz [9] indicated the optimal $e/(e-1+a)$-competitive randomized algorithm in the two states setting. They also studied the multislope ski rental [10] and gave an optimal randomized strategy with an $e/(e-1)$ competitive ratio in the additive model and with an e competitive ratio in the non-additive model. Later, Levi and Patt-Shamir [8] improved the competitive ratio for the non-additive model with two slopes. Fujiwara et al. [4] explored the upper and lower bounds for the competitive ratio in the k slope setting for small values k.

As discussed, Dinitz et al. [3] is the first to consider controlling the tail risk for the classical ski rental problem. They gave the optimal randomized strategy and analyzed the structure of the optimal solution, which is significantly different from the no tail constraint case. Subsequently, Christianson et al. [1] introduced a related type of tail bound called the *conditional* value-at-risk (as opposed to the *absolute* tail bounds of [3]). This type of tail bound turns out to be more amenable to analysis, and they were able to derive a number of interesting bounds and algorithms.

1.3 Paper Outline

We begin in Sect. 2 by formally defining the problem and important concepts. We expand upon this in the full version [2], where we prove important properties of zero intervals and zero suffixes that will be useful for our main proofs. We then explore the structure of the optimal purchase distributions in Sect. 3, proving that the greedy algorithm gives an optimal solution (Theorem 1) in Sect. 3.1 and then argue about the structure of optimal solutions in the two different regimes of δ (Theorems 2 to 5) in Sect. 3.2. In [2], we give our two algorithms for computing optimal purchase distributions (Theorem 6 and 7) and discuss the trade-off between the algorithms. Finally, we show different structures by exploring different parameters experimentally in [2].

2 Preliminaries

We restate our model here. We consider the ski rental problem with tail constraints in the two states setting (defined in Sect. 1). State 1, where we call it the rental state, is to pay one unit of rental cost per time unit. To leave state 1 and reach state 2, one needs to pay a one-time cost $1-a$ and then the rental rate drops to a, for some $0 \leq a < 1$. If we see the geometric interpretation of the model, it consists of two intersecting lines with the intersection point $(1,1)$. If the adversary chooses the stopping time $t \leq 1$, then the optimal strategy is to stay in state 1 from the beginning, with an optimal cost t. If the adversary chooses $t > 1$ to stop, then the optimal strategy is to switch to state 2 at the beginning, with an optimal cost $1-a+at$. After one moment of thinking, the best deterministic algorithm is $(2-a)$-competitive. Our task is to determine when to switch to state 2 by giving the purchase distribution, at the same time not violate the tail constraints (γ, δ) where generally we have $\gamma \geq 2-a$ and $0 \leq \delta \leq 1$.

2.1 Bad Interval

Let $\{f_t\}_{t \geq 0}$ be the purchase distribution function, where f_∞ is the probability that we never buy. Note that $x, t \in \mathbb{R}^+ \cup \{\infty\}$. Let $\alpha(t, x)$ be the competitive ratio of the algorithm when adversary chooses for x to be the last time of skiing and we buy at the time t, and a is the slope of the second option. Then we have

$$\alpha(t,x) = \begin{cases} \frac{t+1-a+a(x-t)}{x}, & t \leq x \text{ and } 0 \leq x \leq 1 \\ 1, & t > x \text{ and } 0 \leq x \leq 1 \\ \frac{t+1-a+a(x-t)}{1-a+ax}, & t \leq x \text{ and } x > 1 \\ \frac{x}{1-a+ax}, & t > x \text{ and } x > 1. \end{cases}$$

Together with the tail constraint (γ, δ) where $\gamma \geq 2 - a$ and $0 \leq \delta \leq 1$, we define the bad interval of a point x.

Definition 1. *The bad interval $I_\gamma(x)$ consists of all time t such that if we buy at time t and adversary chooses time x, then the competitive ratio is larger than γ. In other words, $I_\gamma(x) = \{t : \alpha(t,x) > \gamma\}$. Specifically, we have*

$$I_\gamma(x) = \begin{cases} \{t \leq x : \frac{t+1-a+a(x-t)}{x} > \gamma\}, & 0 \leq x \leq 1 \\ \{t \leq x : \frac{t+1-a+a(x-t)}{1-a+ax} > \gamma\} \cup \{t > x : \frac{x}{1-a+ax} > \gamma\}, & x > 1 \end{cases}$$

For every point x, we want to simplify and explicitly write down the formula of its bad interval. We note that bad intervals may have different forms. For sake of completeness, we denote two bound points $I_3 := \frac{(\gamma-1)(1-a)}{1-a\gamma}$ and $L_b = \frac{\gamma-1}{1-a\gamma}$, where we show in the full version [2] that I_3 is the last point whose bad interval is empty and L_b is the largest point not contained in any bad interval that stretches out to ∞. We defer the calculation and more discussions (bounds of different bad intervals) to [2].

2.2 Feasible Solution and Optimal Solution

Definition 2. *A distribution f is feasible if it satisfies the constraint with respect to (γ, δ) where $\gamma \geq 2 - a$ and $0 \leq \delta \leq 1$. We call f a feasible solution if the distribution is feasible.*

The following lemma gives a way to check the feasibility of a distribution.

Lemma 1. *f is feasible if and only if $\int_{t \in I_{\gamma(x)}} f_t \, dt \leq \delta$ for all $x \in \mathbb{R}^+ \cup \{\infty\}$.*

Proof. Follows directly from the definition.

Now, we explore if a feasible solution f is an optimal solution. Given a feasible solution f, let $\alpha_f(x)$ be the expectation of $\alpha(t, x)$ where t is drawn from f.

If $0 \leq x \leq 1$,

$$\alpha_f(x) = \int_{t=0}^{\infty} \alpha(t,x) f_t \, dt = \int_{t=0}^{x} \frac{t+1-a+a(x-t)}{x} f_t \, dt + \left(\int_{t=x}^{\infty} f_t \, dt + f_\infty \right).$$

If $x \geq 1$,

$$\alpha_f(x) = \int_{t=0}^{x} \frac{t+1-a+a(x-t)}{1-a+ax} f_t\, dt + \left(\int_{t=x}^{\infty} \frac{x}{1-a+ax} f_t\, dt + \frac{x}{1-a+ax} f_\infty \right).$$

It is not hard to see that $\alpha_f(x)$ is continuous.

We say that a feasible solution is optimal if it gives the best worst-case competitive ratio. Let f^* be an optimal solution. By Lemma 1, f^* is the distribution f that

$$\begin{aligned} \min \quad & max_x\ \alpha_f(x) \\ \text{s.t} \quad & \int_{t \in I_{\gamma(x)}} f_t\, dt \leq \delta. \end{aligned}$$

For some point x, if the above inequality constraint is an equality constraint, then it is called tight. See Definition 3, 4 for the formal definition of tightness.

2.3 Tightness of the Mass Constraint and the Competitive Ratio Constraint

Let f be a feasible solution. Denote $OPT{:=}\max_x \alpha_f(x)$. Note that we have a closed analytical form of OPT if we have $OPT = \alpha_f(\infty)$. In this case, OPT can be computed as

$$OPT = \alpha_f(\infty) = 1 \cdot (1 - f_\infty) + \frac{1}{a} \cdot f_\infty = 1 + \left(\frac{1}{a} - 1\right) f_\infty.$$

A useful observation is that both the competitive ratio and mass at time ∞ are functions only of the mass at ∞. Now we define both the tightness of the mass constraint and the competitive ratio (CR) constraint.

Definition 3. *We say that the mass constraint is tight at x if $\int_{t \in I_{\gamma(x)}} f_t\, dt = \delta$. In short, we say that it is mass tight at time x (or, x is mass tight).*

Definition 4. *We say that the competitive ratio constraint is tight at x if $\alpha_f(x) = OPT$. In short, we say that it is CR tight at time x (or, x is CR tight).*

2.4 Zero Suffix and Zero Interval

There are two special structures in a feasible solution f that have nice properties, which we call the zero suffix and the zero interval. We explore such properties in the full version [2].

3 The Structure of Optimal Solutions

In this section, we will prove the structure of optimal solutions. We assume that the true OPT is given and answer the structural question we mentioned in Sect. 1 for the optimal solutions. In Sect. 3.1 we will argue that there is an optimal solution by greedily placing mass without violating any mass or CR constraints, as argued in Theorem 1. Moreover, we prove that every optimal solution must have the same prefix. In Sect. 3.2 we will explore structures of optimal solutions in large and small δ regimes, and prove the corresponding structure theorems, as stated in Theorem 2 to 5.

3.1 Optimal Solution by the Greedy Algorithm

We prove that at time 1 the mass constraint cannot be tight (since we show that its bad interval is empty) in the full version [2] and that the CR constraint at time 1 must be tight (Theorem 8) in any optimal solution. A consequence of this result is that either the CR constraint or the mass constraint must be tight in $[0, 1]$ (Theorem 9). It turns out that the greedy algorithm, which places one unit of mass to make either CR or mass constraint tight at each point until it is not allowed to do so, gives an optimal solution to the problem (Theorem 1). Due to space constraints, full proofs of the lemmas and theorems stated in this section are deferred to [2].

Lemma 2. *Let $T \geq 0$ be some finite point that is not CR tight. Then for any feasible solution f, if time $t < \infty$ is the first CR tight point after T, then there is nonzero mass at time t.*

Based on Lemma 2, we explore the tightness at time 1 in any optimal solution.

Theorem 8. *Let f be any optimal solution. Then the competitive ratio constraint is tight at time 1.*

Now, we can use Theorem 8 to show that in any optimal solution, either the mass constraint or the CR constraint is tight in $[0, 1]$.

Theorem 9. *Let f be any optimal solution. Then for any point in $[0, 1]$, either the competitive ratio constraint or the mass constraint is tight.*

Given Theorem 9, we can use the same technique (by simple induction) as [3] to prove the uniqueness in $[0, 1]$ for optimal distributions. This is because for every optimal solution, $[0, 1]$ is always either CR or mass tight by Theorem 9 and we use the fact that the bad interval of $x \in [0, 1]$ is a (sub)interval of its prefix.

Theorem 10. *If f and f' are optimal solutions, then $f_t = f'_t$ for all $0 \leq t \leq 1$.*

After we show the structure in $[0, 1]$ of an optimal solution, we examine its structure after time 1. The next lemma shows that CR is non-increasing over zero intervals in any optimal solution.

Lemma 3. *In any optimal solution, in any zero interval, the competitive ratio is non-increasing.*

Now, let us introduce the definition of a tight prefix.

Definition 5. *We say that* $[0, t]$ *is a tight prefix of a solution* f *if every point in* $[0, t]$ *is either* CR *tight or mass tight.*

Thus, $[0, 1]$ is a tight prefix for any optimal solution f. One natural way to obtain a tight prefix is to use the *greedy* algorithm—informally, we iterate through the times starting at time 0 and greedily place mass to make either the CR constraint or mass constraint tight at each time step. Technically, we do a special check at time right after point 1. For a formal definition of the greedy algorithm, see the full version [2].

Then, a natural question is to ask if the greedy algorithm gives an optimal distribution. First, one might worry that by placing probability mass greedily in $[0, 1]$, there is actually some time $T > 1$ where the CR already exceeds OPT. Fortunately, we show that this cannot happen.

Lemma 4. *By placing mass greedily in* $[0, 1]$*, there is no point* $T > 1$ *where the competitive ratio already exceeds* OPT.

Let $\tau > 0$ be a sufficiently small time step to discretize the domain of the problem. As a corollary, we have the following.

Corollary 1. *After mass is assigned greedily in* $[0, 1]$*, placing a zero mass at time* $1 + \tau$ *will not increase the competitive ratio.*

Corollary 1 states that after greedily placing mass in $[0, 1]$, if we place a zero mass at $1 + \tau$, the CR either stays flat or decreases. Since we know that the CR constraint is tight at time 1 (Theorem 8), this means that we can either place nonzero mass at time $1 + \tau$ (if the CR would decrease if we place mass 0) or we must place 0 mass at time $1 + \tau$ (if the CR stays flat when placing 0 mass).

Definition 6. *After placing mass greedily in* $[0, 1]$*, we say that we are in* world 1 *if one has to place a zero mass at time* $1 + \tau$ *and we are in* world 2 *if one can place some nonzero mass at time* $1 + \tau$.

Note that world 1 and world 2 partition the solution space. In the following lemma, we show that in different world, CR behaves differently in zero intervals after time 1.

Lemma 5. *Let* f *be any distribution and let* I *be any zero interval after time 1. In world 1, the competitive ratio is flat in* I*. In world 2, the competitive ratio decreases in* I.

Since solutions in world 1 and world 2 have different structures, the remaining part of Sect. 3.1 shows that the greedy algorithm gives an optimal solution separately. The easier case is when we are in world 1, since we are left with no further choices. So we have the following lemma.

Lemma 6. *In world 1, the greedy algorithm gives an optimal solution.*

When we are in world 2, it is not trivial to say that the greedy algorithm is optimal because we need to rule out the case that when the greedy algorithm has already placed a total of δ mass in the bad interval suffix $(L_b, \infty]$ (we show that in the full version [2] there is at most δ mass in this suffix $(L_b, \infty]$ to get a feasible solution), the total mass is still less than 1. Therefore, to prove the final Theorem 1, we first prove the following lemma in world 2.

Lemma 7. *In world 2, there exists an optimal solution of the following form:*
—Either competitive ratio or mass constraint is tight for some prefix ending at time T.
—Possibly nonzero mass is at time $T + \tau$.
—zero mass after time $T + \tau$.

This lemma indicates that some optimal solutions are in an equivalence class, namely that they have the same length of tight prefix (which is the longest tight prefix among all optimal solutions), but after the tight prefix there might be a semi-flexible way to place mass by just not violating any constraint at any point outside the tight prefix.

We note that Theorem 1 is not trivial by Lemma 7. For example, after we make a tight prefix $[0, x]$ where x is the first point whose bad interval is of the form $I_\gamma(x) = (\cdot, \infty]$, if there is still remaining mass to place, it could be the case that the longest tight prefix solution puts mass to ∞. Thus, every point in the zero suffix could be mass tight. However, greedy algorithm will place mass greedily, resulting in another optimal solution. We will show that the greedy algorithm constructs the solution which coincides with the solution in Lemma 7 from 0 to L_b, and hence gives an optimal solution to the problem (Theorem 1).

Definition 7. *We call the optimal solution constructed by the greedy algorithm in Theorem 1 the* greedy solution.

3.2 General Structure of Optimal Solutions

In this section, we will use properties in Sect. 3.1 to examine the structures of optimal solutions. We show that there is a threshold such that the optimal solution is unique and has a zero suffix when δ is larger than that threshold (Theorem 2), and there is not a unique optimal solution when δ is smaller than the threshold (Theorem 3). In a small δ regime, we further show the structure of the greedy solution (Theorem 4). Finally, we prove that there is an optimal solution with a finite support $[0, L_b] \cup \{\infty\}$ (Theorem 5). Full proofs of the lemmas and theorems stated in this section are deferred to the full version [2].

To explore the general structure of optimal solutions, we first show that CR must be tight in $[1, I_3]$ in any optimal solution.

Theorem 11. *For any optimal solution f, the competitive ratio constraint must be tight in $[1, I_3]$.*

Next, we show that the structure in $(1, I_3]$ of any optimal solution is particularly simple, even when the optimal solution is in world 2. Let $\tau > 0$ be a sufficiently small time step (again we discretize the problem). The following lemma calculates the amount of mass at each point to keep CR tight before I_3 in any optimal solution in world 2.

Lemma 8. *To keep CR tight in $(1, I_3]$, for any $k \geq 1, k \in \mathbb{N}$, we have $f_{1+k\tau} = (1+\tau)^{k-1} f_{1+\tau}$, where $f_{1+\tau}$ is calculated in [2].*

Then, Theorem 11 and Lemma 8 lead to the following Corollary.

Corollary 2. *Every optimal solution in the same world (world 1 or world 2) is the same in $[0, I_3]$. Moreover, in world 2, the distribution in $[1, I_3]$ of every optimal solution is the same exponential function.*

Now, we determine whether the optimal solution lies in world 1 or world 2 as δ varies. We first show that, in world 2, any optimal solution must have a δ mass in the bad interval suffix $(L_b, \infty]$.

Theorem 12. *For any optimal solution, if the competitive ratio is strictly decreasing over zero intervals after time 1 (i.e., world 2), then there is exactly δ mass in the bad interval suffix $(L_b, \infty]$.*

Then, we show that when $\delta < \frac{OPT-1}{\frac{1}{a}-1}$, every optimal solution must be in world 2. It is worth mentioning that this case could happen when δ is small enough because $OPT \in [\frac{e}{e-1+a}, 2-a]$ and then the implicit expression $\frac{OPT-1}{\frac{1}{a}-1}$ is bounded below by some value which is independent of δ.

Theorem 13. *If $\delta < \frac{OPT-1}{\frac{1}{a}-1}$, then every optimal solution must be in world 2.*

Now we know that when δ lies in $[0, \frac{OPT-1}{\frac{1}{a}-1})$ (the small δ regime) we are in world 2, and when $\delta = 1$ (no tail constraint) the classical solution is in world 1. We will prove that when $\frac{OPT-1}{\frac{1}{a}-1} \leq \delta \leq 1$ (the large δ regime), we are in world 1. That is, we prove Theorem 2.

Thus, in the large δ regime, the difference between the two-state with tail constraints setting and pure buying with tail constraints setting [3] is just that extra mass is added at ∞. However, when δ is decreasing and passing the threshold $\frac{OPT-1}{\frac{1}{a}-1}$, the structure of the optimal solution is fundamentally changed. Since there are multiple optimal solutions in the small δ regime, we explore the structure of the *greedy solution* in the small δ regime.

By Theorem 11 and Lemma 8, we know that in the greedy solution, after time 1 the solution first has an exponential function to keep CR tight. Then there are two possibilities: adding mass exponentially until L_b does not violate any mass constraint (thus there is no mass tight point inside $[1, L_b]$), or at some point we must add less amount of mass to make the corresponding point mass tight. In the simulation (see [2]) one can see that both cases could happen. We show in Theorem 4 that in the greedy solution if the latter case happens, once some point has to become mass tight, it will keep mass tight until L_b.

Note that although Theorem 4 states that the greedy solution always has finite support, the support can vary depending on different inputs a, δ and γ. However, based on the theorem, one can easily find another optimal solution with a finite support $[0, L_b] \cup \{\infty\}$ which is closely related to the greedy solution, by placing all the mass in $(L_b, \infty]$ to ∞. Together with the fact that in the large δ regime, there is a unique optimal solution with support $[0, 1] \cup \{\infty\}$, we have a unified structure result for all δ (Theorem 5).

In the full version [2], we leverage the structure theorems to design two algorithms for computing optimal purchase distributions and then present visualizations of the solution structures under various parameter settings.

Acknowledgments. Supported in part by NSF award 2228995.

Disclosure of Interests. The authors have no competing interests to declare that are relevant to the content of this article.

References

1. Christianson, N., Sun, B., Low, S., Wierman, A.: Risk-sensitive online algorithms (extended abstract). In: Agrawal, S., Roth, A. (eds.) Proceedings of Thirty Seventh Conference on Learning Theory. Proceedings of Machine Learning Research, 30 Jun–03 Jul, vol. 247, pp. 1140–1141. PMLR (2024)
2. Cui, Q., Dinitz, M.: Controlling tail risk in two-slope ski rental (2025). arXiv:2508.06809
3. Dinitz, M., Im, S., Lavastida, T., Moseley, B., Vassilvitskii, S.: Controlling tail risk in online ski-rental. In: Proceedings of the 2024 Annual ACM-SIAM Symposium on Discrete Algorithms (SODA), pp. 4247–4263. SIAM (2024)
4. Fujiwara, H., Kitano, T., Fujito, T.: On the best possible competitive ratio for multislope ski rental. In: Asano, T., Nakano, S., Okamoto, Y., Watanabe, O. (eds.) ISAAC 2011. LNCS, vol. 7074, pp. 544–553. Springer, Heidelberg (2011). https://doi.org/10.1007/978-3-642-25591-5_56
5. Karlin, A.R., Kenyon, C., Randall, D.: Dynamic tcp acknowledgement and other stories about e/(e-1). In: Proceedings of the Thirty-Third Annual ACM Symposium on Theory of Computing, STOC 2001, pp. 502–509. Association for Computing Machinery, New York (2001)
6. Karlin, A.R., Manasse, M.S., McGeoch, L.A., Owicki, S.: Competitive randomized algorithms for non-uniform problems. In: Proceedings of the First Annual ACM-SIAM Symposium on Discrete Algorithms, SODA 1990, pp. 301–309. Society for Industrial and Applied Mathematics, USA (1990)
7. Karlin, A.R., Manasse, M.S., Rudolph, L., Sleator, D.D.: Competitive snoopy caching. In: 27th Annual Symposium on Foundations of Computer Science, pp. 244–254 (1986)

8. Levi, A., Patt-Shamir, B.: Non-additive two-option ski rental. Theoretical Computer Science **584**, 42–52 (2015), special Issue on Structural Information and Communication Complexity
9. Lotker, Z., Patt-Shamir, B., Rawitz, D.: Ski rental with two general options. Inf. Process. Lett. **108**(6), 365–368 (2008)
10. Lotker, Z., Patt-Shamir, B., Rawitz, D.: Rent, lease, or buy: randomized algorithms for multislope ski rental. SIAM J. Discret. Math. **26**(2), 718–736 (2012)

Incremental–Decremental Maximization

Yann Disser[1], Max Klimm[2(✉)], Annette Lutz[1], and Lea Strubberg[2]

[1] TU Darmstadt, Darmstadt, Germany
{disser,lutz}@mathematik.tu-darmstadt.de
[2] TU Berlin, Berlin, Germany
{klimm,strubberg}@math.tu-berlin.de

Abstract. We introduce a framework for incremental–decremental maximization that captures the gradual transformation or renewal of infrastructures. In our model, an initial solution is transformed one element at a time and the utility of an intermediate solution is given by the sum of the utilities of the transformed and untransformed parts. We propose a simple randomized and a deterministic algorithm that both find an order in which to transform the elements while maintaining a large utility during all stages of transformation, relative to an optimum solution for the current stage. More specifically, our algorithms yield competitive solutions for utility functions of bounded curvature and/or generic submodularity ratio, and, in particular, for submodular functions, and gross substitute functions. Our results exhibit that incremental–decremental maximization is substantially more difficult than incremental maximization.

1 Introduction

Incremental maximization provides a mathematical model to study situations where a new infrastructure is constructed over time. The infrastructure consists of a set of elements E. The benefit for having built a subset of elements is defined by a set function $f\colon 2^E \to \mathbb{R}_{\geq 0}$ that maps each partial solution $S \subseteq E$ to its benefit $f(S)$. Depending on the application, the elements may model, e.g., pipes of a pipe system, wires in an electric network, or machines in a workshop. The benefit $f(S)$ models, e.g., averaged annual returns from operating the partial infrastructure S. Given such a situation, it is a natural question in which order the elements should be constructed. This question is addressed with the theory of incremental maximization problems.

Formally, an incremental solution is an order $\pi = (e_1, \dots, e_n)$ of the elements of E where $n = |E|$. The quality of an incremental solution is measured by a competitive analysis. To this end, for $k \in \{1, \dots, n\}$, let

$$\mathrm{OPT}_k := \max\bigl\{f(S) : S \subseteq E \text{ with } |S| = k\bigr\}$$

be the value of an optimal solution with k elements. For an incremental solution $\pi = (e_1, \dots, e_n)$ and $k \in \{1, \dots, n\}$, let $S_k := \{e_1, \dots, e_k\}$ denote the prefix

J. Matuschke and J. Verschae (Eds.): WAOA 2025, LNCS 16077, pp. 97–110, 2026.
https://doi.org/10.1007/978-3-032-06706-7_7

of the first k elements of π. Then, the competitive ratio of an incremental solution π is defined as $\rho := \inf\{r \in \mathbb{R}_{\geq 0} : r\, f(S_k) \geq \mathrm{OPT}_k \text{ for all } k \in \{1, \dots, n\}\}$, and one is interested in finding incremental solutions with a low competitive ratio. In this framework, upper and lower bounds on the best-possible competitive ratio have been derived for various classes of set functions f [1,7,8,20].

While the incremental maximization framework is a suitable model for situations where a new infrastructure needs to be constructed from scratch, it fails to capture scenarios where an existing infrastructure is upgraded to a new technological standard. As an example, suppose that E is a set of cell phone towers, and let $g(S)$ be the total amount of data transferrable with a subset $S \subseteq E$ of the towers using an old technological standard. The technology of the cell phone towers is to be updated to a new and non-interoperable standard; let $h(S)$ be the total amount of data transferable with a subset $S \subseteq E$ of the towers using the new standard.[1] In this context it is natural to ask in which order the cell towers should be upgraded in order to ensure that the amount of data transferrable in total across both standards is high at every point in time. To phrase and answer this question formally, in this paper, we introduce a novel *incremental–decremental* optimization framework that generalizes incremental maximization.

In analogy to the incremental maximization framework, we call an ordering $\pi = (e_1, \dots, e_n)$ of the elements an incremental–decremental solution. Such an order increments the set of elements using the new technology while decrementing the set of elements using the old technology, hence the name. For two arbitrary set functions $h, g \colon 2^E \to \mathbb{R}_{\geq 0}$, let $f \colon 2^E \to \mathbb{R}_{\geq 0}$ be defined as $f(S) := h(S) + g(E \setminus S)$. Note that in the application above $f(S)$ is equal to the total amount of data transferable using both standards whence the subset $S \subseteq E$ of cell phone towers has been upgraded. As before, let $\mathrm{OPT}_k := \max\{f(S) : S \subseteq E \text{ with } |S| = k\}$. The *competitive ratio* of the incremental–decremental solution π is then defined as

$$\rho := \inf\{r \in \mathbb{R}_{\geq 0} : r f(S_k) \geq \mathrm{OPT}_k \text{ for all } k \in \{1, \dots, n\}\}.$$

The competitive ratio of an algorithm is the supremum over the competitive ratios of its solutions. Likewise the competitive ratio of a problem is the infinum over the competitive ratios of an algorithm for the problem. We say that an algorithm or a solution is *ρ-competitive* if its competitive ratio is at most ρ.

Since the notions of an incremental–decremental solution for h and g, and that of an incremental solution for f defined by $f(S) = h(S) + g(E \setminus S)$ coincide, one may wonder why it is insightful to introduce incremental–decremental solutions in the first place. The reason for this stems from the fact that the incremental–decremental framework allows to impose natural properties of the

[1] In a similar vein, one may consider a pipeline network being upgraded from operation with natural gas (g) to operation with hydrogen (h); this is our motivation to use g for the function evaluating the benefit from the old standard and h for the function evaluating the benefit from the new standard.

functions h and g that may not be phrased so conveniently for the combined function f. For instance, in the above example of upgrading the technology of cell phone towers, it is natural to assume that both the function h and g are monotone. However, $f(S) = h(S) + g(E \setminus S)$ is then generally non-monotone and thus known results for incremental maximization (that all require monotonicity) do not apply. As a consequence, we will see that the best-possible competitive ratio in the incremental–decremental setting is strictly worse than in the incremental setting for monotone submodular functions. A general treatment of the incremental maximization problem with non-monotone objective function f however does not yield meaningful results as it is unbounded even if f is submodular.

1.1 Our Results

We give polynomial-time algorithms computing an incremental–decremental solution whose performances are parametrized by the generic submodularity ratio $\gamma \in (0, 1]$ and the curvature $c \in (0, 1]$ of the functions g and h; these properties of set functions are formally defined in Definition 1 and 2.

As a warm-up, we examine a simple randomized algorithm that takes uniformly at random the greedy order of the elements in E regarding h or the reverse greedy order with respect to g. For this randomized algorithm, an upper bound of $2c\frac{e^{c\gamma}}{e^{c\gamma}-1}$ on the competitive ratio follows rather straightforwardly from known results for incremental maximization.

Table 1. Our lower and upper bounds on the competitive ratio of incremental–decremental maximization for different classes of functions h and g with curvature c and generic submodularity ratio γ; results with shaded background are for randomized algorithms. Known upper bounds on the competitive ratios for incremental maximization (corresponding to the case $g \equiv 0$) are for comparison.

	incremental–decremental				**incremental**	
	lower bound		upper bound		upper bound	
general	$1 + \frac{c}{2-c}$	(Theorem 3)	$2c\frac{e^{c\gamma}}{e^{c\gamma}-1}$	(Corollary 2)	$c\frac{e^{c\gamma}}{e^{c\gamma}-1}$	[2]
	$\frac{2}{3} + \frac{1}{3\gamma}$	(Theorem 3)	$\frac{1}{\gamma}(1 + c\frac{e^{c}}{e^{c}-1})$	(Theorem 1)		
submodular	2	(Theorem 3)	$1+\frac{e}{e-1} \approx 2.58$	(Corollary 1)	$\frac{e}{e-1}$	[17]
gross substitute	5/4	(Theorem 3)	2	(Theorem 2)	1	[18]

As our main contribution, we then introduce a deterministic algorithm that can be viewed as a way to derandomize the randomized approach. Instead of following the greedy order either only for g or only for h, the algorithm builds both greedy orders partially at the same time depending on which promises a larger incremental gain in the current step. For this deterministic algorithm that we term *double-greedy*, we prove the following competitive ratio which improves on the randomized competitive ratio for large range of values of the curvature $c \in (0,1]$ and the generic submodularity ratio $\gamma \in (0,1]$. Specifically, it provides a better competitive ratio as long as $\gamma \geq 0.54$ which includes in particular all submodular functions.

Theorem 1. *For monotone functions $g, h\colon 2^E \to \mathbb{R}_{\geq 0}$ with generic submodularity ratio $\gamma \in (0,1]$ and curvature $c \in (0,1]$, the double-greedy algorithm is $\frac{1}{\gamma}\big(1 + c\frac{\mathrm{e}^c}{\mathrm{e}^c-1}\big)$-competitive.*

Using that for a submodular function, we have $\gamma = 1$, $c \in (0,1]$, and that the function $c \mapsto c\frac{\mathrm{e}^c}{\mathrm{e}^c-1}$ is non-decreasing in c, we obtain the following direct corollary of Theorem 1.

Corollary 1. *For monotone submodular functions g, h, the double-greedy algorithm is $\big(1 + \frac{\mathrm{e}}{\mathrm{e}-1}\big)$-competitive.*

We further consider gross substitute functions, a subclass of submodular functions. For this class of functions, we show the following result.

Theorem 2. *For monotone and gross substitute functions $g, h : 2^E \to \mathbb{R}_{\geq 0}$ the double-greedy algorithm is 2-competitive.*

We complement our results by the following lower bounds.

Theorem 3. *Let $g, h\colon 2^E \to \mathbb{R}_{\geq 0}$ be monotone functions. Then no algorithm has a better competitive ratio than*

- $1 + \frac{c}{2-c}$ *if g and h are submodular with curvature $c \in [0,1]$;*
- $\frac{2}{3} + \frac{1}{3\gamma}$ *if g and h have generic submodularity ratio $\gamma \in (0,1]$;*
- $5/4$ *if g and h are gross substitute.*

For an overview of known results as well as our contributions, we refer to Table 1.

1.2 Related Work

Our work on incremental–decremental maximization directly relates to literature on incremental maximization where the latter corresponds to the special case that $g \equiv 0$. The natural greedy algorithm for cardinality constrained maximization inherently produces an incremental solution. Its analysis for monotone submodular functions by Nemhauser et al. [17] translates to a competitive ratio of $\frac{\mathrm{e}}{\mathrm{e}-1}$. This approximation guarantee is best-possible among all polynomial algorithms, unless $\mathsf{P} = \mathsf{NP}$, as shown by Feige [10]. It is interesting to note that in a setting where both h and g are monotone submodular, our results imply that the best-possible competitive ratio is in the interval $[2, 1 + \frac{\mathrm{e}}{\mathrm{e}-1}]$ thus separating the competitive ratios achievable in the incremental and the incremental–decremental settings. For monotone gross substitute functions, a result of Paes Leme [18] implicitly yields an incremental solution that is 1-competitive. In contrast, our results imply that in a setting where both g and h are gross substitute, the best-possible competitive ratio for incremental–decremental maximization is in the interval $[5/4, 2]$ exhibiting that also for this class of functions incremental–decremental maximization is strictly more difficult than incremental maximization. For submodular functions with curvature c, Conforti et al. [5] provided a more fine-grained analysis that yields a competitive ratio of $c\frac{\mathrm{e}^c}{\mathrm{e}^c-1}$ for the greedy algorithm. For functions with submodularity ratio $\widetilde{\gamma}$, Das et al. [6] showed that the greedy algorithm yields a $\frac{\mathrm{e}^{\widetilde{\gamma}}}{\mathrm{e}^{\widetilde{\gamma}}-1}$-approximation. Bian et al. [2] combined the analysis of Conforti et al. and Das et al. and showed that for functions of curvature c and submodularity ratio $\widetilde{\gamma}$, the greedy algorithm yields a $c\frac{\mathrm{e}^{c\widetilde{\gamma}}}{\mathrm{e}^{c\widetilde{\gamma}}-1}$-approximation. Disser et al. [8] considered the more general setting of $\widetilde{\gamma}$-α-augmentable functions and showed that the greedy algorithm is a $\frac{\alpha}{\widetilde{\gamma}} \cdot \frac{\mathrm{e}^{\alpha}}{\mathrm{e}^{\alpha}-1}$-approximation. This class of functions contains all functions with submodularity ratio $\widetilde{\gamma}$ as well as all α-augmentable functions. While the results above are stated in terms of the submodularity ratio $\widetilde{\gamma}$, they also hold for the generic submodularity ratio γ since $\widetilde{\gamma} \geq \gamma$.

Incremental maximization beyond the greedy algorithm has been considered by Bernstein et al. [1]. For accountable functions, they gave a $(\varphi + 1)$-competitive solution where $\varphi \approx 1.62$ is the golden ratio. Here, a function $f \colon 2^E \to \mathbb{R}_{\geq 0}$ is called accountable if for all $S \subseteq E$ there is $s \in S$ such that $f(S \setminus \{s\}) \geq f(S) - f(S)/|S|$. They further showed that no deterministic algorithm can provide a competitive ratio better than 2.18 in this setting. Disser et al. [7] gave an improved lower bound of 2.24 and gave a randomized upper bound of 1.77.

Goemans and Unda [12] considered a different setting of incremental maximization where, instead of the worst-case ratio over all time steps, the ratio of the sum over all time steps is to be maximized, and obtained bounds for special functions containing rank functions of independence systems.

The double-greedy algorithm that we consider is reminiscent of double-greedy algorithms for non-monotone submodular maximization. Buchbinder et al. [3] used this approach and obtained a deterministic 3-approximation and a randomized 2-approximation for this problem. Feige et al. [11] showed earlier that no better approximation than 2 is possible with polynomially many oracle calls, implying that the randomized 2-approximation of Buchbinder et al. cannot be improved. Non-montone submodular function maximization has also been considered under knapsack, matroid, and other constraints [4,9,15,16].

2 A Simple Randomized Algorithm

To begin with, we state a first observation that leads to a naive randomized algorithm for the incremental–decremental problem. For a randomized algorithm the returned incremental–decremental solution π and, hence, all prefixes S_k are random variables; such an algorithm is called ρ-competitive if $\rho\,\mathbb{E}[f(S_k)] \geq \mathrm{OPT}_k$ for all k. The proof of the following proposition and further proofs are deferred to the full version of the paper [19].

Proposition 1. *Let* $g, h\colon 2^E \to \mathbb{R}_{\geq 0}$. *If there is a* ρ*-competitive algorithm both for the incremental problem for* g *and for the incremental problem for* h*, then there is a* 2ρ*-competitive randomized algorithm for the incremental–decremental maximization problem for* $f\colon 2^E \to \mathbb{R}_{\geq 0}$, $S \mapsto h(S) + g(E \setminus S)$.

To obtain concrete bounds on the competitiveness of this algorithm, we use the upper bounds on the competitive ratio of incremental maximization for functions with bounded curvature $c \in (0, 1]$ and bounded generic submodularity ratio $\gamma \in (0, 1]$ listed in Table 1. The generic submodularity ratio is a way to relax submodularity and allows to bound the competitive ratio for functions that are in some sense close to submodular. In the following definitions, we set $\min \emptyset := 1$. For the marginal increase in $f\colon 2^E \to \mathbb{R}_{\geq 0}$ of an element $e \in E$ or of a set $T \subseteq E$ with respect to $S \subseteq E$, we introduce the notations $f(e \mid S) := f(\{e\} \cup S) - f(S)$ and $f(T \mid S) := f(T \cup S) - f(S)$.

Definition 1 (Gong et al. [13]). *A monotone set function* $f\colon 2^E \to \mathbb{R}_{\geq 0}$ *has* generic submodularity ratio

$$\gamma := \min\left\{ \frac{f(e \mid A)}{f(e \mid A \cup B)} : A, B \in 2^E, e \in E \textit{ with } f(e \mid A \cup B) > 0 \right\}.$$

Intuitively, the generic submodularity ratio interpolates between monotone functions with $\gamma = 0$ and submodular functions with $\gamma = 1$. To see this, note that for a submodular function, we have $f(e \mid A) \geq f(e \mid A \cup B)$ for all $A, B \in 2^E$ and $e \in E$, so that the minimum is attained for $B = \emptyset$ and the submodularity ratio is $\gamma = 1$. Das and Kempe [6] introduced the stronger notion of the *submodularity ratio*

$$\widetilde{\gamma} := \min\left\{ \frac{\sum_{b \in B} f(b \mid A)}{f(B \mid A)} : A, B \in 2^E \text{ with } f(B \mid A) > 0 \right\}$$

and showed that the greedy algorithm is $\frac{e^{\widetilde{\gamma}}}{e^{\widetilde{\gamma}}-1}$-competitive for $g \equiv 0$. Observe that the generic submodularity ratio γ is always smaller or equal to $\widetilde{\gamma}$, since for all subsets $B = \{b_1, \ldots, b_k\}, A \subseteq E$ with $f(B \mid A) > 0$,

$$\frac{\sum_{b \in B} f(b \mid A)}{f(B \mid A)} = \frac{\sum_{i=1}^k f(b_i \mid A)}{\sum_{i=1}^k f(b_i \mid A \cup \{b_1, \ldots, b_{i-1}\})} \geq \frac{\sum_{i=1}^k f(b_i \mid A)}{\gamma^{-1} \sum_{i=1}^k f(b_i \mid A)} = \gamma.$$

Thus, their bound implies also $\frac{e^{\gamma}}{e^{\gamma}-1}$-competitiveness in terms of the generic submodularity ratio in the incremental setting.

The curvature $c \in [0, 1]$ measures how close a function is to being supermodular. A function is supermodular if it has curvature $c = 0$.

Definition 2 (Bian et al. [2]). *A monotone set function* $f\colon 2^E \to \mathbb{R}_{\geq 0}$ *has* curvature

$$c := 1 - \min\left\{ \frac{f(e \mid A \cup B)}{f(e \mid A)} : A, B \in 2^E, e \in E \setminus B \textit{ with } f(e \mid A) > 0 \right\}.$$

Bian et al. [2] showed that the greedy algorithm is $c\frac{e^{c\tilde{\gamma}}}{e^{c\tilde{\gamma}}-1}$-competitive for $g \equiv 0$. For submodular functions the notion of curvature was already introduced by Conforti and Cornuéjols [5]. In this case, the curvature simplifies to

$$1 - \min_{e \in E} \frac{f(e \mid E \setminus \{e\})}{f(\{e\})} \tag{1}$$

because of $f(e \mid A \cup B) \geq f(e \mid E \setminus \{e\})$ and $f(e \mid A) \leq f(\{e\})$. We observe that submodular functions with curvature zero are modular since they are both submodular and supermodular.

The upper bound on the competitive ratio of the incremental problem together with Proposition 1 translates to a first upper bound on the competitive ratio of incremental–decremental maximization.

Corollary 2. *Let $g, h \colon 2^E \to \mathbb{R}_{\geq 0}$ be monotone with generic submodularity ratio $\gamma \in (0,1]$ and curvature $c \in (0,1]$. The algorithm that returns uniformly at random the greedy order of E with respect to h or the reverse greedy order with respect to g is $2c\frac{e^{c\tilde{\gamma}}}{e^{c\tilde{\gamma}}-1}$-competitive.*

3 A Deterministic Algorithm

The double-greedy algorithm (see below) can be seen as a way to derandomize the randomized solution. Instead of only considering the elements in the greedy-order defined by g or the inverse greedy-order defined by h, it instead builds partial greedy orders for both functions from both ends simultaneously. Initially both the considered prefix and the considered suffix of the solution are empty. In each step, an element is selected that maximizes the maximum of the increment relative to the current prefix for h and the increment relative to the current suffix for g. It is then appended from the right to the current prefix if the maximum is attained for h, and from the left to the current suffix if the maximum is attained for g. This procedure is visualized in Fig. 1.

In the following, we denote by $F_k := G_k \cup H_k$ the set of all elements that are added to the solution by the double-greedy algorithm up to step k. Also, we denote the complement of a set S in E by $S^{\mathrm{c}} := E \setminus S$ and write $S_k^{\mathrm{c}} := (S_k)^{\mathrm{c}}$ for indexed sets. Further, we generally let $n := |E|$ and let $f \colon 2^E \to \mathbb{R}_{\geq 0}$ denote the objective function of the incremental–decremental maximization problem, i.e., $f(S) := h(S) + g(S^{\mathrm{c}})$. The double-greedy algorithm is symmetric with respect to the roles of g and h with the slight complication that the handling of ties between g and h is inherently asymmetric. In order to still exploit symmetry, we parameterize the algorithm by a comparison operator $\prec \in \{\leq, <\}$ that governs whether it favors adding elements to G_i or H_i in case of a tie and consider both variants simultaneously. The following proposition makes precise how we can exploit symmetry in our proofs. In particular, it suffices to show competitiveness of the solution up to cardinality $|H_n|$.

Algorithm: double-greedy algorithm

input : $g, h\colon 2^E \to \mathbb{R}_{\geq 0}$ monotone, $\prec \in \{\leq, <\}$
output : ordering $\pi = (e_1, \ldots, e_n)$ of the elements in E

$G_0 \leftarrow \emptyset$; $H_0 \leftarrow \emptyset$
for $i \leftarrow 1, \ldots, n$**:**
  $e_i^* \leftarrow \arg\max_{e \in E \setminus (G_{i-1} \cup H_{i-1})} \{\max\{g(e \mid G_{i-1}), h(e \mid H_{i-1})\}\}$
  if $h(e_i^* \mid H_{i-1}) \prec g(e_i^* \mid G_{i-1})$**:**
    $G_i \leftarrow G_{i-1} \cup \{e_i^*\}$
    $H_i \leftarrow H_{i-1}$
    $e_{n-|G_i|+1} \leftarrow e_i^*$
  else
    $H_i \leftarrow H_{i-1} \cup \{e_i^*\}$
    $G_i \leftarrow G_{i-1}$
    $e_{|H_i|} \leftarrow e_i^*$
return $\pi \leftarrow (e_1, \ldots, e_n)$

Proposition 2. *Let $\mathcal{C} \subseteq \{h\colon 2^E \to \mathbb{R}_{\geq 0}$ monotone$\}$. Assume that for all functions $g, h \in \mathcal{C}$, for all $\prec \in \{\leq, <\}$, for all $k \leq n$ and for all subsets $S \subseteq E$ with $|S| = |H_k|$ the double-greedy algorithm ensures that*

$$\rho f(H_k) \geq f(S), \tag{2}$$

for some $\rho \geq 1$. Then, the double-greedy algorithm is ρ-competitive for all functions $g, h \in \mathcal{C}$.

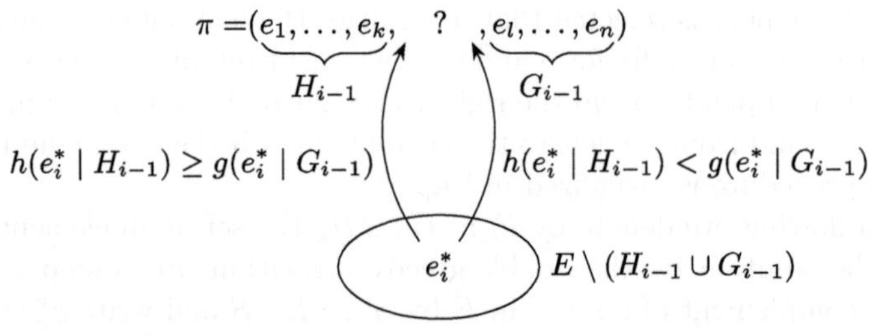

Fig. 1. In step i of the double-greedy algorithm (with $\prec = <$), the element e_i^* maximizing $\max\{g(e \mid G_{i-1}), h(e \mid H_{i-1})\}$ among the remaining elements $e \in E \setminus (G_{i-1} \cup H_{i-1})$ is added to the left or right depending on which marginal increase, $g(e_i^* \mid G_{i-1})$ or $h(e_i^* \mid H_{i-1})$, is higher.

From now on, we assume that $h(\emptyset) = g(\emptyset) = 0$. This assumption is without loss of generality, since it suffices to show ρ-competitiveness for the functions $g'(S) := g(S) - g(\emptyset)$ and $h'(S) := h(S) - h(\emptyset)$, because of

$$\begin{aligned} f(S) &:= h(S) + g(S^{\mathrm{c}}) = h'(S) + h(\emptyset) + g'(S) + g(\emptyset) \\ &\leq \rho(h'(H_k) + g'(H_k^{\mathrm{c}})) + h(\emptyset) + g(\emptyset) \leq \rho(h'(H_k) + h(\emptyset) + g'(H_k^{\mathrm{c}}) + g(\emptyset)) = \rho f(H_k), \end{aligned}$$

for all $k \leq n$ and $|S| = |H_k|$, which proves competitiveness by Proposition 2.

We conclude this subsection with a lemma which we will use to prove our upper bounds.

Lemma 1. *Let $g, h\colon 2^E \to \mathbb{R}_{\geq 0}$ be monotone. After iteration k of the double-greedy algorithm we have $f(H_k) \geq \sum_{i=1}^{k} \max\{h(e_i^* \mid H_{i-1}), g(e_i^* \mid G_{i-1})\}$.*

3.1 Bounded Generic Submodularity Ratio and Curvature

In this section, we consider the case where g and h have bounded curvature c and bounded generic submodularity ratio γ which we have defined in Sect. 2. Both parameters, the generic submodularity ratio and the curvature characterize how much the marginal increase of an element is allowed to change on different subsets. A function has generic submodularity ratio γ if the marginal values of each element increase at most by a factor of γ, i.e., we have by Definition 1 that

$$f(e \mid S) \geq \gamma f(e \mid T) \qquad \text{for } S \subseteq T \subseteq E, e \in E. \tag{3}$$

Likewise, the curvature measures how much the marginal increase decreases at most, i.e., Definition 2 implies

$$(1-c) f(e \mid S) \leq f(e \mid T) \qquad \text{for } S \subseteq T \subseteq E, e \in E \setminus T. \tag{4}$$

We proceed to analyze the double-greedy algorithm for functions with fixed generic submodularity ratio and curvature. Recall, that $F_m = H_m \cup G_m$ is the set of elements that are added to the solution up to step m. Let $\varphi_i := \max\{g(e_i^* \mid G_{i-1}), h(e_i^* \mid H_{i-1})\}$ be the increment governing the choice of e_i^* in each step of the algorithm. We obtain a lower bound on φ_i by considering the gap to the value of a larger solution $S \subseteq E$. In particular, we will use this bound with S being an optimum solution for f of size $|S|$.

Lemma 2. *Let $g, h\colon 2^E \to \mathbb{R}_{\geq 0}$ be monotone with generic submodularity ratio $\gamma \in (0,1]$ and curvature $c \in (0,1]$. For $S \subseteq E$ and $0 \leq m := |S|$ the double-greedy algorithm ensures that*

$$\varphi_{m+1} \geq \frac{c}{|S \setminus F_m|}\left(\gamma h(S) - \sum_{i=1}^{m} \varphi_i\right) + \frac{1-c}{|S \setminus F_m|}\left(\gamma h(S) - \sum_{i: e_i^* \in F_m \cap S} \varphi_i\right).$$

We combine the bounds derived in Lemma 2 on the increments φ_{m+1} in each step of the algorithm to bound the value of every solution by the sum of those increments. An optimum solution of each size is particularly bounded as follows.

Lemma 3. *Let $g, h\colon 2^E \to \mathbb{R}_{\geq 0}$ be monotone with generic submodularity ratio $\gamma \in (0,1]$ and curvature $c \in (0,1]$. For $S \subseteq E$ and $k := |S|$, the double-greedy algorithm ensures that $\frac{1}{c}\left(1 - \frac{1}{e^c}\right)\gamma h(S) \leq \sum_{i=1}^{k} \varphi_i$.*

With this at hand, we are ready to prove our main result.

Theorem 1. *For monotone functions $g, h\colon 2^E \to \mathbb{R}_{\geq 0}$ with generic submodularity ratio $\gamma \in (0,1]$ and curvature $c \in (0,1]$, the double-greedy algorithm is $\frac{1}{\gamma}\big(1 + c\frac{\mathrm{e}^c}{\mathrm{e}^c - 1}\big)$-competitive.*

Proof. Let $k \leq n$ and $S \subseteq E$ with $\ell := |H_k| = |S|$. By Proposition 2, it suffices to show that

$$\gamma^{-1}\left(1 + c\frac{\mathrm{e}^c}{\mathrm{e}^c - 1}\right) f(H_k) \geq f(S). \tag{5}$$

Since $\varphi_i \geq 0$ by monotonicity of g and h, and since $k \geq \ell$, we obtain

$$f(H_k) \overset{\text{Lem. 1}}{\geq} \sum_{i=1}^{k} \varphi_i \geq \sum_{i=1}^{\ell} \varphi_i \overset{\text{Lem. 3}}{\geq} \frac{1}{c}\left(1 - \frac{1}{\mathrm{e}^c}\right)\gamma h(S). \tag{6}$$

By monotonicity of g, we have

$$g(S^{\mathrm{c}}) \leq g(E) = g(H_k \cup H_k^{\mathrm{c}}) = \sum_{i: e_i^* \in H_k} g(e_i^* \mid H_{i-1} \cup H_k^{\mathrm{c}}) + g(H_k^{\mathrm{c}}) \tag{7}$$

Further, using $G_{i-1} \subseteq H_{i-1} \cup H_k^{\mathrm{c}}$ and the definition of the algorithm, we derive

$$\begin{aligned}
g(S^{\mathrm{c}}) &\overset{(7)}{\leq} \sum_{i: e_i^* \in H_k} g(e_i^* \mid H_{i-1} \cup H_k^{\mathrm{c}}) + g(H_k^{\mathrm{c}}) \\
&\overset{(3)}{\leq} \gamma^{-1} \sum_{i: e_i^* \in H_k} g(e_i^* \mid G_{i-1}) + g(H_k^{\mathrm{c}}) \\
&\overset{\text{alg.}}{\leq} \gamma^{-1} \sum_{i: e_i^* \in H_k} h(e_i^* \mid H_{i-1}) + g(H_k^{\mathrm{c}}) \\
&= \gamma^{-1} h(H_k) + g(H_k^{\mathrm{c}}) \\
&\overset{\gamma \leq 1}{\leq} \gamma^{-1} f(H_k).
\end{aligned} \tag{8}$$

Together with (6), we obtain the desired approximation (5)

$$\begin{aligned}
f(S) = h(S) + g(S^{\mathrm{c}}) &\overset{(6),(8)}{\leq} \gamma^{-1} c \left(\frac{\mathrm{e}^c}{\mathrm{e}^c - 1}\right) f(H_k) + \gamma^{-1} f(H_k) \\
&= \gamma^{-1}\left(1 + c\frac{\mathrm{e}^c}{\mathrm{e}^c - 1}\right) f(H_k).
\end{aligned}$$

We now show the lower bound on the competitive ratio for the case that g and h are submodular with curvature c. This proves the first part of Theorem 3.

Proposition 3. *For $g, h\colon 2^E \to \mathbb{R}_{\geq 0}$ monotone submodular functions with curvature $c \in [0,1]$ no algorithm has a better competitive ratio than $1 + \frac{c}{2-c}$.*

We also give a lower bound in terms of the generic submodularity ration; this proves the second part of Theorem 3.

Proposition 4. *For monotone functions $g, h\colon 2^E \to \mathbb{R}_{\geq 0}$ with generic submodularity ratio $\gamma \in (0,1]$ no algorithm has a better competitive ratio than $\frac{2}{3} + \frac{1}{3\gamma}$.*

4 Gross Substitute Functions

The gross substitute property originates from the economics literature [14]. In this context, given a value function $f\colon 2^E \to \mathbb{R}_{\geq 0}$ and a modular price function $p\colon 2^E \to \mathbb{R}$, one is interested in sets S maximizing the utility $f(S) - p(S)$. A function f is gross substitute if, for every price function and for every subset S maximizing the utility before increasing some of those prices, all elements of S with non-increased price are still part of a utility-maximizing set after the prices are increased.

Definition 3 (Kelso and Crawford [14]**).** *A function $f\colon 2^E \to \mathbb{R}$ is* gross substitute *if for all modular price functions $p, p'\colon 2^E \to \mathbb{R}$ with $p \leq p'$ and $S \in \arg\max_{S \subseteq E}\{f(S) - p(S)\}$ there is $S' \in \arg\max_{S \subseteq E}\{f(S) - p'(S)\}$ with $S \cap \{e \mid p(\{e\}) = p'(\{e\})\} \subseteq S'$.*

Paes Leme [18] surveys several equivalent ways to characterize gross substitute functions. We will work with the following characterization.

Lemma 4 (Paes Leme [18]**).** *A monotone set function $f\colon 2^E \to \mathbb{R}_{\geq 0}$ is* gross substitute *if and only if it is submodular and for all $A, B \subseteq E$ with $A \cap B = \emptyset$, $|B| \geq 2$ and all $b \in B$ there is $b' \in B \setminus \{b\}$ with*

$$f(b \mid A) + f(B \setminus \{b\} \mid A) \leq f(b' \mid A) + f(B \setminus \{b'\} \mid A).$$

The greedy algorithm is optimal for the incremental maximization problem with gross substitute objective function. Together with Proposition 1 we obtain the following result for the randomized algorithm discussed in Sect. 2 that returns uniformly at random the greedy order with respect to h or the reverse greedy order with respect to g.

Corollary 3. *For $g, h\colon 2^E \to \mathbb{R}_{\geq 0}$ monotone and gross substitute the randomized algorithm is 2-competitive.*

However, we obtain the same bound for gross substitute functions with the deterministic double-greedy algorithm. To show this, we need the following observation, which even holds for submodular functions. Recall that $f(S) = h(S) + g(S^{\mathrm{c}})$.

Lemma 5. *For $g, h\colon 2^E \to \mathbb{R}_{\geq 0}$ submodular and $\prec \in \{\leq, <\}$, the double-greedy algorithm ensures that*

$$g(E) = f(H_0) \leq \cdots \leq f(H_n) = f(G_n^c) \geq \cdots \geq f(G_0^c) = h(E).$$

With this, we are able to prove our second main result.

Theorem 2. *For monotone and gross substitute functions $g, h : 2^E \to \mathbb{R}_{\geq 0}$ the double-greedy algorithm is 2-competitive.*

Remark 1. The analysis of the double-greedy algorithm for gross substitute functions is tight. Even for modular functions, the algorithm is not better than 2-competitive. Consider the instance $E = \{a, b\}$, $g, h\colon 2^E \to \mathbb{R}_{\geq 0}$ modular, given by $g(\{a\}) = h(\{a\}) = h(\{b\}) = 1$ and $g(\{b\}) = 0$, and $\prec = <$. Then, the double-greedy algorithm can choose $e_1^* = a$ computing $\pi = (a, b)$. The solution of size 1 has value $f(\{a\}) = h(\{a\}) + g(\{b\}) = 1$. But the optimum value for a solution of size 1 is $f(\{b\}) = h(\{b\}) + g(\{a\}) = 2$. Thus, the algorithm is no better than 2-competitive.

We finally state that no algorithm can be better than 5/4-competitive which proves the third part of Theorem 3.

Proposition 5. *For $g, h\colon 2^E \to \mathbb{R}_{\geq 0}$ monotone and gross substitute functions no algorithm has a better competitive ratio than 5/4.*

5 Discussion

In this work, we introduced the incremental–decremental maximization framework and established several competitive ratios for randomized and deterministic algorithm. While our results provide new insights, some open questions remain.

Concerning lower bounds for the double-greedy algorithm, for monotone and submodular functions h and g, Theorem 1 establishes $(1 + \frac{\mathrm{e}}{\mathrm{e}-1})$-competitiveness. In particular, the algorithm cannot perform better than that. We do not know, whether the general upper bound of $\frac{1}{\gamma}(1 + c\frac{\mathrm{e}^c}{\mathrm{e}^c-1})$ for the setting where h and g have curvature c and generic submodularity ratio γ is tight and leave this question as an open problem.

For the special case that $g(E) = h(E)$ alongside submodularity and monotonicity, the double-greedy algorithm guarantees a 2-competitive solution, which is optimal.

Proposition 6. *For monotone and submodular set functions $g, h\colon 2^E \to \mathbb{R}_{\geq 0}$ with $g(E) = h(E)$, the double-greedy algorithm is exactly 2-competitive.*

It would be interesting to see similar results which exploit dependencies between h and g.

Another open question is whether Theorem 1 can be strengthened in settings where g and h have different curvatures $c_g \neq c_h$ and different generic submodularity ratios $\gamma_g \neq \gamma_h$. By defining $c := \max\{c_g, c_h\}$ and $\gamma := \min\{\gamma_g, \gamma_h\}$ one can adapt the proofs leading to Theorem 1 obtaining $\frac{1}{\gamma}(1 + c\frac{e^c}{e^c-1})$-competitiveness of the double-greedy algorithm. Note that the function space $\mathcal{C}$ in Proposition 2 is then given by all monotone set functions on the ground set E with curvature at most c and generic submodularity ratio at least γ.

While the simple randomized algorithm stated in Corollary 2 provides in some cases a better competitive analysis, it is outperformed by the double-greedy algorithm for submodular functions. Buchbinder et al. [3] introduced a randomized version of the double-greedy algorithm improving the approximation factor for unconstrained submodular maximization from 3 to 2. Their improved analysis, however, does not directly lead to a better competitiveness for a randomized version of the double-greedy for incremental–decremental maximization as the worst case ratio is not necessarily attained at the position where k equals the cardinality of a maximum set of f. We leave the design of randomized algorithms with better competitive ration as an interesting problem for future research.

Finally, another open problem is to narrow or even close the gaps identified in Table 1 for deterministic algorithms, in particular, the gap between 2 and $1+\frac{e}{e-1}$ for submodular h and g, as well as the gap between 1.25 and 2 in the gross substitute setting.

Acknowledgments. We thank Martin Knaack, Jannik Matuschke, and David Weckbecker for helpful discussions on functions of bounded curvature, gross substitute functions and incremental maximization, respectively. We also acknowledge David Weckbecker, Martin Knaack, Julia Baligacs and Nils Mosis for taking part in an initial exchange about the topic. This work is funded by the DFG via the SFB TRR 154 subprojects A07 and A09.

Disclosure of Interests. The authors have no competing interests to declare that are relevant to the content of this article.

References

1. Bernstein, A., Disser, Y., Groß, M., Himburg, S.: General bounds for incremental maximization. Math. Program. **191**(2), 953–979 (2022)
2. Bian, A.A., Buhmann, J.M., Krause, A., Tschiatschek, S.: Guarantees for greedy maximization of non-submodular functions with applications. In: Proceedings of the 34th International Conference on Machine Learning (ICML), pp. 498–507 (2017)
3. Buchbinder, N., Feldman, M., Naor, J., Schwartz, R.: A tight linear time (1/2)-approximation for unconstrained submodular maximization. SIAM J. Comput. **44**(5), 1384–1402 (2015)
4. Chekuri, C., Vondrák, J., Zenklusen, R.: Submodular function maximization via the multilinear relaxation and contention resolution schemes. SIAM J. Comput. **43**(6), 1831–1879 (2014)
5. Conforti, M., Cornuéjols, G.: Submodular set functions, matroids and the greedy algorithm: tight worst-case bounds and some generalizations of the rado-edmonds theorem. Discret. Appl. Math. **7**(3), 251–274 (1984)

6. Das, A., Kempe, D.: Approximate submodularity and its applications: subset selection, sparse approximation and dictionary selection. J. Mach. Learn. Res. **19**(3), 1–34 (2018)
7. Disser, Y., Klimm, M., Schewior, K., Weckbecker, D.: Incremental maximization via continuization. In: Proceedings of the 50th International Colloquium on Automata, Languages and Programming (ICALP), p. 47(17) (2023)
8. Disser, Y., Weckbecker, D.: Unified greedy approximability beyond submodular maximization. SIAM J. Discret. Math. **38**(1), 348–379 (2024)
9. Fadaei, S., Fazli, M.A., Safari, M.A.: Maximizing non-monotone submodular set functions subject to different constraints: combined algorithms. Oper. Res. Lett. **39**(6), 447–451 (2011)
10. Feige, U.: A threshold of $\ln n$ for approximating set cover. J. ACM **45**(4), 634–652 (1998)
11. Feige, U., Mirrokni, V.S., Vondrák, J.: Maximizing non-monotone submodular functions. In: Proceedings of the 48th Annual IEEE Symposium on Foundations of Computer Science (FOCS), pp. 461–471 (2007)
12. Goemans, M.X., Unda, F.: Approximating incremental combinatorial optimization problems. In: Proceedings of the 20th/21st International Workshop on Approximation, Randomization, and Combinatorial Optimization. Algorithms and Techniques. (APPROX/RANDOM), vol. 81, pp. 6:1–6:14 (2017)
13. Gong, S., Nong, Q., Sun, T., Fang, Q., Dingzhu, D., Shao, X.: Maximize a monotone function with a generic submodularity ratio. Theoret. Comput. Sci. **853**, 16–24 (2021)
14. Kelso Jr, A.S., Crawford, V.P.: Job matching, coalition formation, and gross substitutes. Econometrica J. Econom. Soc. 1483–1504 (1982)
15. Kulik, A., Shachnai, H., Tamir, T.: Approximations for monotone and nonmonotone submodular maximization with knapsack constraints. Math. Oper. Res. **38**(4), 729–739 (2013)
16. Lee, J., Mirrokni, V.S., Nagarajan, V., Sviridenko, M.: Maximizing nonmonotone submodular functions under matroid or knapsack constraints. SIAM J. Discret. Math. **23**(4), 2053–2078 (2010)
17. Nemhauser, G.L., Wolsey, L.A., Fisher, M.L.: An analysis of approximations for maximizing submodular set functions—I. Math. Program. **14**, 265–294 (1978)
18. Leme, R.P.: Gross substitutability: an algorithmic survey. Games Econom. Behav. **106**, 294–316 (2017)
19. Disser, Y., Klimm, M., Lutz, A., Strubberg, L.: Incremental-decremental maximization. arXiv: 2508.14516 (2025). https://arxiv.org/abs/2508.14516
20. Disser, Y., Weckbecker, D.: Incremental maximization for a broad class of objectives. In: Proceedings of the 33rd Annual European Symposium on Algorithms (ESA), pp. 91:1–91:14 (2025)

Robust Scheduling on Uniform Machines

New Results Using a Relaxed Approximation Guarantee

Hauke Brinkop[1], David Fischer[2(✉)], and Klaus Jansen[1]

[1] Algorithms and Complexity Group, Kiel University, Kiel, Germany
{hab,kj}@informatik.uni-kiel.de

[2] Institute for Management Science and Operations Research, Helmut Schmidt University, Hamburg, Germany
fischerd@hsu-hh.de

Abstract. We consider $Q||C_{\max}$, the problem of scheduling n jobs on m uniform machines while minimizing the makespan, in an online setting with migration of jobs. In this online setting, the jobs are inserted or deleted over time, and at each step, the goal is to compute a near-optimal solution while reassigning some jobs, such that the overall processing time of reassigned jobs, called migration, is bounded by some factor β times the processing time of the job added or removed. We propose an Efficient Polynomial Time Approximation Schemes (EPTAS) with an additional load error of $\mathcal{O}(\varepsilon p_{\max})$ and constant amortized migration factor β, where $p_{\max}$ is the maximum processing time in the instance over all steps. As an intermediate step, we obtain an Efficient Parameterized Approximation Scheme (EPAS), a $(1+\varepsilon)$-competitive algorithm parameterized by $p_{\max}$ and the number of different processing times d in an instance, with β bounded in a function of $p_{\max}$, d and ε.

Keywords: Scheduling · Approximation · FPT · Online Algorithms

1 Introduction

In this work, we mainly consider the *online* version of the classical scheduling problem on uniform machines $Q||C_{\max}$, which consists of a set $\mathcal{J}$ of n jobs and m uniform machines, where each job J has a processing time $p_J \in \mathbb{N}_{>0}$, and each machine i has a speed $s_i \in \mathbb{Q}_{>0}$. A job J needs time $\frac{p_J}{s_i}$ to be completed on a machine i, and the goal is to find a schedule σ such that the makespan, the maximum completion time of any machine in σ, is minimized. In the online versions of these problems, the set of jobs to be scheduled is not fixed, but instead extended or reduced in discrete steps q. At the start $(q = 0)$, we are given an initial job set $\mathcal{J}_0$, and at any step thereafter $(q \geq 1)$, a single job is added and its processing time revealed, or a single job is removed from the job

Funded by the Deutsche Forschungsgemeinschaft (DFG, German Research Foundation) – Project Number 453769249

The original version of the chapter has been revised. The funding acknowledgement has been corrected. A correction to this chapter can be found at
https://doi.org/10.1007/978-3-032-06706-7_16

J. Matuschke and J. Verschae (Eds.): WAOA 2025, LNCS 16077, pp. 111–126, 2026.
https://doi.org/10.1007/978-3-032-06706-7_8

set. We call this setting for online scheduling problems (fully) dynamic. One may also restrict the setting to be permanent, where jobs may only be added. The run time of algorithms for online problems is usually measured as the time it takes to process one step, e.g. add or remove a single job. In the classical online setting, as first considered by Graham [16], as soon as a job is added to or removed from the instance, one must assign it to a machine, without changing the assignment of any other jobs. It is impossible to maintain an optimal solution in this setting already for $P||C_{\max}$, where all machines have the same speed, even if jobs are only inserted, unless P = NP [26]. We are thus interested in algorithms with small competitive ratio α. For $Q||C_{\max}$, it is known that no better competitive ratio than 2.564 can be achieved without reassignment of jobs [11]. If one does allow a (small) number of jobs to be reassigned at each step however, the goal now becomes to design an algorithm with small competitive ratio α, while simultaneously bounding this reassignment in some measure. The bound on the reassignment is often done in terms of the processing time of the added and removed jobs, as first introduced by Westbrook [29], and called *migration.*

Let Δ denote the difference in value of a variable, set or function compared to itself from one specified step to another, and let $\Delta\mathcal{J}$ be the difference of the job set $\mathcal{J}$ between steps q and $q+1$. An online algorithm has *bounded amortized migration* with amortized migration factor β, if for any step, the overall processing time ξ of jobs moved (from one machine to another) satisfies $\xi \leq \beta\left(\sum_{j\in\Delta\mathcal{J}} p_j + \Delta\Phi\right)$ for some non-negative potential function Φ that is 0 at the beginning. If $|\Delta\Phi|$ is upper bounded by some value β', we say that the algorithm has bounded reassignment potential β'. This amortized notion of the migration factor is e.g. used by Skutella and Verschae [27]. If the reassignment potential is not bounded, this can lead to large changes at a single step, as e.g. is the case for the Polynomial-Time Approximation Scheme (PTAS) for bin packing in the permanent setting by Ivković and Lloyd [19], where sometimes the whole instance has to be repacked. If $\beta' = 0$, then the algorithm has *bounded migration* with *migration factor* β, as used e.g. by Sanders *et al.* [26]. This restricted definition implies that $\xi \leq \beta \sum_{j\in\Delta\mathcal{J}} p_j$ for any step.

Related Work. It is well known that both $Q||C_{\max}$ and $P||C_{\max}$ are strongly NP-hard already in the offline setting, but both problems do admit offline Efficient Polynomial-Time Approximation Schemes (EPTASS) [5,20]. This hardness also holds for the related problems $Q||C_{\min}$ and $P||C_{\min}$, which have the same input parameters as their $C_{\max}$ counterparts, but the the goal is to find a schedule σ such that the minimum completion time of any machine in σ is maximized. While there however exists an EPTAS for $P||C_{\min}$ [27], for $Q||C_{\min}$, currently only a PTAS is known [2]. The first to consider approximation schemes in the migration framework were Sanders *et al.* [26], who gave an $(1+\varepsilon)$-competitive algorithm for permanent $P||C_{\max}$, for all $\varepsilon > 0$, with migration factor $\beta \in 2^{\mathcal{O}(\varepsilon^{-1}(\log \varepsilon^{-1})^2)}$ and run time $2^{2^{\mathcal{O}(\varepsilon^{-1}(\log \varepsilon^{-1})^2)}} + \mathcal{O}(n)$, with the run time later being improved to $2^{\mathcal{O}(\varepsilon^{-2}(\log \varepsilon^{-1})^3)} + \mathcal{O}(n)$ by Ehlers and Jansen [12]. Due to the $(1+\varepsilon)$-competitiveness, this algorithm constitutes an online EPTAS with migration factor β. Aiming to bound the migration factor of an online PTAS in $f(\varepsilon)$ for some computable function f seems natural, as then, for fixed ε, not only is the run time polynomial, but the migration factor is constant. To this end, Sanders *et al.* [26] introduce the term of *robustness*. An online scheduling algorithm is called robust if it has a constant (amortized) migration factor for any fixed $\varepsilon > 0$. Skutella and Verschae [27] managed to expand this result to both dynamic $P||C_{\max}$ and dynamic $P||C_{\min}$ by allowing some amortization with bounded reassignment potential. After showing that there exist no PTASS for general instances of dynamic $P||C_{\min}$ with constant (non-amortized) migration factor, the authors give a robust EPTAS for dynamic $P||C_{\max}$ and dynamic $P||C_{\min}$ with implicitly given amortized migration factor $\beta \in 2^{\mathcal{O}(\varepsilon^{-3}(\log \varepsilon^{-1})^3)}$, reassignment potential β' bounded by an ε fraction of the optimal objective value, and run time $2^{\mathcal{O}(\varepsilon^{-3}(\log \varepsilon^{-1})^3)} + \mathcal{O}(n)$. Gálvez et al. [15] recently gave a robust algorithm for permanent $P||C_{\min}$ with polynomial run time and migration factor, but relaxed competitive ratio of $(4/3+\varepsilon)$. The latest migration-related result for dynamic $Q||C_{\max}$ is by Maack [24], using a strategy employed in previous results without reassignment together with the migration framework to achieve a $(8/3+\varepsilon)$-competitive robust algorithm with amortized migration factor in $\mathcal{O}(\varepsilon^{-1})$, and a $(4+\varepsilon)$-competitive robust algorithm with non-amortized migration factor in $\mathcal{O}(\varepsilon^{-1})$, the latter for $\varepsilon \leq 0.8$. To the best of our knowledge however, there are no known results on robust PTASS for either $Q||C_{\max}$ nor $Q||C_{\min}$ on uniform machines, neither in the permanent nor in the dynamic setting. A way to tackle such gaps that has become popular in recent years are so-called Efficient Parameterized Approximation Schemes (EPASS), $(1+\mathcal{O}(\varepsilon))$-approximation algorithms whose run time is in $f(k,\varepsilon)\,\mathrm{poly}(|I|)$, for some parameter k and the encoding length of the instance $|I|$. Such algorithms can thus be seen as fixed-parameter tractable (FPT) EPTASS, parameterized in k. As Abbasi *et al.* [1] remark, these algorithms are not directly comparable to PTASS. Other notable ways to measure reassignment in an online setting are in the number of jobs reassigned per step (see e.g. [14]), fixed movement costs per item (see e.g. [13]) or job-individual assignment costs (see e.g. [4]).

Our Contribution. Let $p_{\max}$ be the maximum processing time of any job occurring in a scheduling instance. We assume to know $p_{\max}$ a priori, a common assumption in online scheduling [8], which can also be viewed as a particular kind of advice, see e.g. [25]. As our main contribution, we achieve a robust EPTAS with amortized migration factor and bounded reassignment potential, but an additional load error of $\mathcal{O}(\varepsilon p_{\max})$ on the solution, for dynamic $Q||C_{\max}$. An additional load error on a solution means that the loads, i.e. the total processing time of jobs on a machine, of the solution deviate by at most this error from $(1+\mathcal{O}(\varepsilon))$-approximate solutions. The algorithm can thus also be considered an *additive approximation scheme* as defined by Buchem *et al.* [7] with parameter $h = \max(\mathrm{OPT}^\star, p_{\max})$ and optimal makespan $\mathrm{OPT}^\star$.

Theorem 1. *There exists a robust* EPTAS *for dynamic* $Q||C_{max}$ *with additional load error* $\mathcal{O}(\varepsilon p_{\max})$, *run time* $2^{\mathcal{O}(\varepsilon^{-1}\log(\varepsilon^{-1})\log\log(\varepsilon^{-1}))} + \mathcal{O}(|I|)$, *amortized migration factor* $2^{\mathcal{O}(\varepsilon^{-1}\log(\varepsilon^{-1})\log\log(\varepsilon^{-1}))}$, *and reassignment potential bounded by* $3\varepsilon p_{\max}$.

For specific instances (instances with no jobs $< \varepsilon p_{\max}$), the result is a true robust EPTAS with constant migration factor.

Theorem 2. *For instances of dynamic* $Q||C_{max}$ *without jobs of processing time less than* $\varepsilon p_{\max}$, *there exists a robust* EPTAS *with run time* $2^{\mathcal{O}(\varepsilon^{-1}\log(\varepsilon^{-1})\log\log(\varepsilon^{-1}))} + \mathcal{O}(|I|)$, *and migration factor* $2^{\mathcal{O}(\varepsilon^{-1}\log(\varepsilon^{-1})\log\log(\varepsilon^{-1}))}$.

While no true robust EPTASs for general instances, to the best of our knowledge, these results are the first in the direction of robust approximation schemes for both problems on uniform machines. This addresses an open questions posed by Verschae [28] and Maack [24] on the existence of such algorithms for this machine setting. It is also important to note that the need for amortization and the additional load error of Theorem 1 on the load are purely due to the small jobs with processing time $< \varepsilon p_{\max}$. To handle the small jobs in the dynamic setting of general instances, we develop a general framework on how to place and migrate these jobs in an online setting, given a robust algorithm for large jobs ($p_j \geq \varepsilon p_{\max}$). This framework for small jobs is similar to the widely used offline approach [17], and thus might be of independent interest for online scheduling.

Theorem 3 (Dynamic Grouping). *Given a robust algorithm for dynamic* $Q||C_{max}$ *with run time* g *for jobs of size* $\geq \varepsilon p_{\max}$, *where migration is bounded by a function* $\beta(p_J)$ *for each job* J, *there exists a robust algorithm for dynamic* $Q||C_{max}$ *for all jobs with amortized migration bounded by a function* $\bar{\beta}(p_J) = 2 \cdot \beta(3p_J)$, *bounded reassignment potential* $3\varepsilon p_{\max}$, *run time at most* $\mathcal{O}(g \cdot \mathrm{poly}(\varepsilon^{-1}))$, *and with additional load error at most* $\varepsilon p_{\max}(4 + \lceil \frac{3}{m} \rceil) \in \mathcal{O}(\varepsilon p_{\max})$.

As an intermediate step, we also give a robust EPAS for dynamic $Q||C_{\max}$, parameterized in the maximum processing time $p_{\max}$ and the number of different processing times d occurring in the instance. We hereby extend the notion of robustness to EPASS by allowing the (amortized) migration factor to be parameterized.

Lemma 1 (EPAS for Dynamic $Q||C_{max}$). *There exists a robust* EPAS *for dynamic* $Q||C_{max}$ *with parameters* $p_{\max}$ *and* d, *run time* $2^{\mathcal{O}((d+\varepsilon^{-1}\log(\varepsilon^{-1}p_{\max}))\cdot\log(\varepsilon^{-1}p_{\max}))} + \mathcal{O}(|I|)$, *and parameterized migration factor* $2^{\mathcal{O}(d\log(\varepsilon^{-1}p_{\max}))}$.

In the full version of this work [6], we manage to apply all these aforementioned results to $Q||C_{\min}$ as well. This in particular justifies our use of amortization, which is in fact necessary for $Q||C_{\min}$, as Skutella and Verschae [27] have shown that already $P||C_{\min}$ does not admit a robust PTAS with constant, non-amortized migration factor.

The general idea behind obtaining the algorithms stated in Theorem 2 and Theorem 1 is to first round the input values between $\varepsilon p_{\max}$ and $p_{\max}$ in such a way that $p_{\max}, d \in \mathcal{O}(\varepsilon^{-c})$ for some constant $c \in \mathbb{N}$, and afterwards, to apply the EPAS as stated in Lemma 1, directly implying the results. This rounding is done via a more sophisticated version of the classical rounding scheme for scheduling problems [18], as introduced by Berndt *et al.* [3], in order to get the claimed run times and migration factors. The basic concept behind the EPAS for $Q||C_{\max}$ is to first guess the optimal makespan $\mathrm{OPT}^\star$ at each step q. Then, all machines in an instance whose capacity (speed times $\mathrm{OPT}^\star$) is large enough to allow an additive error of $\Omega(p_{\max})$ can be covered in an ILP formulation by a single knapsack-like constraint. Their actual assignment can now be constructed by a simple greedy algorithm, essentially a robust version of the well-known list-scheduling by Graham [16]. Machines whose capacities are not large enough then have a capacity bounded by a polynomial in $p_{\max}$ and ε^{-1}, which allows the efficient usage of a configuration ILP based approach. The main difficulties to overcome here are the combination of both greedy and configurations based approaches for different machines in the same step, and the transitioning of machines whose capacity either becomes large enough, or too small, to be handled by the knapsack-like constraint through the addition/removal of jobs from the instance. This is the main technical contribution. Due to space constraints, in this work, we focus on giving an intuition on the proofs for and techniques behind our main results. All formal proofs can be found in the full version of this paper [6].

2 Preliminaries

We use migration, (amortized) migration factor and reassignment factor as defined in the introduction, where the (amortized) migration factor is called "parameterized" if β is dependent on some parameter k. For minimization problems, a robust EPTAS is an online algorithm, such that for any $\varepsilon > 0$, the algorithm computes an $(1 + \mathcal{O}(\varepsilon))$-approximate solution with (amortized) migration factor dependent only on ε, and whose run time is bounded by $f(\varepsilon) \cdot \mathrm{poly}|I|$, for some computable function f. Similarly, for minimization problems, a robust EPAS is an online algorithm, such that for any $\varepsilon > 0$, the algorithm computes an $(1 + \mathcal{O}(\varepsilon))$-approximate solution with parameterized (amortized) migration factor dependent only on ε and k, and whose run time is bounded by $f(k, \varepsilon) \cdot \mathrm{poly}|I|$, for some parameter k and some computable function f. We assume wlog. for any occurring $\varepsilon > 0$ that ε^{-1} is integral. Let $\mathbb{e}^{(j)}$ denote the j-th unit vector, i.e. $\mathbb{e}^{(j)}_j = 1$ and $\mathbb{e}^{(j)}_{j'} = 0$ for all $j' \neq j$. We use the Bourbaki interval notation, e.g. write $[0; 1[$ for $\{x | 0 \leq x < 1\}$. We extend this notation to integer intervals, e.g. write $]0 \,..\, 4]$ for $\{1, 2, 3, 4\}$. As shorthand, we use $[x] := [1 \,..\, x]$. We use the operator $\nabla^{(q)}$ to denote a variable, set or function at a specific step q. Usually when Δ is used for two different steps q and q', these steps are clear from context.

Let $p \in [p_{\min} \,..\, p_{\max}]^d$ be the vector of (different) job processing times and $\nu \in \mathbb{N}_0^d$ the corresponding multiplicity vector, i.e. for $j \in [d]$ there are ν_j jobs of processing time p_j in the instance. For this encoding, also referred to as a *high-multiplicity* encoding, we formally define a schedule σ to be a map $\sigma\colon [m] \times [d] \to \mathbb{N}$, mapping a machine $i \in [m]$ and a job type $j \in [d]$ to the number of jobs of type j that are scheduled on a machine i, such that $\sum_{i\in[m]} \sigma(i,j) = \nu_j$ for all $j \in [d]$. Note that for this definition, jobs of the same type have become indistinguishable. If not stated otherwise, from now on we exclusively deal with high-multiplicity instances and schedules as redefined above. Adding a job to or removing it from the job set $\mathcal{J}$ is akin to adding/removing a job of this type to/from the multiplicity vector ν. The minimal makespan over all schedules for some multiplicity vector $\nabla^{(q)}\nu$ at step q is called $\mathrm{OPT}^\star(\nabla^{(q)}\nu)$, where we may omit either step, job set, or both if they are evident from context. For some given makespan T, we call $s_i T$ the *capacity* of machine i. For some schedule σ, the *load* of a machine i refers to the total processing time of all jobs assigned to i by σ.

We make use of known results for N-fold ILPs, a specific type of block structured ILPs with an arbitrary objective vector c and constraint matrix A of size $R + SH \times NH$ for $R, S, H, N \in \mathbb{N}$ consisting of R arbitrary rows, with a block diagonal matrix below where each block has size $S \times H$. Such ILPs can be solved in time $2^{\mathcal{O}(RS^2)}\,(RS\|A\|_\infty)^{\mathcal{O}(R^2S+S^2)}\,(NH)^{1+o(1)}$ using the algorithm of Cslovjecsek *et al.* [10]. We also formally define "configurations", i.e. all ways to place jobs with total processing time at most some upper bound u, which are a helpful tool in formulating ILPs for the considered problems.

Definition 1 (Configurations)

$$\mathcal{C}\colon \mathbb{Q}_{\geq 0} \to \mathbb{P}(\mathbb{N}_0^d)\ , \qquad u \mapsto \{c \in \mathbb{N}_0^d | p^\intercal c \leq u\}\ .$$

Robust algorithms often make use of *ILP sensitivity* of a subproblem. Formally, sensitivity measures how much a solution has to change in relation to a change of the input, usually dependent on the right-hand side (RHS) of the ILP. A very strong tool that is often applied in the design of such algorithms (as in the case of the aforementioned robust algorithms for $P||C_{\max}$) is the classical sensitivity result by Cook *et al.* [9] for general ILPs. In this work however, we use a special sensitivity result for N-fold ILPs that admits a smaller bound in the block parameters.

Lemma 2 (Jansen *et al.* [21], Theorem 16). *Let x be an optimal solution of an N-fold ILP with right-hand side b, in particular, $Ax = b$. If the right-hand side changes to b' and the N-fold ILP still has a finite optimum, then there exists and optimal solution x' for b' ($Ax' = b'$) with $\|x - x'\|_1 \leq \mathcal{O}(RS\|A\|_\infty)^{RS} \cdot \|b - b'\|_1$.*

It is well known that there exists an ILP formulation for (offline) $Q||C_{\max}$ that has N-fold structure [22], the so-called Assignment-ILP. This formulation can be adapted to represent dynamic $Q||C_{\max}$ via simple and straightforward modifications incorporating *lexicographically minimal solutions* as used by Skutella and Verschae [27] that particularly ensure that the changes of the RHS are bounded.[1] Using the algorithm by Cslovjecsek *et al.* [10], one can then solve

[1] We thank an anonymous reviewer for pointing out this approach and the necessary modifications to us.

both problems optimally in time $(p_{\max})^{\mathcal{O}(d^2)}\mathcal{O}(n)$, and with migration factor $(p_{\max})^{\mathcal{O}(d)}$ by using Lemma 2. However, directly applying a standard rounding scheme (e.g. [17]) on this approach in order to obtain an EPTAS, as was successfully done before for the offline problems, suffers from the same problems as other ILP formulations, even with the special sensitivity: The sensitivity bound is dependent not only on ε^{-1}, and the resulting algorithm therefore not robust.

Our results now stem from a new N-fold-ILP formulation for online $Q||C_{\max}$ with additionally overall small dimension and coefficient, and a non-standard rounding scheme depending on $p_{\max}$ instead of T. This allows us to achieve robustness when using Lemma 2 to accommodate changes of the solution, together with carefully designed techniques to change the RHS in robust terms of the size of the inserted/removed job. Moreover, the small overall dimensions allow us to use a state-of-the-art algorithm for general ILPs, which yields better run time in the parameters in the approximate parameterized setting (Lemma 1) compared to the exact approach.

3 Construction of the Parameterized Robust Algorithm

In this section, we aim to give a high-level idea on the proof of Lemma 1, i.e. the EPAS for dynamic $Q||C_{\max}$. We show the Lemma by iteratively constructing an algorithm and proving all the claimed properties for it; we refer to this algorithm simply as "the robust parameterized algorithm".

By definition of $\mathrm{OPT}^\star$, any solution where the load of any machine i is smaller or equal to its capacity $s_i\,\mathrm{OPT}^\star$ is an optimal solution. The construction of the parameterized robust algorithm is subdivided as follows: First, we show how to obtain such an algorithm if we are ensured that each machine exceeds a given minimum capacity. Then, we show how to do this if all machines subceed this very threshold. Eventually we show how we can combine these approaches to get rid of assumptions about machine capacities, which is by far the most difficult part. The upcoming (sub-)sections make use of the threshold

$$\ell := (1+\varepsilon)^{1+\lceil \log_{1+\varepsilon}(2\varepsilon^{-1}p_{\max})\rceil} \; .$$

Note that

$$\varepsilon(1+\varepsilon)^{-1}\ell \geq \varepsilon\ell \geq 2p_{\max} \; , \qquad \text{and } \ell \leq \mathrm{poly}(\varepsilon^{-1}, p_{\max}) \; .$$

In the following, we assume that machine speeds are all powers of $(1+\varepsilon)$. We can achieve this property by rounding each machine speed up to the next power of $(1+\varepsilon)$. This process introduces an additional multiplicative error of at most $(1+\varepsilon)$ on the solution. Let τ denote the number of different machine speeds after rounding. Let $s \in \mathbb{Q}^\tau$ be the vector of rounded machine speeds and $\mu \in \mathbb{N}_0^\tau$ the corresponding multiplicity vector.

3.1 Case 1: All Machines (always) Have Large Capacity

In this subsection, we assume that at all times, every machine has a capacity of *at least* ℓ w.r.t $\mathrm{OPT}^\star$. To compute approximate solutions at step q, here, we use a simple parameterized robust adaptation of the list-scheduling algorithm [16] that allows for insertion and deletion of jobs with migration bounded in $p_j + p_{\max}$ for insertion/deletion of a job j, returning a solution at every step q where each machine has completion time at most $\frac{p^\intercal \nu}{s^\intercal \mu} + p_{\max} \leq \mathrm{OPT}^\star + p_{\max}$. Since $\varepsilon\ell \geq p_{\max}$, and all machines of each type t have capacity $s_t \,\mathrm{OPT}^\star \geq \ell$ by assumption, this is a $(1+\varepsilon)$–approximate solution at every step q due to

$$(1+\varepsilon)s_t \,\mathrm{OPT}^\star \geq s_t \,\mathrm{OPT}^\star + \varepsilon\ell \geq s_t \,\mathrm{OPT}^\star + p_{\max} \;,$$

with migration factor at most $p_{\max} + 1$.

3.2 Case 2: All Machines (always) have Small Capacity

For this section, we assume that at all times, every machine has a capacity of *at most* ℓ w.r.t $\mathrm{OPT}^\star$. Then, all we need is the fact that $\ell \leq \mathrm{poly}(\varepsilon^{-1}, p_{\max})$, i.e. ℓ is bounded by some fixed polynomial in ε^{-1} and $p_{\max}$. Consider the following, configuration-based ILP formulation for $Q||C_{\max}$ and some makespan guess T.

<u>Configuration-ILP</u>

$$\sum_{t\in[\tau]} \sum_{c\in\mathcal{C}(s_t T)} x_{t,c} \cdot c = \nu$$

$$\sum_{c\in\mathcal{C}(s_t T)} x_{t,c} = \mu_t \qquad \forall t \in [\tau]$$

<u>with variables</u>

$\mathtt{var}\; x_{t,c} \in \mathbb{N}_0 \qquad \forall t \in [\tau], c \in \mathcal{C}(s_t T)$

To obtain a solution at step q, we could simply guess the makespan T, discard machine types where $s_t T < 1$, resulting in at most $1 + \log_{1+\varepsilon} \ell \in \mathcal{O}(\varepsilon^{-1} \log(\varepsilon^{-1} p_{\max}))$ machine types that can be assigned a non-zero configuration, as all capacities $s_t T$ are bounded by ℓ by definition of the case. We could then solve this using a standard ILP algorithm parameterized by the number of constraints ($\leq d + 1 + \log_{1+\varepsilon} \ell$) and the largest coefficient ($\leq \ell + 1$). However, doing this makes going from step q to $q+1$ – in contrast to the case where all machines have large capacities – not that straightforward. This is only due to removal: If we were only allowed to add jobs, we could simply increase T when necessary, possibly include some newly available configurations into the ILP, convert our old solution to a solution of the new ILP by adding "0"'s for the new entries (those that correspond to newly introduced configurations) and then, as this ILP has N-fold structure, use Lemma 2 to find a solution nearby. However, if we can remove jobs, we cannot convert our old solution to one with smaller T in general, as we might have to change the assignment of $\Theta(m)$ machines, if they are assigned a configuration that has to be removed from the ILP due to the decrease of T.

This problem also occurs for identical machines, and Skutella and Verschae [27] solved it in the following way: They list a superset of necessary configurations and use a linear objective $\varphi^{(\text{lex})}$ that encodes "Minimizing the makespan" via *lexicographically minimal solutions.* We can proceed similarly here as – due to the assumption that capacities never exceed ℓ – we can simply list all configurations whose load does not exceed ℓ. We additionally need to generalize the construction of the linear objective such that it can be applied to uniform machines with a few machine types, which is rather straightforward. Recall that at most $1+\log_{1+\varepsilon}\ell \in \mathcal{O}(\varepsilon^{-1}\log(\varepsilon^{-1}p_{\max}))$ machines can be assigned a non-zero configuration at each step, and if we list only machine types that – before or after the operation – can actually be assigned a job, this number stays $\mathcal{O}(\varepsilon^{-1}\log(\varepsilon^{-1}p_{\max}))$. We then achieve the desired result using the sensitivity bound as stated in Lemma 2. Note that the objective constructed leads to unique optimal solutions. To compute a new solution it thus suffices to simply *solve* the ILP, while only implicitly bounding the changes of variables.

3.3 Combining both Cases

Let us call the algorithm for only large capacities "blue algorithm", and the one for only small capacities "red algorithm". Our goal is now to partition the jobs and machines into red and blue jobs and machines at each step q, respectively, and use the blue algorithm for blue jobs and blue machines, and the red algorithm for red jobs and machines to compute feasible approximate solutions for both sets in the partition, and by that, for the whole instance. Let $\mu^{(\text{blue})} \in \mathbb{N}^{\tau}$ be the vector of blue machines, $\nu^{(\text{blue})} \in \mathbb{N}^{d}$ be the vector of blue jobs at some step q. Consider the following ILP that is parameterized by ν, $\mu^{(\text{blue})}$ and some value $\alpha \in \mathbb{Q}_{\geq 0}$. This ILP amounts to a modified version of the Configuration-ILP (with upper bound ℓ on the configurations of each machine type), extended by a knapsack-like constraint with capacity α dependent on the total capacity of the blue machines. The total capacity on the blue machines is itself dependent on the current number of blue machines $\mu^{(\text{blue})}$ and the current makespan guess T. The parameter ν gives us the current multiplicity of each job type in the instance. The objective function $\varphi^{(\text{lex})}$ again encodes "Minimizing the makespan" on the given set of configurations.

(ILP^{dynamic})

$$\min \varphi^{(\text{lex})}(x)$$

s.t.

$$p^{\mathsf{T}}\nu^{(\text{blue})} \leq \alpha \qquad (1)$$

$$\sum_{c\in\mathcal{C}(\ell)} x_{t,c} = \mu_t - \mu_t^{(\text{blue})} \qquad \forall t \in [\tau] \qquad (2)$$

$$\sum_{\substack{t\in[\tau]\\ c\in\mathcal{C}(\ell)}} x_{t,c}\cdot c + \nu^{(\text{blue})} = \nu \qquad (3)$$

<u>with variables</u>

var $x_{t,c} \in \mathbb{N}_0 \qquad \forall t \in [\tau], c \in \mathcal{C}(\ell)$

var $\nu^{(\text{blue})} \in \mathbb{N}_0^d$

<u>with parameters</u>

param $\nu \in \mathbb{N}_0^d$

param $\mu^{(\text{blue})} \in \mathbb{N}_0^{\tau}$

param $\alpha \in \mathbb{Q}_{\geq 0}$

Variables $x_{t,c}$ represent the configurations of red machines, and thus encode all jobs placed on these machines. Variables $\nu^{(\text{blue})}$ represent all jobs placed on blue machines instead. Constraint 1 ensures that the total load of jobs on blue machines represented by $\nu^{(\text{blue})}$ does not exceed the currently available capacity on blue machines α. Constraint 2 guarantees that the number of configurations of each machine type t in any solution matches the current number of red machines $\mu_t - \mu_t^{(\text{blue})}$ of this type. Finally, Constraint 3 makes sure that each job is assigned via a configuration to a red machine, or assigned to the blue machines via $\nu^{(\text{blue})}$.

Let T be a guess on the optimal makespan at some step q. We are now able to compute an approximate solution for dynamic $Q||C_{\max}$ at step q by setting $\mu_t^{(\text{blue})} := \mu_t$ if $s_t T \geq \ell$ and $\mu_t^{(\text{blue})} := 0$ otherwise, for each $t \in [\tau]$, and setting $\alpha := T s^\intercal \mu^{(\text{blue})}$, i.e. the total capacity of the blue machines. Using the red algorithm on such a modified ILP then computes a solution with makespan T on the red machines for some subset of the jobs $\nu - \nu^{(\text{blue})}$, such that the load of jobs not placed on the red machines $p^\intercal \nu^{(\text{blue})}$ is smaller or equal to the total capacity on the blue machines $T s^\intercal \mu^{(\text{blue})}$, if it exists. If a solution exists, using the blue algorithm to compute an assignment of $\nu^{(\text{blue})}$ on $\mu^{(\text{blue})}$ then returns a solution with makespan at most $(1+\varepsilon)T$ by its definition. Combining these solutions gives us a schedule with a makespan of at most $(1+\varepsilon)T$ over all machines. If no solutions exists to the ILP with makespan T, we can correctly conclude there exists no solution with this makespan.

However, to go from step q to $q+1$, we need to repartition the set of machines into blue and red machines depending on the possibly different optimal makespan, a process we call *recolouring.* This recolouring affects $\mu^{(\text{blue})}$, changing the RHS of (ILP^{dynamic}), and thus impacting the bound on the sensitivity and the migration factor. In fact, setting up (ILP^{dynamic}) in such a way that, at each step q, using the parameterized red and blue algorithms leads to the computation of *robust* solutions turns out to be quite challenging. The two main problems are (i) that we might have to recolour $\Omega(m)$ machines, if their capacities cross the threshold ℓ due to changes in the optimal makespan; and (ii) the RHS of the knapsack constraint is highly dependent on the optimal makespan: even a tiny change in the optimum can trigger a $\omega_{\varepsilon,p_{\max},d}(1)$ change. Both scenarios would directly lead to unbounded migration.

Our approach to tackle this is twofold: (i) we discretize the makespan guarantees in order to not be forced to react to small changes all the time; and (ii) we distribute the work that needs to be done when the makespan guarantee changes over the preceding steps, such that in the step where this change occurs, the remaining necessary changes are properly bounded. The latter means that we recolour machines over time such that when our guarantee changes, there are at most a few machines left to recolour; and analogously we change the RHS of the knapsack constraint over time. First, let us capture the restriction on the machine colouring by an invariant:

Invariant 3.A (Machine Colouring Invariant). *Given a bound $T \in \mathbb{Q}_{\geq 0}$ on the objective value, for any blue machine of type t we have $s_t T \geq \ell$; and for any red machine of type t we have $s_t T \leq \ell$.*

Since our ILP is parameterized by the value α that represents the part of the overall capacity on blue machines that may be used by the ILP, our next step is to formalize this interaction, i.e. ensure that we restrict ourselves to values of α that do not exceed the overall total capacity of the blue machines, for which we actually allow another additional multiplicative error of $(1+\varepsilon)$.

Invariant 3.B (Compatible Colouring Invariant)

$$0 \leq \alpha \leq (1+\varepsilon)Ts^{\intercal}\mu^{(\text{blue})} .$$

┘

For fixed T and $\mu^{(\text{blue})}$, we will also call $(1+\varepsilon)Ts^{\intercal}\mu^{(\text{blue})}$ the "available blue area (for T and $\mu^{(\text{blue})}$)". As mentioned before, the direct application of Lemma 2 does not lead to a sufficiently small sensitivity bound, due to possible large differences in the RHS b and b' – this is the main issue to overcome. Given a (approximate) makespan T, we require the red schedule to not exceed T, while for the blue schedule we only require a makespan of $(1+\varepsilon)T$. This will allow us to bound the number of machine recolourings by some kind of witness argument.

One property of the objective $\varphi^{(\text{lex})}$ is that, if applied to all variables, optimal solutions are unique as long as feasible solutions can be bounded by some fixed value. This property also applies to (ILP^{dynamic}). Our main focus thus does not rely on finding solutions that meet the requirements on the migration, but to adjust the parameters in such a way that the unique optimal solution to these parameters does so. Thus, instead of classifying solutions, we classify triples of parameter values, called "states", calling those "T-valid" where the optimal solutions meets our requirements (for makespan T):

Definition 2 (States, associated solutions, ILP-makespan, T-validity). *A triple of the ILP parameters* $(\alpha, \mu^{(\text{blue})}, \nu)$ *is called* state. *The (unique) optimal solution* $(x, \nu^{(\text{blue})})$ *of* (ILP^{dynamic}) *for these parameters is called* the associated solution. *The* (ILP) makespan of a state *is the (ILP) makespan of the associated solution, i.e.* $C_{\max}(x)$. *If the ILP is infeasible, we define the makespan to be* ∞.

A state S is T-valid *iff Invariant 3.A and Invariant 3.B both hold for T; and the associated solution of S has (ILP) makespan at most T.*

Note that if there is a T-valid state, there is also T'-valid state for any $T' \geq T$. Solving (ILP^{dynamic}) for a T-valid state directly yields a solution with makespan at most $(1+\varepsilon)^2T$, with the error being introduced by the additional εT error on the blue area, and the actual placement of blue jobs on blue machines by the greedy algorithm. As we want to solve the ILP for (approximately) the optimal makespan OPT*, we introduce the following definition.

Definition 3 (Optimal ILP Makespan). *Given a job vector* $\nu \in \mathbb{N}_0^d$, $\text{OPT} \in \mathbb{Q}_{\geq 0}$ *is the smallest number such that there is a* OPT*-valid state S. Moreover, we define* $\text{OPT}^{\downarrow}$ *to be the largest power of* $(1+\varepsilon)$ *strictly smaller than* OPT; *and* $\text{OPT}^{\uparrow} := (1+\varepsilon)\,\text{OPT}^{\downarrow}$.

Definition 4 (Valid States). *A state S is valid if it is* $\mathrm{OPT}^{\uparrow}$*-valid.*

Note that there is not necessarily just one, but often times many valid states. We will exploit this heavily, by only converting a valid state to another valid state with "nice" properties, i.e. bounded migration. By the definition of OPT and Invariant 3.B, $\mathrm{OPT}^{\star} \geq \mathrm{OPT}$. Thus, computing a solution to the ILP for any valid state always returns an at most $(1+\varepsilon)^3 \in (1+\mathcal{O}(\varepsilon))$-approximate solution. Now, we can discuss the general idea of achieving bounded migration using this framework of ($\mathrm{ILP}^{\mathrm{dynamic}}$), states and associated solutions. Let $S = (\alpha, \mu^{(\mathrm{blue})}, \nu)$ be a valid state and $\bar{\nu}$ a changed job set. We assume wlog. that $\bar{\nu}$ differs from ν by exactly one job. Our goal is to always find a valid state $\bar{S} = (\bar{\alpha}, \bar{\mu}^{(\mathrm{blue})}, \bar{\nu})$ such that $|\Delta\alpha| + \|\Delta\mu^{(\mathrm{blue})}\|_1 \leq \|\Delta b\|_1 + 1 \leq \mathrm{poly}(p_{\max}, \varepsilon^{-1})$. Plugging our ILP parameters into Lemma 2 then directly ensures that the unique associated solutions have properly bounded ℓ_1-distance, and thus we obtain bounded migration when solving the ILP. Since we require $\bar{S}$ to be valid, the associated solution to this state is then again a $(1+\mathcal{O}(\varepsilon))$-approximate solution to the new job set, and the algorithm therefore is a robust EPAS. We construct such an $\bar{S}$ by carefully recolouring machines and changing α from S to $\bar{S}$. To do this, we only consider machines of capacity exactly ℓ for recolouring in each step: Due to the discretization of the makespan to $\mathrm{OPT}^{\uparrow}$, these are the only types of machines that are "critical", i.e. they may be forced to be either red (if $\mathrm{OPT}^{\uparrow}$ decreases) or blue (if $\mathrm{OPT}^{\uparrow}$ increases) at the next makespan change, but may not be appropriately coloured yet before this change. At the same time, these machines are the only ones allowed to be coloured either red or blue for $\mathrm{OPT}^{\uparrow}$ by Invariant 3.A.

First, let us consider the case where $\Delta\,\mathrm{OPT}^{\uparrow} = 0$, i.e. the makespan guarantee does not change. When some job j is added, we only increase the value of α if necessary – thus, we act some kind of "lazy". If an increase is necessary but we cannot increase α due to Invariant 3.B, we can conclude that there is some red machine that can be recoloured blue – otherwise $\mathrm{OPT}^{\uparrow}$ would have increased. By doing so, due to the makespan guarantee on blue machines being only $(1+\varepsilon)\,\mathrm{OPT}^{\uparrow}$, we gain an additional space of at least $\varepsilon\ell \geq p_{\max}$ that can be used for the necessary increase of α. Thus, recolouring a single machine suffices in this case. When some job j is removed, as long as there is a blue machine that must not be blue after $\mathrm{OPT}^{\uparrow}$ decreases, we recolour a single machine if possible without ending up in an invalid state, to prepare for an eventual decrease of $\mathrm{OPT}^{\uparrow}$. If there is no such machine, we decrease α by $\mathcal{O}(p_{\max})$ if possible (and the maximal possible value otherwise), while ensuring that α does not subceed a specific threshold. The latter is necessary to ensure validity once has $\mathrm{OPT}^{\uparrow}$ dropped, i.e. just after a drop we want α to be the maximal possible value that it may take under the new $\mathrm{OPT}^{\uparrow}$-value and its implied blue area.

An increase of $\mathrm{OPT}^{\uparrow}$ only occurs when a job is added and no more critical machines can be recoloured blue. Since an increase of $\mathrm{OPT}^{\uparrow}$ increases the available blue area on every blue machine by at least $\varepsilon\,\mathrm{OPT}^{\uparrow} \geq \varepsilon\ell$, we can simply increase α by as much. A decrease of $\mathrm{OPT}^{\uparrow}$ implies the removal of a job, and we recolour the machines that we need to recolour (some of those that are currently blue but may not be allowed to be blue under the new value of $\mathrm{OPT}^{\uparrow}$), and reduce α to ensure that it does not exceed its maximal value in respect to the new, reduced $\mathrm{OPT}^{\uparrow}$-value. Due to our continuous blue $\rightarrow$ red recolouring at each job removal, we can show that this affects only a small number of machines and a small change in α at the point of $\mathrm{OPT}^{\uparrow}$ decrease, as otherwise, they could have (and would have) been recoloured before, or were never coloured blue for this $\mathrm{OPT}^{\uparrow}$-value in the first place. Crucial for proving this is the fact that, by the continuous recolouring, α always uses almost all available blue area, implying that if the recolouring and α changes were too big, already the job set previous to the removal would have been valid for a decreased $\mathrm{OPT}^{\uparrow}$-value – a contradiction to the validity of this state. We thus show that we go from valid state to valid state when adding or removing jobs, while changing the state parameters only in terms of $\mathrm{poly}(\varepsilon^{-1}, p_{\max})$. As the parameters are the only changes in the RHS of ($\mathrm{ILP}^{\mathrm{dynamic}}$), besides the job set being changed by 1 for the inserted or removed job, this then directly implies a bound $\mathrm{poly}(\varepsilon^{-1}, p_{\max})$ on the change of the RHS.

With this result in hand, and the fact that ($\mathrm{ILP}^{\mathrm{dynamic}}$) has N-fold structure, we can use the sensitivity result stated in Lemma 2 to get our bound on the sensitivity, and thus, migration factor. As ($\mathrm{ILP}^{\mathrm{dynamic}}$) has only a relatively small number of constraints in total, the most efficient way to compute the (unique) solutions at each step q is via an algorithm for solving ILPs with small number of constraints by Jansen and Rohwedder [20]. Details on the sensitivity and ILP solver can be found in the full version. This is also where the run time and migration factor of the algorithms come from, as solving the ILP and bounding the change in parameters is the bulk of the computational work.

4 Getting Rid of the Parameterization

With the EPASS in hand, we can now describe the necessary modifications made to any instance in order to achieve the EPTAS results as stated in Theorem 2 and Theorem 1. For this, we generally use the standard approach of reducing the complexity of an instance by preprocessing, i.e. rounding and simplifying, the input at the cost of an additional error on the solutions. In contrast to the usual approaches however, which do the rounding dependent on a guess of the optimal objective value $\mathrm{OPT}^{\star}$, our rounding depends on $p_{\max}$ instead, as $\mathrm{OPT}^{\star}$ itself is not fixed but dependent on the current job set at a specific step.

Our preprocessing follows a recent improvement over the well known rounding and scaling scheme for scheduling problems. This improvement, as first described by Berndt *et al.* [3], aims to reduce the largest absolute entry in the configurations necessary to describe a scheduling problem from (ε^{-1}) to $\log(\varepsilon^{-1})$ after rounding. They successfully apply this technique to reduce the ℓ_1-norm of the columns, and thus the hereditary discrepancy of the matrix, from ε^{-1} to $\log(\varepsilon^{-1})$ in a Configuration-ILP formulation for offline $P||C_{\max}$. Solving this ILP with the algorithm by Jansen and Rohwedder [20], whose run time depends on the hereditary discrepancy, then yields an EPTAS with run time $2^{\mathcal{O}(\varepsilon^{-1}\log(\varepsilon^{-1})\log\log(\varepsilon^{-1}))}$.

As ($ILP^{dynamic}$) is an extension of Configuration-ILP, and we use the same algorithm by Jansen and Rohwedder [20] to solve the ILP, we also make use of this improved rounding scheme to not only reduce $p_{\max}$ and d to values only dependent on ε^{-1}, but also to reduce the ℓ_1-norm of the configuration part of the ILP. Additionally however, we have to reduce the largest coefficient in the other constraints of ($ILP^{dynamic}$) to $\log(\varepsilon^{-1})$ as well, in order to get the proper bound on the ℓ_1-norm of the columns. Luckily, after applying the rounding above, the only other time where a coefficient larger than $\log(\varepsilon^{-1})$ appears in ($ILP^{dynamic}$) is in the (single) knapsack constraint. To deal with this, we use a technique introduced by Knop *et al.* [23] called "Coefficient Reduction" to reduce the largest coefficient of this constraint to 1, while introducing an additional $\log \varepsilon^{-1}$ constraints and variables as a trade-off. This such rounded and reduced ILP we call (${ILP^{dynamic}}_{decr}$). As our rounding scheme is dependent on $p_{\max}$, we first consider only instances with processing time values in $[\varepsilon p_{\max}, p_{\max}]$. Instances that include processing times $< \varepsilon p_{\max}$ are dealt with using our dynamic grouping technique as stated in Theorem 3.

For the former type of instances, any solution to (${ILP^{dynamic}}_{decr}$) then is a $(1+\mathcal{O}(\varepsilon))$-competitive solution for the original instance, and all relevant parameters for solving the ILP are now bound in terms of ε^{-1} only. Again, using the same functions for inserting and removing jobs as in the parameterized case, we transition from step q to $q+1$. Here, this keeps the change of the RHS bounded in terms of $\mathrm{poly}(\varepsilon^{-1})$. Thus, by again using Lemma 2 to bound sensitivity and the algorithm by Jansen and Rohwedder [20] to solve the ILP, we achieve a run time of $2^{\mathcal{O}(\varepsilon^{-1}\log(\varepsilon^{-1})\log\log(\varepsilon^{-1}))} + \mathcal{O}(|I|)$, matching the currently fastest EPTAS for offline scheduling of jobs on identical machines [3], and a migration factor of $2^{\mathcal{O}(\varepsilon^{-1}\log(\varepsilon^{-1})\log\log(\varepsilon^{-1}))}$. Note that by modifying a standard rounding scheme (e.g. [17]) to be dependent on $p_{\max}$ instead of T, results with the same approximation guarantees can also be achieved by applying this rounding to the exact approach. However, this results in an algorithm with a worse run time of $2^{\mathcal{O}(\varepsilon^{-2}\log^3(\varepsilon^{-1}))} + \mathcal{O}(|I|)$, and a worse migration factor $2^{\mathcal{O}(\varepsilon^{-1}\log^2(\varepsilon^{-1}))}$.

References

1. Abbasi, F., et al.: Parameterized approximation schemes for clustering with general norm objectives. In: Proceedings of FOCS 2023, pp. 1377–1399 (2023)
2. Azar, Y., Epstein, L.: Approximation schemes for covering and scheduling in related machines. In: Jansen, K., Rolim, J. (eds.) APPROX 1998. LNCS, vol. 1444, pp. 39–47. Springer, Heidelberg (1998). https://doi.org/10.1007/BFb0053962
3. Berndt, S., Brinkop, H., Jansen, K., Mnich, M., Stamm, T.: New support size bounds for integer programming, applied to makespan minimization on uniformly related machines. In: Proceedings of ISAAC 2023. LIPIcs, pp. 13:1–13:18. Schloss Dagstuhl - Leibniz-Zentrum für Informatik (2023)
4. Berndt, S., Deppert, M.A., Jansen, K., Rohwedder, L.: Load balancing: the long road from theory to practice. In: Proceedings of ALENEX 2022, pp. 104–116 (2022)

5. Berndt, S., Eberle, F., Megow, N.: Online load balancing with general reassignment cost. Oper. Res. Lett. **50**(3), 322–328 (2022)
6. MISC
7. Buchem, M., Rohwedder, L., Vredeveld, T., Wiese, A.: Additive approximation schemes for load balancing problems. In: Proceedings of ICALP 2021. LIPIcs, pp. 42:1–42:17. Schloss Dagstuhl – Leibniz-Zentrum für Informatik (2021)
8. Chekuri, C., Khanna, S., Zhu, A.: Algorithms for minimizing weighted flow time. In: Proceedings of STOC 2001, pp. 84–93. ACM (2001)
9. Cook, W.J., Gerards, A.M.H., Schrijver, A., Tardos, É.: Sensitivity theorems in integer linear programming. Math. Program. **34**(3), 251–264 (1986)
10. Cslovjecsek, J., Eisenbrand, F., Hunkenschröder, C., Rohwedder, L., Weismantel, R.: Block-structured integer and linear programming in strongly polynomial and near linear time. In: Proceedings of SODA 2021, pp. 1666–1681. SIAM (2021)
11. Ebenlendr, T., Sgall, J.: A lower bound on deterministic online algorithms for scheduling on related machines without preemption. TOCS **56**(1), 73–81 (2015)
12. Ehlers, T., Jansen, K.: Online-scheduling on identical machines with bounded migration. In: Proceedings of SYNASC 2013, pp. 361–366. IEEE Computer Society (2013)
13. Feldkord, B., et al.: Fully-dynamic bin packing with little repacking. In: Proceedings of ICALP 2018. LIPIcs, pp. 51:1–51:24. Schloss Dagstuhl - Leibniz-Zentrum für Informatik (2018)
14. Foussoul, A., Goyal, V., Kumar, A.: Fully-dynamic load balancing. In: Proceedings of IPCO 2024, pp. 182–195. Springer, Heidelberg (2024)
15. Gálvez, W., Soto, J.A., Verschae, J.: symmetry exploitation for online machine covering with bounded migration. TALG **16**(4) (2020)
16. Graham, R.L.: Bounds on multiprocessing timing anomalies. SIAP **17**(2), 416–429 (1969)
17. Hochbaum, D.S.: Approximation algorithms for NP-hard problems. SIGACT News **28**(2), 40–52 (1997)
18. Hochbaum, D.S., Shmoys, D.B.: Using dual approximation algorithms for scheduling problems theoretical and practical results. J. ACM **34**(1), 144–162 (1987)
19. Ivković, Z., Lloyd, E.L.: Fully dynamic bin packing. In: Fundamental Problems in Computing: Essays in Honor of Professor Daniel J. Rosenkrantz, pp. 407–434. Springer, Heidelberg (2009). https://doi.org/10.1007/978-1-4020-9688-4_15
20. Jansen, K., Lassota, A., Rohwedder, L.: Near-linear time algorithm for n-fold ILPs via color coding. SIDMA **34**(4), 2282–2299 (2020)
21. Jansen, K., Rohwedder, L.: On integer programming, discrepancy, and convolution. Math. Oper. Res. (2022)
22. Knop, D., Koutecký, M.: Scheduling meets n-fold integer programming. J. Sched. **21**(5), 493–503 (2018)
23. Knop, D., Pilipczuk, M., Wrochna, M.: Tight complexity lower bounds for integer linear programming with few constraints. ACM Trans. Comput. Theory **12**(3) (2020)
24. Maack, M.: Online load balancing on uniform machines with limited migration. Oper. Res. Lett. **51**(3), 220–225 (2023)
25. Renault, M.P., Rosén, A., van Stee, R.: Online algorithms with advice for bin packing and scheduling problems. Theor. Comput. Sci. **600**, 155–170 (2015)
26. Sanders, P., Sivadasan, N., Skutella, M.: Online scheduling with bounded migration. Math. Oper. Res. **34**(2), 481–498 (2009)

27. Skutella, M., Verschae, J.: Robust polynomial-time approximation schemes for parallel machine scheduling with job arrivals and departures. Math. Oper. Res. **41**(3), 991–1021 (2016)
28. Verschae, J.: The Power of Recourse in Online Optimization: Robust Solutions for Scheduling, Matroid and MST Problems. PhD thesis, Technical University of Berlin (2012)
29. Westbrook, J.: Load balancing for response time. J. Algor. **35**(1), 1–16 (2000)

Max-Min and 1-Bounded Space Algorithms for the Bin Packing Problem

Hiroshi Fujiwara[1(✉)], Rina Atsumi[2], and Hiroaki Yamamoto[1]

[1] Shinshu University, Nagano, Japan
fujiwara@shinshu-u.ac.jp
[2] Mitsubishi Electric Digital Innovation Corporation, Tokyo, Japan

Abstract. In the (1-dimensional) bin packing problem, we are asked to pack all the given items into bins, each of capacity one, so that the number of non-empty bins is minimized. Zhu [Chaos, Solitons & Fractals 2016] proposed an approximation algorithm MM that sorts the item sequence in a non-increasing order by size at the beginning, and then repeatedly packs, into the current single open bin, first as many of the largest items in the remaining sequence as possible and then as many of the smallest items in the remaining sequence as possible. In this paper we prove that the asymptotic approximation ratio of MM is at most 1.5. Next, focusing on the fact that MM is at the intersection of two algorithm classes, max-min algorithms and 1-bounded space algorithms, we comprehensively analyze the theoretical performance bounds of each subclass derived from the two classes. Our results include a lower bound of 1.25 for the intersection of the two classes. Furthermore, we extend the theoretical analysis over algorithm classes to the cardinality constrained bin packing problem.

Keywords: bin packing problem · online algorithm · approximation algorithms analysis

1 Introduction

The bin packing problem is a basic combinatorial optimization problem in which, given a sequence of items, each of size at most one, the number of non-empty bins is minimized under the constraint that each item is contained in some bin of capacity one. The bin packing problem is known to be NP-complete [13]. Many approximation algorithms that run in polynomial time and assign bins as close to the minimum number as possible have been studied (see [12,15] for example).

Zhu proposed an algorithm MM that sorts the item sequence in a non-increasing order by size, and then repeatedly packs at the beginning, into the current single open bin, first as many of the largest items in the remaining item sequence as possible and then as many of the smallest items in the remaining item sequence as possible [20]. His paper shows through several computational experiments that algorithm MM has good practical performance.

In this paper we conduct a theoretical performance analysis of algorithm MM as well as the relevant classes of algorithms (see Fig. 1). We refer to an

J. Matuschke and J. Verschae (Eds.): WAOA 2025, LNCS 16077, pp. 127–141, 2026.
https://doi.org/10.1007/978-3-032-06706-7_9

algorithm as a *max-min* algorithm if it first sorts the given item sequence in a non-increasing order by size and then repeatedly packs either the head or tail item in the currently remaining item sequence into a bin, without seeing other items. From the above explanation, MM is obviously a max-min algorithm.

There is a closely related class of algorithms. We refer to an algorithm as a *pre-sorted online* algorithm if it first sorts the given item sequence in a non-increasing order by size and then repeatedly packs the head item in the currently remaining item sequence into a bin, without seeing other items. Algorithm NFD (Next Fit Decreasing) [2] and algorithm FFD (First Fit Decreasing) [14] are typical examples and have been studied for a long time. In the context of priority algorithms [7,19], a pre-sorted online algorithm can be viewed as a *fixed priority* algorithm with a non-increasing order by size. Any pre-sorted online algorithm is also a max-min algorithm, in the sense that it is allowed to choose the head or tail item but only the head item is chosen.

Our research is further motivated by the potential of max-min algorithms as heuristics for high-dimensional packing problems to which pre-sorted online algorithms are often applied (see [8,17] for example). How useful will it be to pack even the smallest items?

Another aspect is the space complexity. During the execution of an algorithm, a bin is called *open* if some item has already been packed in it and the algorithm may pack further items into it. The algorithm keeps track of the sum of the sizes of the items in each open bin. An algorithm is called a *bounded space* algorithm if there exists a constant B such that the algorithm maintains simultaneously at most B open bins (for example, algorithm $Harmonic$ [16]). We use the terminology a B-*bounded space* algorithm to specify the maximum number of open bins. An algorithm with no such constant B is called an *unbounded space* algorithm (for example, algorithm FF (First Fit) [14]). Since algorithm MM maintains a single open bin, MM is a 1-bounded space algorithm.

The performance of approximation algorithms is often measured by the *asymptotic approximation ratio* [12,14]. Let $ALG(I)$ denote the number of non-empty bins by an algorithm ALG for an item sequence I. Let $OPT(I)$ denote the minimum number of bins that are needed to contain all items in I. The asymptotic approximation ratio of ALG is defined as

$$R_{ALG} = \lim_{m\to\infty} \sup\left\{ \frac{ALG(I)}{OPT(I)} \;\middle|\; OPT(I) \geq m \right\}. \tag{1}$$

For the k-*cardinality constrained* bin packing problem, which is a variant with an additional constraint that at most k items can be packed in each bin, the asymptotic approximation ratio of ALG is defined as,

$$R_{ALG} = \lim_{m\to\infty} \sup\left\{ \frac{ALG(I)}{OPT_k(I)} \;\middle|\; OPT_k(I) \geq m \right\}, \tag{2}$$

where $OPT_k(I)$ denotes the minimum number of bins that are needed to contain all items in I so that at most k items are packed in each bin. From now on, we will refer to the bin packing problem that we have argued so far as the *classic* bin packing problem to distinguish the problem from the k-cardinality constrained bin packing problem.

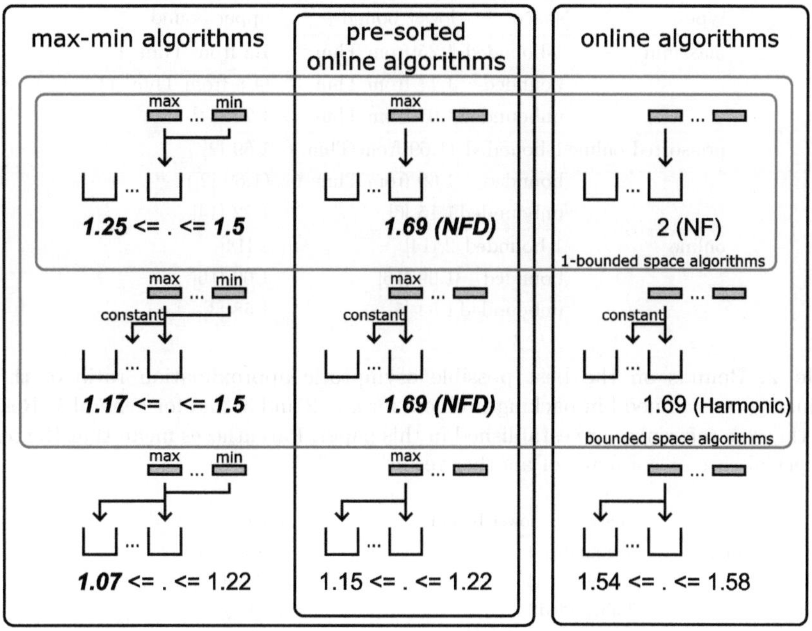

Fig. 1. Illustration of algorithm classes and bounds on the best possible asymptotic approximation ratio for the classic bin packing problem. Bold italic font indicates bounds established in this paper

1.1 Our Contribution

It was a breakthrough that for the classic bin packing problem, Lee and Lee clarified a theoretical performance gap between online bounded space algorithms and online unbounded space algorithms [16], where an *online* algorithm is an algorithm that repeatedly packs the head item in the currently remaining item sequence into a bin, without seeing other items. In this paper, extending this perspective, we comprehensively analyze theoretical performance bounds for algorithm classes (see Table 1).

We make some preliminary definitions to express the value of the asymptotic competitive ratio. Define the sequence $\{\pi_i\}_{i=1}^{\infty}$ by $\pi_1 = 2$ and $\pi_{i+1} = \pi_i(\pi_i - 1) + 1$ for $i \geq 1$. Let $\gamma = \sum_{i=1}^{\infty} \frac{1}{\pi_i - 1}$, which is known to converge to approximately 1.69 [2]. Define the sequence $\{\lambda_j\}_{j=1}^{\infty}$ by $\lambda_j = \sum_{i=1}^{j} \max\{\frac{1}{\pi_i - 1}, \frac{1}{j}\}$ for $j \geq 1$. It is known that $\lim_{j\to\infty} \lambda_j = \gamma + 1 \approx 2.69$ [11]. Throughout this paper, numerical rounding is always done to the nearest value. The theorems in Sects. 2 and 3,

Table 1. Bounds on the best possible asymptotic approximation ratio of the classic bin packing problem. Results with theorem references are established in this paper. Parentheses mean that the result is a special case or corollary of another result

type	space	lower bound	upper bound
max-min	1-bounded	1.25 from Thm. 2	1.5 from Thm. 1
	bounded	1.17 from Thm. 3	(1.5 from Thm. 1)
	unbounded	1.07 from Thm. 4	1.22 [14]
pre-sorted online	1-bounded	(1.69 from Thm. 6)	1.69 [2]
	bounded	1.69 from Thm. 6	(1.69 [2])
	unbounded	1.15 [6]	1.22 [14]
online	1-bounded	2 [14]	2 [14]
	bounded	1.69 [16]	1.69 [16]
	unbounded	1.54 [5]	1.58 [3]

Table 2. Bounds on the best possible asymptotic approximation ratio of the k-cardinality constrained bin packing problem for $k = 2$ and 3, and for general k. Results with theorem references are established in this paper. Parentheses mean that the result is a special case or corollary of another result

k	type	space	lower bound	upper bound
2	max-min	1-bounded	1	1 from Thm. 7
		bounded	1	(1 from Thm. 7)
		unbounded	1	(1 [18])
	pre-sorted online	1-bounded	(1.5 [11])	(1.5 from Thm. 11)
		bounded	(1.5 [11])	(1.5 [11])
		unbounded	1	1 [18]
	online	1-bounded	(2 from Thm. 14)	2 (trivial)
		bounded	(1.5 [11])	(1.5 [11])
		unbounded	1.43 [4]	1.45 [1]
3	max-min	1-bounded	1.33 from Thm. 9	(1.5 from Thm. 8)
		bounded	(1.07 from Thm. 4)	(1.5 from Thm. 8)
		unbounded	(1.07 from Thm. 4)	(1.5 from Thm. 8)
	pre-sorted online	1-bounded	(1.83 [11])	(1.83 from Thm. 11)
		bounded	(1.83 [11])	(1.83 [11])
		unbounded	1.14 [12]	(1.75 [11])
	online	1-bounded	(2.33 from Thm. 14)	(2.33 from Thm. 13)
		bounded	(1.83 [11])	(1.83 [11])
		unbounded	1.56 [4]	1.75 [11]
k	max-min	1-bounded	(1.5 for $k \geq 6$ from Thm. 10)	$\lambda_k - \frac{1}{k} \to 2.69$ from Thm. 8
		bounded	1.5 for $k \geq 6$ from Thm. 10	($\lambda_k - \frac{1}{k} \to 2.69$ from Thm. 8)
		unbounded	(1.07 from Thm. 4)	($\nu_k \to 2$ [1])
	pre-sorted online	1-bounded	($\lambda_k \to 2.69$ [11])	$\lambda_k \to 2.69$ from Thm. 11
		bounded	$\lambda_k \to 2.69$ [11]	$\lambda_k \to 2.69$ [11]
		unbounded	1.15 for $k \geq 5$ [6]	($\nu_k \to 2$ [1])
	online	1-bounded	$3 - \frac{2}{k} \to 3$ from Thm. 14	$3 - \frac{2}{k} \to 3$ from Thm. 13
		bounded	$\lambda_k \to 2.69$ [11]	$\lambda_k \to 2.69$ [11]
		unbounded	$\mu_k \to 2$ [4]	$\nu_k \to 2$ [1]

each of which states an inequality on the number of bins, provide the following bounds on the asymptotic approximation ratio:

(I) We obtain the following results for the classic bin packing problem (see Fig. 1 and Table 1). We prove $R_{MM} \le \frac{3}{2} = 1.5$ (Theorem 1). Compare with $R_{NFD} = \gamma \approx 1.69$ [2]. NFD can be seen as a pre-sorted online 1-bounded space algorithm. Our result explains the extent to which performance can be improved by packing also the smallest items in the remaining sequence.

On the other hand, we show $R_{ALG} \ge \frac{5}{4} = 1.25$ for any max-min 1-bounded space algorithm ALG (Theorem 2). Compare with $R_{FFD} = \frac{11}{9} \approx 1.22$ [14]. FFD is a pre-sorted online unbounded space algorithm. Our result reveals that with only 1-bounded space, any max-min algorithm cannot achieve better performance than FFD. In addition, we provide a lower bound of $\frac{7}{6} \approx 1.17$ for max-min bounded space algorithms (Theorem 3), and that of $\frac{16}{15} \approx 1.07$ on max-min unbounded space algorithms (Theorem 4).

Furthermore, we give a lower bound for pre-sorted online bounded space algorithms of $\gamma \approx 1.69$ (Theorem 6). Combining with $R_{NFD} = \gamma$ introduced above [2], we conclude that the best possible performance of pre-sorted online algorithms is not affected by a single open bin available or a constant number of open bins available. This result contrasts with the gap between the asymptotic competitive ratios of the best possible online 1-bounded space algorithm and the best possible online bounded space algorithm, which are 2 [14] and $\gamma \approx 1.69$ [16].

(II) Our results for the k-cardinality constrained bin packing problem are as follows (see Table 2). For the k-cardinality-constrained version of algorithm MM, denoted by MM_k, we show $R_{MM_k} \le \lambda_k - \frac{1}{k}$ for each $k \ge 2$ (Theorems 7 and 8), which breaks the lower bound barrier of λ_k for pre-sorted online bounded space k-cardinality constrained algorithms [11]. For example, the values of $\lambda_k - \frac{1}{k}$ are: $\frac{3}{2} = 1.5$ for $k = 3$, $\frac{7}{4} = 1.75$ for $k = 4$, and $\frac{19}{10} = 1.9$ for $k = 5$. As $k \to \infty$, we have $\lambda_k - \frac{1}{k} \to \gamma + 1 \approx 2.69$. In particular, for $k = 2$ and 3, MM_k also breaks the lower bounds of $\frac{10}{7} \approx 1.43$ and 1.56 [4], respectively, for online unbounded space k-cardinality constrained algorithms.

We provide a lower bound for max-min 1-bounded space algorithms: $\frac{4}{3} \approx 1.33$ for $k = 3$ (Theorem 9), $\frac{4}{3} \approx 1.33$ for $k = 4$, $\frac{17}{12} \approx 1.42$ for $k = 5$, and $\frac{3}{2} = 1.5$ for $k \ge 6$ (Theorem 10). We show that the k-cardinality-constrained version of algorithm NFD is a best possible pre-sorted online bounded space algorithm (Theorem 11), by combining with the lower bound by Epstein [11]. We prove that the k-cardinality-constrained version of algorithm NF (Next Fit) is a best possible online 1-bounded space algorithm (Theorems 13 and 14).

1.2 Previous Results

(I) In the literature we can find the following results for the classic bin packing problem: While a tight bound of $\gamma \approx 1.69$ for online bounded space algorithms, achieved by algorithm *Harmonic*, is established by Lee and Lee [16], a tight bound for online unbounded space algorithms is unknown so far. The current best upper and lower bounds are 1.58 [3] and $\frac{1363-\sqrt{1387369}}{120} \approx 1.54$ [5], respectively.

For online 1-bounded space algorithms, it is known that a tight bound of 2 is achieved by NF (Next Fit) [14].

Pre-sorted online algorithms such as NFD [2] and FFD [14] have been studied by many researchers for a long time. The refinement of the analysis of FFD has attracted great interest [9,10]. Balogh et al. gave a lower bound of $\frac{54}{47} \approx 1.15$ for pre-sorted online unbounded space algorithms [6].

(II) The results below are known for the k-cardinality constrained bin packing problem. Epstein provided a tight bound of λ_k for online bounded space algorithms, achieved by algorithm CCH_k (Cardinality Constrained Harmonic) [11]. This is the counterpart to the work of Lee and Lee mentioned above [16]. Since the lower bound is proved by a non-increasing item sequence, the tight bound also applies to pre-sorted online bounded space algorithms.

Babel et al. designed an online unbounded space algorithm with an asymptotic approximation ratio of $\nu_k = 2 + \frac{k+\eta_k(\eta_k-3)}{(k-\eta_k+1)\eta_k}$, where $\eta_k = \lceil \frac{-k+\sqrt{k^3-2k}}{k-2} \rceil$ for each k [1]. For $k = 2$ and 3, there are better algorithms: one with an asymptotic approximation ratio of $1+1/\sqrt{5} \approx 1.45$ for $k = 2$ [1] and one with an asymptotic approximation ratio of $\frac{7}{4} = 1.75$ for $k = 3$ [11]. Balogh et al. also gave a lower bound for online unbounded space algorithms [4]: $\frac{10}{7} \approx 1.43$ for $k = 2$, 1.56 for $k = 3$, and $\mu_k = \frac{2k+k^2-k^3+\sqrt{k^6-2k^5-3k^4+12k^3-12k^2+4k}}{2}$ for general k.

The lower bound of $\frac{54}{47} \approx 1.15$ for pre-sorted online unbounded space algorithms by Balogh et al. [6] is valid for $k \geq 5$. The book [12] introduces a lower bound of $\frac{8}{7} \approx 1.14$ for pre-sorted online unbounded space algorithms by Csirik, Galambos, and Turán with their proof, which is still the best each for $k = 3$ and 4. Tinhofer showed that the 2-cardinality-constrained version of FFD achieves an asymptotic approximation ratio of 1 [18].

2 The Classic Bin Packing Problem

We formulate the *classic* bin packing problem. (We have added the term classic to distinguish the problem from the k-cardinality constrained version.) Let $\mathbb{N}$ be the set of all positive integers. The input is an item sequence $I = (a_1, a_2, \ldots, a_n) \in (0,1]^n$, where each item i is of size a_i. We refer to an item of size a as an *a-item.* An item sequence $I = (a_1, a_2, \ldots, a_n) \in (0,1]^n$ is said to be *sorted* if $a_1 \geq a_2 \geq \cdots \geq a_n$.

A solution to the classic bin packing problem is an assignment $f : \{1, 2, \ldots, n\} \to \mathbb{N}$ which maps each item index to a bin index and satisfies the constraint that for all bin indices $j \in \mathbb{N}$, $\sum_{i:f(i)=j} a_i \leq 1$ holds. The constraint means that the sum of the sizes of the items in each bin does not exceed 1. If bin j is assigned to item i, i.e., $f(i) = j$, we say that bin j *contains* item i. An item is said to be a *sole* item if a bin is assigned exclusively to the item.

Given an item sequence I and an assignment f for I, the number of non-empty bins is the number of elements in the range of assignment f, denoted by $|f(I)|$. The goal of the classic bin packing problem is to minimize $|f(I)|$.

An algorithm for the classic bin packing problem is to output an assignment $f : \{1, 2, \ldots, n\} \to \mathbb{N}$. If an algorithm assigns bin j to item i, we interchangeably say that the algorithm *packs* item i into bin j. Without loss of generality, we assume that when an algorithm is about to assign one of the bins that have not previously had any items assigned to it (often referred to as *opening* the bin), the algorithm always chooses the bin of the *smallest* index. We refer to a bin with a smaller (respectively, larger) index as an *earlier* (respectively, a *later*) bin.

We write the number of non-empty bins of an assignment obtained by an algorithm ALG for an item sequence I as $ALG(I)$, without specifying the assignment itself. We denote by $OPT(I)$ the minimum number of bins that are assigned to all items in I. We also denote by OPT an algorithm that for a given I, outputs such an assignment that the number of non-empty bins is $OPT(I)$.

The definition of the asymptotic competitive ratio of an algorithm ALG, denoted by R_{ALG}, is given by (1). The statements of the theorems in this section all take the form of an inequality for $ALG(I)$ and $OPT(I)$. Bounds on the asymptotic competitive ratio are obtained from the coefficients of $OPT(I)$.

2.1 A Max-Min 1-Bounded Space Algorithm

Zhu proposed the following algorithm MM [20]: First, sort the item sequence in a non-increasing order. Then, repeatedly assign the current single open bin first to as many of the head items in the remaining sequence as possible and next to as many of the tail items in the remaining sequence as possible. Open a new bin only when the current single open bin cannot accept the next tail item. If algorithm MM has assigned a bin to the head (respectively, tail) item in the currently remaining item sequence, then in the resulting assignment we call that item a *head* (respectively, *tail*) item.

The theorem below proves $R_{MM} \leq \frac{3}{2} = 1.5$. The proof is done either by an analysis of levels of bins or by an analysis using weights as used in [2,11,14], depending on whether an item of size at most $\frac{1}{3}$ is contained in the latest bin in the assignment.

Theorem 1. *It holds that* $MM(I) \leq \frac{3}{2} \cdot OPT(I) + 1$ *for all item sequences* I.

Without loss of generality, let $I = (a_1, a_2, \ldots, a_n) \in (0, 1]^n$ be a sorted item sequence and assume that I does not change by sorting by algorithm MM. In this proof, we call an item of size $(\frac{1}{2}, 1]$ a *class 1* item, an item of size $(\frac{1}{3}, \frac{1}{2}]$ a *class* 2 item, and an item of size $(0, \frac{1}{3}]$ a *class* 3 item.

(i) The case where the latest bin in the assignment by algorithm MM for I contains at least one class 3 item (see Fig. 2). Since I is sorted, throughout the execution of algorithm MM, the tail item in the currently remaining item sequence must always be a class 3 item. Therefore, the sum of the sizes of the items of each bin, except the latest one, must exceed $\frac{2}{3}$. (If there were a bin which does not satisfy this, MM would have packed one of the tail class 3 items additionally into the bin.) Hence, by summing up the sizes of the items for all the bins other than the latest one, we get $\sum_{i=1}^{n} a_i > \frac{2}{3} \cdot (MM(I) - 1)$. Since $\sum_{i=1}^{n} a_i \leq OPT(I)$, we have $MM(I) < \frac{3}{2} \cdot OPT(I) + 1$.

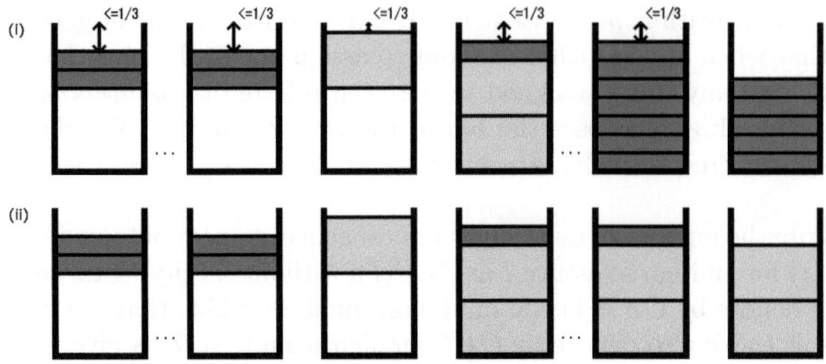

Fig. 2. The assignment by MM. The white, light grey, and dark grey items are of classes 1, 2, and 3, respectively (Color figure online)

(ii) The case where the latest bin in the assignment by algorithm MM for I does not contain any class 3 item (see Fig. 2). The following weight function assigns a weight to each item according to the class:

$$w_1(x) = \begin{cases} 1, & x \in (\frac{1}{2}, 1]; \\ \frac{1}{2}, & x \in (\frac{1}{3}, \frac{1}{2}]; \\ 0, & x \in (0, \frac{1}{3}]. \end{cases} \quad (3)$$

Since I is sorted, throughout the execution of algorithm MM, the head item in the currently remaining item sequence must always be a class 1 or class 2 item. Hence, there is no bins that contain only class 3 items in the assignment. Focus on the head items in the assignment. Ignore the latest bin. Then, due to the behavior of MM, each of the remaining bins turns out to be a bin containing one head class 1 item whose weight is 1, or a bin containing two head class 2 items whose total weight is $\frac{1}{2} + \frac{1}{2} = 1$. We thus obtain that the sum of the weights of the items in each of those bins is at least 1. Therefore, $\sum_{i=1}^{n} w_1(a_i) \geq MM(I) - 1$ holds in total.

On the other hand, the sum of the weights of items that fit in a single bin is at most $\frac{3}{2}$, where the maximum is achieved by a class 1 item and a class 2 item. Therefore, $\sum_{i=1}^{n} w_1(a_i) \leq \frac{3}{2} \cdot OPT(I)$ holds. We thus have $MM(I) \leq \frac{3}{2} \cdot OPT(I) + 1$.

2.2 A Lower Bound for Max-Min Bounded Space Algorithms

A Lower Bound for 1-Bounded Space Algorithms. The next theorem states that $R_{ALG} \geq \frac{5}{4} = 1.25$ holds for any max-min 1-bounded space algorithm ALG. Compare with algorithm FFD (First Fit Decreasing) that achieves $R_{FFD} = \frac{11}{9} (\approx 1.22)$ [14].

FFD first sorts the given item sequence in a non-increasing order and then runs algorithm FF (First Fit), that is, repeatedly packs the head item in the currently remaining item sequence into the bin with the smallest index that the item can fit in and opens a new bin only when the item cannot fit in any open bin. Thus, *FFD* is a pre-sorted online unbounded space algorithm. The theorem reveals that with only a single open bin, no max-min algorithm can outperform *FFD* (see Table 1). For lack of space, we omit the proof of Theorem 2.

Theorem 2. *For any max-min 1-bounded space algorithm ALG, and any positive integer m divisible by 2, there exists a sorted item sequence I such that* $OPT(I) = m$ *and* $ALG(I) \geq \frac{5}{4} \cdot OPT(I) - \frac{1}{4}$.

A Lower Bound for B -Bounded Space Algorithms. We obtain a weaker lower bound than Theorem 2. More specifically, we present Theorem 3, which states that $R_{ALG} \geq \frac{7}{6} (\approx 1.17)$ holds for any max-min bounded space algorithm *ALG*. The proof is omitted due to space limitations.

Theorem 3. *For any max-min B-bounded space algorithm ALG and any positive integer m that is at least 2B and divisible by 3, there exists a sorted item sequence I such that* $OPT(I) = m$ *and* $ALG(I) \geq \frac{7}{6} \cdot OPT(I) - B$.

2.3 A Lower Bound for Max-Min Unbounded Space Algorithms

We establish, as Theorem 4, an even weaker lower bound for unbounded space algorithms: $R_{ALG} \geq \frac{16}{15} (\approx 1.07)$ holds for any max-min unbounded space algorithm *ALG*. The proof involves a technique of suspending the execution of the algorithm when it packs special items in order to construct another item sequence depending on the behavior of the algorithm. This exploits the weakness of the max-min algorithm in that it can only see the head and tail items of the currently remaining item sequence. The proof is omitted due to lack of space.

Theorem 4. *For any max-min unbounded space algorithm ALG, and any positive integer m, there exists a sorted item sequence I such that* $OPT(I) > m$ *and* $ALG(I) > \frac{16}{15} \cdot (OPT(I) - 1)$.

2.4 A Tight Bound for Pre-Sorted Online Bounded Space Algorithms

Recall the sequence $\{\pi_i\}_{i=1}^{\infty}$ defined by $\pi_1 = 2$ and $\pi_{i+1} = \pi_i(\pi_i - 1) + 1$ for $i \geq 1$ in Sect. 1.1. The values are: $\pi_1 = 2$, $\pi_2 = 3$, $\pi_3 = 7$, $\pi_4 = 43$, $\pi_5 = 1807$, and so on. By induction, we can see that $\{\pi_i\}_{i=1}^{\infty}$ is monotonically increasing and

$$1 - \sum_{i=1}^{j} \frac{1}{\pi_j} = \frac{1}{\pi_{j+1} - 1} \text{ for all } j \geq 1. \tag{4}$$

We also present the following convergence:

$$\sum_{i=1}^{\infty} \frac{1}{\pi_i - 1} = \gamma (\approx 1.69). \tag{5}$$

Lee and Lee provided a tight bound of γ on the asymptotic approximation ratio of online bounded space algorithms [16]. In this section, by combining with an existing result, we provide a tight bound for pre-sorted online bounded space algorithms, which is also equal to γ. This reveals that pre-sorting gives no advantage for online bounded space algorithms.

Algorithm NFD (Next Fit Decreasing) first sorts the given item sequence in a non-increasing order and then runs algorithm NF (Next Fit), that is, repeatedly packs, into the current single open bin, as many of the head items in the remaining item sequence as possible and opens a new bin only when the item cannot fit in the single open bin. It is seen that NFD is a pre-sorted online 1-bounded space algorithm.

Baker and Coffman Jr. proved $R_{NFD} = \gamma$ [2], which we introduce as Theorem 5. Together with Theorem 6, which we will prove soon, we conclude that NFD is best possible. It should be remarked that since NFD is a 1-bounded space algorithm, our bound is tight for pre-sorted online 1-bounded space algorithms as well as for pre-sorted online bounded space algorithms. In other words, pre-sorted online algorithms do not gain from maintaining a large number of open bins, as long as the number is bounded by a constant.

Theorem 6 intuitively follows from the fact that any pre-sorted online bounded space algorithm can not combine items of different sizes if the item sequence contains a sufficiently long sequence of items of the same size. The item sequence used to prove Theorem 6 is essentially the same as that used in the proof of the main theorem in the paper [2]. The reason why the statement of Theorem 6 is nested is to allow K, which is a parameter determined by how close δ is to zero, to be chosen independently of m.

Theorem 5 *([2]). It holds that* $NFD(I) \leq \gamma \cdot OPT(I) + 3$ *for all item sequences* I.

The proof of Theorem 6 is omitted for space reasons.

Theorem 6. *For any* δ *satisfying* $0 < \delta < \gamma$*, there exists an integer* $K \geq 1$ *such that: For any pre-sorted online* B*-bounded space algorithm* ALG *and any positive integer* m*, there exists a sorted item sequence* I *such that* $OPT(I) = m$ *and* $ALG(I) > (\gamma - \delta) \cdot OPT(I) - B(K-1)$.

3 The k-Cardinality Constrained Bin Packing Problem

We consider the *k-cardinality constrained* bin packing problem [11], which is a variant of the classic bin packing problem with an additional constraint that each bin can be assigned to at most k items. Throughout this section, for $j = 1, 2, \ldots, k-1$, we call an item of size $(\frac{1}{j+1}, \frac{1}{j}]$ a *class* j item, and we call an item of size $(0, \frac{1}{k}]$ a *class* k item.

An algorithm is called a *k-cardinality constrained* algorithm if for any item sequence, it outputs an assignment that assigns each bin to at most k items. We denote by $OPT_k(I)$ the minimum number of bins that are assigned to all items in I while ensuring that each bin is assigned to at most k items. We also denote by OPT_k an algorithm that for a given I, outputs such an assignment that the number of non-empty bins is $OPT_k(I)$ and each bin is assigned to at most k items.

The definition of the asymptotic competitive ratio of an algorithm ALG, denoted by R_{ALG}, is given by (2). Similarly to Sect. 2, bounds on the asymptotic competitive ratio are obtained from the coefficients of $OPT_k(I)$ in the inequalities in the statements of the theorems.

3.1 A Max-Min 1-Bounded Space Algorithm

In this section we propose a max-min 1-bounded space k-cardinality constrained algorithm MM_k, which is simply a modified version of algorithm MM in Sect. 2 that opens a new bin every time it has packed k items into a bin. Similarly to algorithm MM, if algorithm MM_k assigns a bin to the head (respectively, tail) item in the currently remaining item sequence, then in the resulting assignment we call that item a *head* (respectively, *tail*) item.

Lemma 1 states a significant property of the assignment by MM_2, which is useful for the analysis not only for $k = 2$ but also for $k \geq 3$. The proof is omitted due to space limitations.

Lemma 1. *Suppose that $I = (a_1, a_2, \ldots, a_n) \in (0, 1]^n$ is a sorted item sequence such that the assignment f by MM_2 for I has at least one bin with a sole class 1 item. Let p be the maximum index of such sole class 1 items in f. Let U be the set of indices of items that are packed with items $1, 2, \ldots, p$ in f, which may be empty. Then, for an arbitrary assignment without cardinality constraints g for I, denoting by W the set of indices of items that are packed with items $1, 2, \ldots, p$ in g, it holds that $W \subseteq U$.*

A Tight Bound for $k = 2$. Tinhofer showed that if FFD, which is a pre-sorted online unbounded space algorithm, is run for the 2-cardinality constrained bin packing problem, it always outputs an assignment with the minimum number of non-empty bins [18]. That is, the asymptotic approximation ratio is 1. The following theorem implies that if we are allowed to employ a max-min algorithm, then a single open bin is enough to enjoy the optimal performance. The proof is omitted due to lack of space.

Theorem 7. *It holds that $MM_2(I) = OPT_2(I)$ for all item sequences I.*

An Upper Bound for $k \geq 3$. Recall the sequence $\{\pi_i\}_{i=1}^{\infty}$ defined by $\pi_1 = 2$ and $\pi_{i+1} = \pi_i(\pi_i - 1) + 1$, and the sequence $\{\lambda_j\}_{j=1}^{\infty}$ defined by $\lambda_j = \sum_{i=1}^{j} \max\{\frac{1}{\pi_i - 1}, \frac{1}{j}\}$ for $j \geq 1$ in Sect. 1.1. For example, the values of $\{\lambda_j\}_{j=1}^{\infty}$ are: $\lambda_1 = 1$, $\lambda_2 = \frac{3}{2} = 1.5$, $\lambda_3 = \frac{11}{6} \approx 1.83$, $\lambda_4 = 2$, and $\lambda_5 = \frac{21}{10} = 2.1$.

For the k-cardinality constrained bin packing problem with $k \geq 3$, it is not so difficult to obtain $R_{MM_k} \leq \lambda_k$ by a standard analysis using weights. In what follows, we conduct a more precise analysis to reveal that $R_{MM_k} \leq \lambda_k - \frac{1}{k}$, breaking the lower bound barrier of λ_k for pre-sorted online bounded space algorithms [11]. For example, the values of $\lambda_k - \frac{1}{k}$ are: $\frac{3}{2} = 1.5$ for $k = 3$, $\frac{7}{4} = 1.75$ for $k = 4$, and $\frac{19}{10} = 1.9$ for $k = 5$.

The main idea of our analysis is to attempt to run MM_2 as well for the item sequence in order to extract a subsequence of items to which an analysis using "discounted" weights can be applied. Lemma 2, proven with the help of Lemma 1, enables us to extract such a subsequence. The proof of Lemma 2 is omitted for space reasons.

Lemma 2. *Let I be a sorted item sequence, and I_1 be a subsequence of I consisting of the sole class 1 items in the assignment f by MM_2 for I. Then, for an arbitrary assignment without cardinality constraints g for I, there exists an assignment without cardinality constraints g' such that $|g(I)| = |g'(I)|$ and every item in I_1 is a sole item in g'.*

We employ the following weight function which assigns a weight to each item according to the class. The weight of 1 minus $\frac{1}{k}$ for a class 1 item will contribute to the performance evaluation smaller by $\frac{1}{k}$.

$$w_2(x) = \begin{cases} 1 - \frac{1}{k}, & x \in (\frac{1}{2}, 1]; \\ \frac{1}{j}, & x \in (\frac{1}{j+1}, \frac{1}{j}] \text{ for } j = 2, \ldots, k-1; \\ \frac{1}{k}, & x \in (0, \frac{1}{k}]. \end{cases} \tag{6}$$

Given $I = (a_1, a_2, \ldots, a_n)$, define $w_2(I) = \sum_{i=1}^{n} w_2(a_i)$. The next theorem bounds from above the sum of the weights of items that fit in a single bin. The proof is given by modifying the proof of the upper bound part of Theorem 3 in [11]. For lack of space, we omit the proof of Theorem 3.

Lemma 3. *For any $k \geq 3$ and any $I = (a_1, a_2, \ldots, a_n) \in (0, 1]^n$ such that $n \leq k$ and $\sum_{i=1}^{n} a_i \leq 1$, it holds that $w_2(I) \leq \lambda_k - \frac{1}{k}$.*

We are now ready to claim that $R_{MM_k} \leq \lambda_k - \frac{1}{k}$. The proof of Theorem 8 is omitted for space reasons.

Theorem 8. *For any $k \geq 3$, it holds that $MM_k(I) \leq \left(\lambda_k - \frac{1}{k}\right) \cdot OPT_k(I) + k$ for all item sequences I.*

3.2 A Lower Bound for Max-Min 1-Bounded Space Algorithms

A Lower Bound for 1-Bounded Space Algorithms. Theorem 9 below states that $R_{ALG} \geq \frac{4}{3} (\approx 1.33)$ holds for any max-min 1-bounded space 3-cardinality constrained algorithm ALG. The proof of Theorem 9 is done in almost the same way as the proof of Theorem 2, except for the cardinality constraint. The proof is omitted due to space limitations.

Theorem 9. *For any max-min 1-bounded space 3-cardinality constrained algorithm ALG, and any positive integer m, there exists a sorted item sequence I such that $OPT_3(I) = m$ and $ALG(I) \geq \frac{4}{3} \cdot OPT_3(I) - \frac{4}{3}$.*

A Lower Bound for B -Bounded Space Algorithms. For $k = 3$, we obtain from Theorem 4 that $R_{ALG} \geq \frac{16}{15}(\approx 1.07)$ holds for any max-min unbounded space 3-cardinality constrained algorithm ALG, since none of the bins appearing in the proof of Theorem 4 contains 4 or more items.

The following theorem states that for each $k \geq 4$, $R_{ALG} \geq \max\{\frac{5}{3} - \frac{1}{k-1}, \frac{3}{2}\}$ holds for any max-min bounded space k-cardinality constrained algorithm ALG. This is the best known lower bound also for 1-bounded space algorithms. The proof of Theorem 10 is similar to that of Theorem 3. The proof is omitted due to lack of space.

Theorem 10. *For any max-min B-bounded space k-cardinality constrained algorithm ALG, and any positive integer m that is at least $(k-1)B$ and divisible by 3 and $k-1$, there exists a sorted item sequence I such that $OPT_k(I) = m$ and: (A) $ALG(I) \geq \left(\frac{5}{3} - \frac{1}{k-1}\right) \cdot OPT_k(I) - \frac{1}{6}B(k+5)$ for each $4 \leq k \leq 6$, (B) $ALG(I) \geq \frac{3}{2} \cdot OPT_k(I) - 2B$ for $k = 7$, and (C) $ALG(I) \geq \frac{3}{2} \cdot OPT_k(I) - \frac{1}{2}B$ for each $k \geq 8$.*

3.3 A Tight Bound for Pre-Sorted Online 1-Bounded Space Algorithms

We denote by NFD_k a k-cardinality constrained version of algorithm NFD (see Sect. 2.4) that opens a new bin every time it has packed k items into a bin. The following Theorems 11 and 12 [11] state that NFD_k is best possible among online bounded space k-cardinality constrained algorithms, achieving $R_{NFD_k} = \lambda_k$, though NFD_k is a 1-bounded space algorithm.

Compare with the fact that algorithm CCH_k (Cardinality Constrained Harmonic) is best possible among online bounded space k-cardinality constrained algorithms, achieving the same asymptotic approximation ratio $R_{CCH_k} = \lambda_k$ [11]. Our result explains a trade-off between algorithm classes: if pre-sorting is allowed, then a single open bin is enough for the same performance.

To analyze the performance of NFD_k, we use the following weight function defined in the paper [11]:

$$w_3(x) = \begin{cases} \frac{1}{j}, & x \in (\frac{1}{j+1}, \frac{1}{j}] \text{ for } j = 1, \ldots, k-1; \\ \frac{1}{k}, & x \in (0, \frac{1}{k}]. \end{cases}$$

Given $I = (a_1, a_2, \ldots, a_n)$, define $w_3(I) = \sum_{i=1}^{n} w_3(a_i)$. The sum of the weights of items that fit in a single bin is known to be bounded from above:

Lemma 4 ([11]). *For any $k \geq 2$ and any $I = (a_1, a_2, \ldots, a_n) \in (0,1]^n$ such that $n \leq k$ and $\sum_{i=1}^{n} a_i \leq 1$, it holds that $w_3(I) \leq \lambda_k$.*

The theorem below proves $R_{NFD_k} \leq \lambda_k$. The proof is done similarly to the analysis of NFD in [2], except for the cardinality constraint. The proof is omitted due to space limitations.

Theorem 11. *For any $k \geq 2$, it holds that $NFD_k(I) \leq \lambda_k \cdot OPT_k(I) + k$ for all item sequences I.*

On the other hand, since Theorem 3 in the paper [11] is proven using a sorted item sequence, a lower bound for pre-sorted online bounded space algorithms is straightforwardly derived.

Theorem 12 ([11]). *For any $k \geq 2$, any pre-sorted online B-bounded space k-cardinality constrained algorithm ALG, and any positive integer m, there exists a sorted item sequence I such that $OPT_k(I) = m$ and $ALG(I) \geq \lambda_k \cdot OPT_k(I) - Bk$.*

3.4 A Tight Bound for Online 1-Bounded Space Algorithms

It has long been known that NF (Next Fit) is a best possible online 1-bounded space algorithm for the classic bin packing problem [14]. Let NF_k denote a k-cardinality constrained version of algorithm NF that opens a new one every time it has packed k items into a bin. In this section, we prove that NF_k is best possible among online 1-bounded space k-cardinality constrained algorithms as well, achieving $R_{NF_k} = 3 - \frac{2}{k}$. For lack of space, we omit the proofs of Theorems 13 and 14.

Theorem 13. *For any $k \geq 2$, it holds that $NF_k(I) \leq \left(3 - \frac{2}{k}\right) \cdot OPT_k(I) + 1$ for all item sequences I.*

Theorem 14. *For any online 1-bounded space k-cardinality constrained algorithm and any positive integer $m \geq 3$, there exists an item sequence I such that $OPT_k(I) = m$ and: (A) $ALG(I) \geq 2 \cdot OPT_k(I) - 2$ for $k = 2$ and (B) $ALG(I) \geq \left(3 - \frac{2}{k}\right) \cdot OPT_k(I) - 5 + \frac{5}{k}$ for each $k \geq 3$.*

4 Future Work

It is interesting to narrow the bounds for each subclass of algorithms. Are MM and MM_k best possible max-min 1-bounded space algorithms? Our conjecture is "yes". Is there a better pre-sorted online unbounded space algorithm than FFD? Furthermore, recall that Theorem 6 claims that for the classic bin packing problem, the tight bounds for pre-sorted online 1-bounded space algorithms and pre-sorted online bounded space algorithms coincide. Then, is there a gap between max-min 1-bounded space algorithms and max-min bounded space algorithms?

Acknowledgments. We are grateful to Toshiki Tsuchiya for useful discussions. We would like to thank Hiromu Hashimoto for pointing out an error in the manuscript. This work was supported by JSPS KAKENHI Grant Numbers 25K14990, 20K11689, 20K11676, 23K11100, and 20K11808.

References

1. Babel, L., Chen, B., Kellerer, H., Kotov, V.: Algorithms for on-line bin-packing problems with cardinality constraints. Discrete Appl. Math. **143**(1–3), 238–251 (2004)
2. Baker, B.S., Coffman, E.G.: A tight asymptotic bound for next-fit-decreasing bin-packing. SIAM J. Algebraic Discrete Methods **2**(2), 147–152 (1981)
3. Balogh, J., Békési, J., Dósa, G., Epstein, L., Levin, A.: A new and improved algorithm for online bin packing. In Proceedings of the ESA 2018, pp. 5:1–5:14 (2018)
4. Balogh, J., Békési, J., Dósa, G., Epstein, L., Levin, A.: Online bin packing with cardinality constraints resolved. J. Comput. Syst. Sci. **112**, 34–49 (2020)
5. Balogh, J., Békési, J., Dósa, G., Epstein, L., Levin, A.: A new lower bound for classic online bin packing. Algorithmica **83**(7), 2047–2062 (2021)
6. Balogh, J., Békési, J., Galambos, G.: New lower bounds for certain classes of bin packing algorithms. Theor. Comput. Sci. **440–441**, 1–13 (2012)
7. Borodin, A., Nielsen, M.N., Rackoff, C.: (Incremental) priority algorithms. Algorithmica **37**(4), 295–326 (2003)
8. Christensen, H.I., Khan, A., Pokutta, S., Tetali, P.: Approximation and online algorithms for multidimensional bin packing: a survey. Comput. Sci. Rev. **24**, 63–79 (2017)
9. Dósa, G.: The tight bound of first fit decreasing bin-packing algorithm is $FFD(I) \leq 11/9OPT(I) + 6/9$. In Proceedings of the ESCAPE 2007, pp. 1–11 (2007)
10. Dósa, G., Li, R., Han, X., Tuza, Z.: Tight absolute bound for first fit decreasing bin-packing: $FFD(L) \leq 11/9OPT(L) + 6/9$. Theor. Comput. Sci. **510**, 13–61 (2013)
11. Epstein, L.: Online bin packing with cardinality constraints. SIAM J. Discrete Math. **20**(4), 1015–1030 (2006)
12. Fiat, A., Woeginger, G.J. (eds.): Online Algorithms, The State of the Art (The Book Grown Out of a Dagstuhl Seminar, June 1996), volume 1442 of LNCS. Springer (1998)
13. Garey, M.R., Johnson, D.S.: Computers and Intractability: A Guide to the Theory of NP-Completeness. W. H. Freeman & Co., New York, NY, USA (1979)
14. Johnson, D.S.: Near-optimal bin packing algorithms. Ph. D thesis. Massachusetts Institute of Technology (1973)
15. Korte, B., Vygen, J.: Combinatorial Optimization: Theory and Algorithms, 5th edn. Springer Publishing Company, Incorporated (2012). https://doi.org/10.1007/978-3-540-71844-4_16
16. Lee, C.C., Lee, D.T.: A simple on-line bin-packing algorithm. J. ACM **32**(3), 562–572 (1985)
17. Mommessin, C., Erlebach, T., Shakhlevich, N.V.: Classification and evaluation of the algorithms for vector bin packing. Comput. Oper. Res. **173**, 106860 (2025)
18. Tinhofer, G.: Bin-packing and matchings in threshold graphs. Discret. Appl. Math. **62**, 279–289 (1995)
19. Ye, Y., Borodin, A.: Priority algorithms for the subset-sum problem. J. Comb. Optim. **16**(3), 198–228 (2008)
20. Zhu, D.: Max-min bin packing algorithm and its application in nano-particles filling. Chaos Solit. Fractals **89**, 83–90 (2016)

Low Recourse Arborescence Forests Under Uniformly Random Arcs

J. Niklas Dahlmeier[1(✉)] and D. Ellis Hershkowitz[2]

[1] RWTH Aachen, Aachen, Germany
dahlmeier@gdm.rwth-aachen.de
[2] Brown University, Providence, USA
delhersh@gmail.com

Abstract. In this work, we study how to maintain a forest of arborescences of maximum arc cardinality under arc insertions while minimizing recourse—the total number of arcs changed in the maintained solution. This problem is the "arborescence version" of max cardinality matching.

On the impossibility side, we observe that even in this insertion-only model, it is possible for m adversarial arc arrivals to necessarily incur $\Omega(m \cdot n)$ recourse, matching a trivial upper bound of $O(m \cdot n)$. On the possibility side, we give an algorithm with expected recourse $O(m \cdot \log^2 n)$ if all m arcs arrive uniformly at random.

1 Introduction

Arborescences (a.k.a. directed spanning trees) are one of the most well-studied objects in algorithmic graph theory. Specifically, given a digraph $G = (V, A)$ an arborescence is a subgraph $T \subseteq A$ where every vertex has in-degree at most 1 and T is a tree if we forget arc directions. Since the formulation of poly-time algorithms for computing a minimum weight arborescence [13,17], arborescences have served as an important case study across areas of algorithms and combinatorial optimization. These include near-linear time, primal-dual, randomized, and approximation algorithms, as well as integral polyhedra/totally unimodular (TU) matrices, edge splitting and tree packing theorems [4,7,16,21,22,32,35].

However, one notable area in which arborescences are poorly understood is the dynamic setting. In dynamic settings, our goal is to maintain a solution as the input changes over discrete time steps. A well-studied notion of the quality of a dynamic algorithm is that of recourse: the total number of changes made to the solution over time. Low-recourse algorithms are desirable both because recourse lower bounds the time to update a solution and because solutions that do not change very much over time are often desirable in their own right. For these reasons, many classic algorithmic problems have been studied in the low recourse setting, including set cover [25,28], matching [1,6,9,12,23,37], load balancing [27,33], minimum spanning and Steiner tree [24,26,30,38], TSP [38], clustering [14,19,34,36] and general convex body chasing [2,3,8,11,39].

J. Matuschke and J. Verschae (Eds.): WAOA 2025, LNCS 16077, pp. 142–156, 2026.
https://doi.org/10.1007/978-3-032-06706-7_10

Very recently, Buchbinder et al. [10] gave low-recourse algorithms for the dynamic version of what we will call the maximum arborescence forest problem.[1]

Definition 1 ((Maximum) Arborescence Forest). *Given a digraph $G = (V, A)$, we say a subgraph $F \subseteq A$ is an arborescence forest if each (weakly) connected component of F is an arborescence. Such a subgraph is maximum if it has maximum arc cardinality among all arborescence forests.*

We refer to any vertex of an arborescence forest with in-degree 0 as a *root*. See Fig. 1. The dynamic version of this problem which we study in this work is as follows.

Definition 2 (Incremental Maximum Arborescence Forest). *In the incremental maximum arborescence forest, we are given a vertex set V and a sequence of arcs $a_1, a_2, \ldots, a_m$ over time steps. In the i-th time step we must output a maximum arborescence forest $F^{(i)}$ for digraph $G^{(i)} := (V, \bigcup_{j \leq i}\{a_j\})$. Our goal is to minimize recourse, defined as*[2]

$$\sum_{i=1}^{m-1} |F^{(i)} \setminus F^{(i+1)}|.$$

Buchbinder et al. [10] showed that $O(n \log^2 n)$ recourse is possible if n vertices—rather than edges—arrive adversarially. That is, if in each time step all in-arcs of a unique vertex arrive, then $O(n \log^2 n)$ recourse is possible after n vertex arrivals. On the other hand, to date nothing is known for this problem if arcs, rather than vertices, arrive over time.

1.1 Our Results

We give results for incremental maximum arborescence forest when arcs arrive over time.

On the hardness side, we observe that if m arcs arrive non-adaptively but adversarially, then it is impossible to achieve recourse $o(m \cdot n)$.

Theorem 1 *Given integral $n \geq 0$, there exists a fixed n vertex instance of incremental maximum arborescence forest with $O(n)$ arc insertion such that every solution has recourse at least $\Omega(n^2)$.*

---

[1] More generally, they give results for low recourse matroid intersection where one matroid is a partition matroid and each part of the partition is incrementally revealed.

[2] We use arc deletions and recourse interchangeably. Recourse is often instead defined as a symmetric difference between solutions. The structure of maximum arborescence forests and the fact that our solutions monotonically increase in size up to $n-1$ means that these two quantities are the same up to a multiplicative 2 and additive $n-1$.

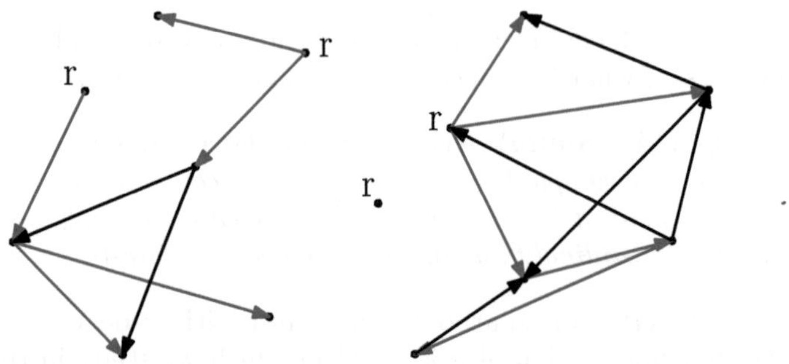

Fig. 1. Red: an arborescence forest where each root is labeled "r". Black: all other arcs. (Color figure online)

Note that it is always possible to achieve recourse $O(n)$ per each arc insertion (as one can fully change the maintained solution with $O(n)$ recourse). Thus, the above impossibility result matches a trivial upper bound of $O(m \cdot n)$. (In the full version of the paper on arXiv we observe that a similar lower bound holds for the problem of dynamically maintaining a minimum-cost arborescence, even when all edges have cost 0 or 1 and parallel arcs are prohibited.)

Since adversarial arc arrivals prohibit nontrivial upper bounds, it becomes natural to study random arc arrivals. On the possibility side, we demonstrate that if m arcs arrive uniformly at random chosen from the complete graph, then expected recourse $O(m \log^2 n)$ is always possible.

Theorem 2 *There is a polynomial-time algorithm for incremental maximum arborescence forest with expected recourse $O(m \log^2 n)$ if all m arcs arrive uniformly at random.*

Similar models where updates happen uniformly at random have been studied by other works in dynamic algorithms. See, for example, work on densest subgraph [18] and subgraph counting [29].

Our results can be seen as a natural complement to existing work in maximum cardinality (bipartite) matching. Specifically, maximum arborescence forest is, in some sense, to arborescence problems as maximum cardinality matching is to matching problems: like maximum arborescence forest, maximum bipartite matching is a special case of finding the largest common independent set of two matroids.[3] Furthermore, like maximum arborescence forest, in maximum bipartite matching $O(n \log^2 n)$ recourse is possible in adversarial vertex arrival models but, similar to our result, $\Omega(n^2)$ recourse is necessary in adversarial edge arrival models [6]. Furthermore, motivated similarly to us, Bernstein and Dudeja [5] studied random edge arrival models for low-recourse bipartite matching, showing that if the graph is chosen adversarially but n edges arrive in a uniformly

[3] For maximum matching this is the intersection of two partition matroids whereas for maximum arborescence forest this is the intersection of a partition matroid and the graphic matroid.

random order, expected recourse of $O(n \log n)$ and $O(n \log^2 n)$ on paths and trees respectively is possible.[4] We show that similar results are possible for maximum arborescence forest on *general graphs* when arcs arrive uniformly at random (among all possible arcs, not an adversarially-chosen subset).

1.2 Challenges and Techniques

Before giving a formal description of our results, we give an overview of challenges and techniques.

Adversarial Recourse Lower Bound. Our recourse lower bound of $\Omega(n^2)$ over the course of $O(n)$ arc insertions follows from a simple "bidirected" path example.

The basic idea is to insert arcs pointing in both directions on a path so that there is always a unique maximum arborescence forest whose nontrivial arborescence either goes from left to right or right to left in the path. We then repeatedly add arcs to the end of this nontrivial arborescence to flip the direction that it points so that once $\Omega(n)$ arcs have been inserted each new arc inserted incurs recourse $\Omega(n)$.

Randomized Recourse Upper Bound. Most of our work focuses on showing our upper bound in the uniformly random arc arrival setting.

Basic Idea of the Algorithm. We begin by discussing the intuition for our algorithm. It is not too hard to see that a forest of arborescences F is maximum in a graph G if and only if there do not exist distinct roots r and r' of F with a path from r to r' in G (Lemma 2).

One natural algorithmic idea, then, is to repeatedly reconfigure one's solution to guarantee that no such paths exist. It is not too hard to see that adding any such path to a solution and appropriately deleting parent arcs of vertices in such a path results in a new arborescence forest of size one larger than the previous (Lemma 5). Furthermore, the result of reconfiguring along such a path from r to r' is that r remains a root but r' is no longer a root; all of the vertices that used to have r' as the root of their arborescence now have r as their root.

A Weak Recourse Bound. However, since our goal is ultimately a low recourse algorithm, we must guarantee that we do not need to perform too many such reconfigurations. Likewise, we must guarantee that the paths between roots along which we update are not too long (and therefore do not result in too many parent arc deletions).

Regarding the former point, one can see that if we already had a maximum arborescence forest, we need only perform at most one such reconfiguration when

[4] This work also shows that recourse $\Omega\left(\frac{n^2}{\log n}\right)$ is necessary for general graphs in this adversarial graph but random edge arrival setting.

a new arc arrives. Intuitively, this is because adding a single arc to a graph increases its maximum arborescence forest cardinality by at most 1 (Lemma 5).

Regarding the latter point, we can show that if there exists some path between two roots, then there always exists a nicely structured one (which we call a "feasible path"; see Definition 4) between some other roots r and r'. These nicely structured paths satisfy the property that, if we update along them, the maximum number of arc deletions we incur is at most the size of the arborescence of r' in the current arborescence forest. Of course, in general, the size of the arborescence of r' may be as large as $\Omega(n)$. Nonetheless, we will leverage this seemingly weak bound on the arborescence to argue $O(\log^2 n)$ amortized recourse.

The key to leveraging the above weak recourse bound will be to observe that a given vertex v must delete its parent arc in our solution only when it is contained in the arborescence of a root r' where we just added a new arc incident to the in-component of r'.[5] Thus, if the in-component of r' is very small, we do not expect v to be forced to delete its parent arc very often.

Of course, in general, the in-component of r' can have size $\Omega(n)$ and so it will not *always* be the case that v is contained in an arborescence whose root has a small in-component.

Using Random Graph Structure. The last piece of our strategy is an amortization argument which shows that, by properties of our random model and the structural properties of our paths, *on average* vertices tend to be in arborescences whose in-components are small and so they need not delete their parent arcs too often.

The first step towards arguing this is to observe that our random model can be equivalently understood in a way that allows us to inherit some structural results from $D(n,p)$, the random directed n-node graph where each arc is independently present with probability p. In particular, if we assign to each arc a uniformly random value in $[0,1]$ and then sequentially add the m arcs with lowest value (in ascending order of value) then this is identical to our model. This then allows us to note that if we take all arcs whose value is less than some p, then our graph at this point is distributed identically to $D(n,p)$ and so we can make use of structural theorems known for $D(n,p)$.

Likewise, we can argue that a desirable property is always exhibited by our graph by either (1) noting that the property is monotone and so proving it for $D(n,p)$ proves it for $p' < p$ or (2) subdividing $[0,1]$ into small intervals each of which contains at most one arc, proving the property for each interval with high probability, and then union bounding over all intervals.

Notably, for p much larger than $1/n$ it is known that $D(n,p)$ has no in-components of size in $[\Omega(\log n), O(n)]$ for a suitable hidden constant (Lemma 11) with high probability. Furthermore, our algorithm guarantees:

---

[5] The in-component of a vertex v in a digraph comprises all vertices that can reach v. The out-component of v comprises all the vertices that v can reach and the strongly connected component of v is the intersection of its in- and out-components.

- The in-component of a root is always fully contained in its arborescence (Lemma 3) and;
- Merging the arborescence of r' into that of r does not change the in-component of r (Lemma 8).

Thus, even if a vertex v is in an arborescence whose root has an in-component of size $\Omega(n)$, it can only be merged into $O(1)$ such arborescences until only arborescences with in-components of size $O(\log n)$ remain (since otherwise its arborescence would have size $> n$). Once v is merged into such an arborescence whose root's in-component has size $O(\log n)$, v will contribute essentially nothing in recourse for the subsequent $\Theta(n/\log n)$ rounds (arc insertions).

More precisely, we divide our recourse analysis into two phases, which correspond to $p < 2/n$ and $p > 2/n$. The two phases use somewhat sophisticated amortization arguments similar to the above to show amortized recourse $O(\log n)$ and $O(\log^2 n)$ respectively.

Deferred proofs throughout our work are available in the full version of our paper on arXiv.

Discussion of Techniques. While our results make use of random graph structure, the key insight of our work is not necessarily only applicable in uniformly random graphs. In particular, the idea that structuring one's solution so that, on average, vertices tend to be in arborescences whose roots have small in-components might feasibly be useful for other randomized models.

For instance, our techniques may prove useful for the above-mentioned model in which the graph is fixed adversarially but arcs arrive in a random order. Notably, it is possible to apply the above insight to our lower bound instance to achieve low recourse in this random model for this particular graph. Whether this is possible in general graphs is an exciting open question for future work. However, some tempering of expectations is likely necessary: low-recourse algorithms have been proven impossible in this model for the closely-related maximum matching problem [5] and so such results may also ultimately be impossible for arborescences.

2 Lower Bound for Adversarial Arc Arrivals

In this section, we briefly explain a simple example that demonstrates that if arcs arrive adversarially, then, in general, recourse $\Omega(m \cdot n)$ is necessary after m arc arrivals for maximum arborescence forest. Notably, this example is non-adaptive in the sense that the adversary can fix all arc insertions ahead of time without knowing what maximum arborescence forests the algorithm outputs. The full example with figure can be found in the full version on arXiv. The idea is to build a bi-directed path of length $\Theta(n)$ such that the root of this path is always on one end. Then a new incoming arc is added to the other side of the path such that all arcs need to get flipped. This newly added arc then get bi-directed as well and this process continues until the graph if strongly connected.

Theorem 1 *Given integral $n \geq 0$, there exists a fixed n vertex instance of incremental maximum arborescence forest with $O(n)$ arc insertion such that every solution has recourse at least $\Omega(n^2)$.*

3 Upper Bound for Uniformly Random Arc Arrivals

In this section, we show how to achieve expected recourse $O(m \log^2 n)$ over the course of m arc insertions. Namely, we prove the following theorem.

Theorem 2 *There is a polynomial-time algorithm for incremental maximum arborescence forest with expected recourse $O(m \log^2 n)$ it if all m it arcs arrive uniformly at random.*

3.1 Algorithm Description

We begin by describing our algorithm for incremental maximum arborescence forest. Given an arborescence forest F and a non-root vertex v, we let $\text{PARENT}(v)$ give the parent arc of v in F (i.e. the unique in-arc of v in the arborescence which contains v in F). To compute $F^{(i)}$ given $F^{(i-1)}$ we check if adding a_i created *some path* between two distinct roots of $F^{(i-1)}$. If doing so does, then we select a *certain structured path* between roots, remove the parent arcs of vertices internal to this path and then add this path to $F^{(i-1)}$ to produce $F^{(i)}$. Note that there could be multiple roots that can reach a root after adding a_i (though the destination root is the same for all) and we specify which one is chosen.

More formally, the way we update along paths is captured by the following update.

Definition 3 (Path Update). *Let F be an arborescence forest of graph G with roots r and r' and let $P = (r, v_1, v_2, \ldots, v_k, r')$ be a path of G from r to r'. Then we define*

$$\text{UPDATE}(F, P) := \left(F \setminus \bigcup_i \text{PARENT}(v_i) \right) \cup P.$$

The forms of the paths along which we update are given by the following notion of feasible paths. Roughly, these are paths that follow one arborescence, take the added arc and then stay inside a single strongly connected component.

Definition 4 (Feasible Paths). *Let F be an arborescence forest of graph G containing two arborescences T and T' with respective roots r and r'. Then we say that a path $P \subseteq G$ from r to r' is feasible if it is of the form $P = P_A \oplus P_V$ where the arcs of P_A are contained in T and the vertices of P_V are contained in T'.*

Using the above definitions, we formally define our algorithm in pseudocode in Algorithm 1.

Algorithm 1. Algorithm for Incremental Maximum Arborescence Forest

Input: Vertex set V and a sequence of arcs $a_1, a_2, \dots a_m$

Output: $F^{(1)}, F^{(2)}, \dots, F^{(m)}$ s.t. each $F^{(i)}$ is a max arborescence forest for $G^{(i)} := (V, \bigcup_{j \le i}\{a_j\})$

```
Let F^(0) = (V, ∅)
for i = 1, ..., m do
    if F^(i-1) has a feasible path P in G^(i) (Definition 4) then
        F^(i) ← UPDATE(F^(i-1), P)
    else
        F^(i) ← F^(i-1)
return F^(0), F^(1), ..., F^(m)
```

3.2 Correctness of Algorithm

We begin by proving that Algorithm 1 is correct; namely, we prove that $F^{(i)}$ is a maximum arborescence forest for $G^{(i)}$ for every i.

We begin by observing that our update results in a forest of arborescences.

Lemma 1. *Let G be a digraph and let $F \subseteq G$ be an arborescence forest with a path $P \subseteq G$ between distinct roots of F. Then* UPDATE(F, P) *is a forest of arborescences.*

Proof. Let $P = (r, v_1, v_2, \dots, v_k, r')$ be the path and let $F' = \text{UPDATE}(F, P)$. First, observe that UPDATE preserves the in-degree of all vertices, except for r' which has its in-degree increased from 0 to 1. Since F was a forest of arborescences, it follows that every vertex in F' has in-degree at most 1. Thus, to show that F' is a forest of arborescences, it suffices to show that F' contains no cycles if we forget about arc directions.

Suppose for the sake of contradiction that such a cycle existed and let P' be a maximal contiguous subpath (arcs do not necessarily point in the same direction) of this cycle where all edges of P' are contained in $F' \setminus P$ (such a subpath must exist since P is itself a path). Let u and v be the first and last vertices of P' with incident arcs a_v and a_u in P'. By the assumption that $a_v, a_u \in F' \setminus P$ we know that a_v and a_u have v and u as their tails (i.e. the arcs point *away* from u and v) since otherwise they would have been the parent arcs of u and v and would have been deleted by UPDATE. However, it follows that some vertex in P' has in-degree at least 2 in F', contradicting the above argument that F' has in-degree at most 1.

We next state a simple helper lemma characterizing the maximum arborescence forests as those which have no paths between roots. This lemma is a special case of more general theorems known regarding matroid intersection—see e.g. [15]—but is considerably simpler in the special case of maximum arborescence forests and so we include a proof for completeness.

Lemma 2. *Given a digraph G, an arborescence forest $F \subseteq G$ is maximum if and only if there do not exist distinct roots r and r' of F where r has a path to r' in G.*

Proof. First, suppose that we have a directed path P from some root r to some other root r' and denote the arborescence starting at r' by T. Now, simply backtrack this path P starting at vertex r' until we hit the first vertex that does not belong to T and call that vertex v. Let r'' be the root of said vertex v. Now r'' can become the new root of the arborescence T that started at r' simply by adding the path from v to r' and deleting the unique incoming arcs of the arborescence T to the vertices in the path P from v to r'. This shows that the arborescence forest was not maximum.

For the other direction, suppose the arborescence forest F is not maximum, then there is a different arborescence forest F' with strictly fewer arborescences that still covers all vertices. By the pigeonhole principle, there must be at least one arborescence T in F' that contains at least two roots r and r' from the original arborescence forest. Denote by r'' the root of this arborescence T. If there is a path from r to r' or vice versa we are done. Else we know that r and r' are not in the strong component of r''. We also know that there must be some root s in the original arborescence forest F whose arborescence contains r''. Note that s is distinct from r and r' since it can reach r'' while the other two cannot. This root s can clearly connect to r and r', meaning that there is a directed path connecting two distinct roots of F.

Lemma 3. *Given a digraph G and a maximum arborescence forest F with a root r, if v can reach r in G then v is in the arborescence of r in F.*

Proof. Suppose F is a maximum arborescence forest and v is a vertex that can reach a root r, however it is not in the arborescence of r. Call r' the root of the arborescence that contains v. By definition of an arborescence, there is a directed path from r' to v and by assumption there is a path from v to r. By Lemma 2 we get a contradiction to the assumption that F is a maximum arborescence forest, since there is a path from root r' to root r.

Lemma 4. *Suppose F is a maximum arborescence forest for digraph $G = (V, A)$. If $G + a$ for some arc $a \notin A$ contains a path between two roots of F, then $G + a$ contains a feasible path for F.*

Proof. Let P be the assumed path in $G + a$ from root r to root r' and let $a = (u, v)$. Let w be the last vertex in P which is not contained in the arborescence of r' and let P_V be the maximal suffix of P which does not contain w. By definition, there exists some root $r'' \neq r'$ with some arborescence T'' of F which has some path $P_A \subseteq T''$ to w. Take as the feasible path $P_A \oplus P_V$.

Lemma 5. *Let $F^{(0)}, F^{(1)}, \ldots, F^{(m)}$ be the output of Algorithm 1 on graphs $G^{(0)}, G^{(1)}, \ldots, G^{(m)}$. Then $F^{(i)}$ is a maximum arborescence forest for $G^{(i)}$ for every i.*

Proof. Our proof is by induction on i, the above lemmas and, in particular, our careful definition of feasible paths (Definition 4). $F^{(0)}$ is trivially a maximum arborescence forest for $G^{(0)}$.

Thus, consider $i > 0$. By our inductive hypothesis, we know that $F^{(i-1)}$ is a maximum arborescence forest for $G^{(i-1)}$. We case on whether adding a_i created a path between roots of $F^{(i-1)}$ or not.

- If there are no paths between any roots of $F^{(i-1)}$ in $G^{(i)}$ then $F^{(i)} = F^{(i-1)}$ is maximum by Lemma 2.
- On the other hand, suppose that there is a path between roots r and r' of $F^{(i-1)}$ in $G^{(i)}$. By Lemma 4, $G^{(i)}$ contains a feasible path P of the form $P_A \oplus P_V$ where the edges of P_A are contained in the arborescence of r in $F^{(i-1)}$ and the vertices of P_V are contained in the arborescence of r' in $F^{(i-1)}$. Furthermore, observe that by the definition of UPDATE and Lemma 1, we have that $F^{(i)}$ is an arborescence forest with roots $R^{(i)} = R^{(i-1)} - r'$ where $R^{(i-1)}$ are the roots of $F^{(i-1)}$.
  We claim that $F^{(i)}$ therefore has no paths between roots in $G^{(i)}$ and so $F^{(i)}$ is maximum in $G^{(i)}$ by Lemma 2. Suppose for the sake of contradiction that such a path existed from root r_0 to root r_1 where $r_0, r_1 \in R^{(i)}$. Since $F^{(i-1)}$ is maximum in $G^{(i-1)}$, it follows by Lemma 2 that such a path must use a_i (since otherwise we would have a path between two distinct roots of $F^{(i-1)}$ in $G^{(i-1)}$, contradicting Lemma 2). Since $r' \notin R^{(i)}$ it must be the case that $r_1 \neq r'$. However, letting v be the head of a_i, we have that v must be reachable from r' in $F^{(i-1)}$ by Lemma 3 and the fact that v can reach r' in $F^{(i-1)}$ (by following a suffix of P). It follows that r_1 has a path to $r' \neq r_1$ in $G^{(i-1)}$, contradicting Lemma 2.

3.3 Recourse of Algorithm

In this section we bound the expected recourse of Algorithm 1 as $O(m \cdot \log^2 n)$. In particular, we prove the following.

Lemma 6. *The expected recourse of Algorithm 1 where arcs are inserted uniformly at random (without replacement) is $O(m \cdot \log^2 n)$.*

For simplicity of presentation, we will assume in our analysis that $n \log n$ arcs are inserted; it is clear from the analysis that terminating the analysis with $m < n \log n$ still yields the same recourse.

We will make use of the following notion of a component. Below, $A(S)$ gives all arcs with both endpoints internal to $S \subseteq V$.

The main fact we will use when bounding recourse is that by our careful choice of paths along which we update—namely feasible paths (Definition 4)—we only pay recourse when we add an arc incident to a root's in-component and when we do so we pay at most the size of that root's arborescence.

Lemma 7. *Suppose Algorithm 1 performs an update in iteration i, namely $F^{(i)} = \text{UPDATE}(F^{(i-1)}, P)$ where P is a feasible path from root r of $F^{(i-1)}$ to root r' of $F^{(i-1)}$. Let $a_i = (u, v)$ and let T' be the arborescence of r' in $F^{(i-1)}$. Then v is in the in-component of r' in $G^{(i-1)}$ and our recourse is at most*

$$|F^{(i-1)} \setminus F^{(i)}| \leq |T'|.$$

Proof. The fact that v is in the in-component of r' in $G^{(i-1)}$ is trivial.

To bound the recourse in terms of $|T'|$, observe that the only arcs we delete are parent arcs of vertices in T' which trivially gives $|F^{(i-1)} \setminus F^{(i)}| \leq |T'|$.

We will also make use of the following lemma which will allow us to argue that after a root r merges with the arborescence of another root r', the in-component of the resulting arborescence is just that of r.

Lemma 8. *Suppose Algorithm 1 performs a merge in the ith iteration; that is, there is some feasible path from root r to root r' of $F^{(i-1)}$ in $G^{(i)}$. Then the in-component of r is the same in $G^{(i-1)}$ and $G^{(i)}$.*

Proof. In particular, let $G = G^{(i-1)}$ be the graph at this point and let $a = (u, v) = a_i$ be the randomly inserted arc so that r has a path to r' in $G + a$ using a. Now suppose for the sake of contradiction that the in-component of r is strictly larger in $G + a$. It follows that there is some w which was not in the in-component of r which now is and this, in turn, can only happen if w has a path to r using a. However, it then follows that v has a path to r in G which by Lemma 3 means that v is in the arborescence of r (before updating). But, since v can reach r' without using a in G then so too can r and so we have a path from r to r' in G which contradicts Lemma 2 and Lema 5.

Our general approach will be to break our analysis into the first, roughly, $2n$ arcs and then all remaining arcs and appeal to Lemma 7. More specifically, we consider the following reformulation of uniformly random arc arrivals. Assign to each directed pair of vertices $(u, v) \in V \times V$ a value ρ_{uv} drawn independently and uniformly at random from $[0, 1]$. Let $a_1, a_2, a_3, \ldots, a_m$ be the first m such pairs after we sort by ρ values. Throughout the rest of this section we will refer to ρ_a as the value of arc a. It is easy to verify that these first m arcs are distributed identically to the m arcs sampled by our uniform random arc arrival model.

What do we gain from such a reformulation? Observe that if we set a fixed threshold $p \in [0, 1]$ and define G_p to consist of all arcs a which satisfy $\rho_a \leq p$ we have an instance of $D(n, p)$, namely the random digraph on n nodes where each arc appears independently at random with probability p. There is a wealth of literature on $D(n, p)$ and so by appropriately setting this threshold p we can hope to leverage some of these results to help bound our recourse. In what follows, we say that the recourse contributed by arc a_i is just $F^{(i-1)} \setminus F^{(i)}$.

Arcs with Value Under $2/n$. We start by bounding the recourse on, roughly, the first $2n$ arcs.

By Lemma 7, the only arborescences that are a problem are those with super-constant size. We prove this theorem by arguing that all arborescences of non-constant size will get merged at most $\log n$ times (in the regime to which this theorem belongs, i.e. during arc insertion of arcs with value under $2/n$). We do this by showing that after $O(\log n)$ merges it is expected that the root of an arborescence is a vertex whose strong component is a single vertex (i.e. the vertex itself). The following result summarizes this.

Lemma 9. *All arcs with a value under $2/n$ contribute a total recourse of $O(n \log n)$ with high probability.*

Arcs with Value Over $2/n$. Having bounded the recourse on the first (approximately) $2n$ arcs, we now bound the recourse on all subsequent arcs. Namely, we show that each arc insertion, where the arc has value at least c/n for constant $c > 1$, gives recourse $O(\log^2 n)$ on average. Note that this regime overlaps with the previous regime but the previous regime for $c > 1$ always contains arcs whose values are not in this regime and so the bounds of the previous section are indeed necessary.

We will make use of a well-known fact showing that $D(n, p)$ is strongly connected whenever p is sufficiently large. In particular, this means that we only have to reason about recourse until this strong connectivity threshold is met (at which point the entire graph is spanned by a single arborescence).

Lemma 10 (*[20]*, Theorem 13.9). *Let $p = 2 \log n/n$. Then $D(n, p)$ is strongly connected with high probability.*

Next, we adapt a known technique to bound the size of components in $D(n, p)$ by using arborescences as witnesses. See [31] and also Lemma 13.3 of [20].

Lemma 11. *Let $p = c/n$ for any $c > 1$. Then with high probability neither the in-component nor out-component of any vertex in $D(n, p)$ has size s for $s \in [10 \log n, \beta n]$ where $\beta = 1 - \frac{1.1}{c} - \frac{\ln c}{c}$.*

The above results apply to a single draw of $D(n, p)$, but we would like it to apply to all of our relevant $G^{(i)}$. We can achieve this by subdividing our value interval and appropriately union bounding over these intervals. Namely, we show the following.

Lemma 12. *With high probability for any $G^{(i)}$ where a_i has a value larger than $1.6/n$ we have that $G^{(i)}$ contains no vertices with an in-component of size in $[10 \log n, \alpha n]$ for some constant $\alpha > 0$.*

Lemma 13. *The expected total recourse from arcs with value larger than $1.6/n$ is $O(n \log^2 n)$.*

Lemma 6 is immediate by combining Lemmas 9 and 13.

3.4 Putting Correctness and Recourse Bounds Together

Our proof of Theorem 2 is immediate by combining the above recourse and correctness bounds.

Theorem 2. *There is a polynomial-time algorithm for incremental maximum arborescence forest with expected recourse $O(m \log^2 n)$ if all m arcs arrive uniformly at random.*

Proof. We use Algorithm 1. Returned solutions are maximum arborescence forests by Lemma 5. The expected recourse is $O(m \cdot \log^2 n)$ by Lemma 6. The runtime is trivial.

4 Conclusion

In this work, we gave the first dynamic algorithms for an arborescence problem with arc insertions—namely, maximum arborescence forest—with amortized expected recourse $O(\log^2 n)$ when arcs are randomly inserted. Likewise, we observed that random arc insertions are, in some sense, necessary for such results, since adversarial insertions can result in amortized recourse $\Omega(n)$.

References

1. Angelopoulos, S., Dürr, C., Jin, S.: Online maximum matching with recourse. J. Comb. Optim. **40**(4), 974–1007 (2020). https://doi.org/10.1007/s10878-020-00641-w
2. Argue, C., Gupta, A., Tang, Z., Guruganesh, G.: Chasing convex bodies with linear competitive ratio. J. ACM (JACM) **68**(5), 1–10 (2021)
3. Bansa, N., Böhm, M., Eliáš, M., Koumoutsos, G., Umboh, S.W.: Nested convex bodies are chaseable. In: Annual ACM-SIAM Symposium on Discrete Algorithms (SODA), pp. 1253–1260 (2018)
4. Bérczi, K., Frank, A.: Packing arborescences (2009)
5. Bernstein, A., Dudeja, A.: Online matching with recourse: random edge arrivals. In: Foundations of Software Technology and Theoretical Computer Science (FSTTCS) (2020)
6. Bernstein, A., Holm, J., Rotenberg, E.: Online bipartite matching with amortized O($log^2 n$) replacements. J. ACM (JACM) **66**(5), 1–23 (2019)
7. Bhalgat, A., Hariharan, R., Kavitha, T., Panigrahi, D.: Fast edge splitting and edmonds' arborescence construction for unweighted graphs. In: Annual ACM-SIAM Symposium on Discrete Algorithms (SODA), pp. 455–464 (2008)
8. Bhattacharya, S., Buchbinder, N., Levin, R., Saranurak, T.: Chasing positive bodies. In: Symposium on Foundations of Computer Science (FOCS) (2023)
9. Bosek, B., Leniowski, D., Sankowski, P., Zych, A.: Online bipartite matching in offline time. In: Symposium on Foundations of Computer Science (FOCS), pp. 384–393 (2014)
10. Buchbinder, N., Gupta, A., Hathcock, D., Karlin, A.R., Sarkar, S.: Maintaining matroid intersections online. In: Annual ACM-SIAM Symposium on Discrete Algorithms (SODA), pp. 4283–4304 (2024)

11. Buchbinder, N., Naor, J.: Online primal-dual algorithms for covering and packing. Math. Oper. Res. **34**(2), 270–286 (2009)
12. Chaudhuri, K., Daskalakis, C., Kleinberg, R.D., Lin, H.: Online bipartite perfect matching with augmentations. In: INFOCOM, pp. 1044–1052 (2009)
13. Chu, Y.J.: On the shortest arborescence of a directed graph. Sci. Sinica **14**, 1396–1400 (1965)
14. Cohen-Addad, V., Hjuler, N.O.D., Parotsidis, N., Saulpic, D., Schwiegelshohn, C.: Fully dynamic consistent facility location. In: Annual Conference on Neural Information Processing Systems (NeurIPS), vol. 32 (2019)
15. Cunningham, W.H.: Improved bounds for matroid partition and intersection algorithms. SIAM J. Comput. **15**(4), 948–957 (1986)
16. Drescher, M., Vetta, A.: An approximation algorithm for the maximum leaf spanning arborescence problem. ACM Trans. Algorithms (TALG) **6**(3), 1–18 (2010)
17. Edmonds, J., et al.: Optimum branchings. J. Res. Natl. Bur. Stand. B **71**(4), 233–240 (1967)
18. Epasto, A., Lattanzi, S., Sozio, M.: Efficient densest subgraph computation in evolving graphs. In: International World Wide Web Conference (WWW), pp. 300–310 (2015)
19. Fichtenberger, H., Lattanzi, S., Norouzi-Fard, A., Svensson, O.: Consistent k-clustering for general metrics. In: Proceedings of the 2021 ACM-SIAM Symposium on Discrete Algorithms (SODA), pp. 2660–2678. SIAM (2021)
20. Frieze, A., Karoński, M.: Introduction to Random Graphs. Cambridge University Press (2015)
21. Gabow, H.N.: A matroid approach to finding edge connectivity and packing arborescences. In: Annual ACM Symposium on Theory of Computing (STOC), pp. 112–122 (1991)
22. Gabow, H.N., Galil, Z., Spencer, T., Tarjan, R.E.: Efficient algorithms for finding minimum spanning trees in undirected and directed graphs. Combinatorica **6**(2), 109–122 (1986)
23. Grove, E.F., Kao, M.-Y., Krishnan, P., Vitter, J.S.: Online perfect matching and mobile computing. In: Akl, S.G., Dehne, F., Sack, J.-R., Santoro, N. (eds.) WADS 1995. LNCS, vol. 955, pp. 194–205. Springer, Heidelberg (1995). https://doi.org/10.1007/3-540-60220-8_62
24. Gu, A., Gupta, A., Kumar, A.: The power of deferral: maintaining a constant-competitive steiner tree online. In: Annual ACM Symposium on Theory of Computing (STOC), pp. 525–534 (2013)
25. Gupta, A., Krishnaswamy, R., Kumar, A., Panigrahi, D.: Online and dynamic algorithms for set cover. In: Proceedings of the 49th Annual ACM SIGACT Symposium on Theory of Computing, pp. 537–550 (2017)
26. Gupta, A., Kumar, A.: Online steiner tree with deletions. In: Annual ACM-SIAM Symposium on Discrete Algorithms (SODA), pp. 455–467 (2014)
27. Gupta, A., Kumar, A., Stein, C.: Maintaining assignments online: matching, scheduling, and flows. In: Annual ACM-SIAM Symposium on Discrete Algorithms (SODA), pp. 468–479 (2014)
28. Gupta, A., Levin, R.: Fully-dynamic submodular cover with bounded recourse. In: Symposium on Foundations of Computer Science (FOCS), pp. 1147–1157 (2020)
29. Henzinger, M., Lincoln, A., Saha, B.: The complexity of average-case dynamic subgraph counting. In: Annual ACM-SIAM Symposium on Discrete Algorithms (SODA), pp. 459–498. SIAM (2022)
30. Imase, M., Waxman, B.M.: Dynamic steiner tree problem. SIAM J. Discret. Math. **4**(3), 369–384 (1991)

31. Karp, R.M.: The transitive closure of a random digraph. Random Structures & Algorithms **1**(1), 73–93 (1990)
32. Korte, B., Vygen, J., Korte, B., Vygen, J.: Spanning trees and arborescences. Combinatorial Optimization: Theory and Algorithms, pp. 133–157 (2018)
33. Krishnaswamy, R., Li, S., Suriyanarayana, V.: Online unrelated-machine load balancing and generalized flow with recourse. In: Annual ACM Symposium on Theory of Computing (STOC), pp. 775–788 (2023)
34. Lacki, J., Haeupler, B., Grunau, C., Jayaram, R., Rozhoň, V.: Fully dynamic consistent k-center clustering. In: Annual ACM-SIAM Symposium on Discrete Algorithms (SODA), pp. 3463–3484 (2024)
35. Laekhanukit, B., Oveis Gharan, S., Singh, M.: A rounding by sampling approach to the minimum size k-arc connected subgraph problem. In: International Colloquium on Automata, Languages and Programming (ICALP), pp. 606–616 (2012)
36. Lattanzi, S., Vassilvitskii, S.: Consistent k-clustering. In: International Conference on Machine Learning (ICML), pp. 1975–1984 (2017)
37. Megow, N., Nölke, L.: Online minimum cost matching on the line with recourse. arXiv preprint arXiv:2001.03107 (2020)
38. Megow, N., Skutella, M., Verschae, J., Wiese, A.: The power of recourse for online MST and tsp. SIAM J. Comput. **45**(3), 859–880 (2016)
39. Sellke, M.: Chasing convex bodies optimally. In: Annual ACM-SIAM Symposium on Discrete Algorithms (SODA), pp. 1509–1518 (2020)

On the FirstFit Algorithm for Online Unit-Interval Coloring

Bob Krekelberg[(✉)] and Alison Hsiang-Hsuan Liu

Utrecht University, Utrecht, The Netherlands
{b.h.a.f.krekelberg,h.h.liu}@uu.nl

Abstract. In this paper, we analyze FirstFit on online unit-length intervals, which can either be open or closed, further investigating FirstFit's true performance. We develop a sophisticated counting method by generalizing the classic neighborhood bound, which limits the color FirstFit can assign an interval by counting the potential intersections. In the generalization, we show that for any interval, there is a critical interval intersecting it that can help reduce overestimation of the number of intersections, and it further helps bound the color an interval can be assigned. The technical challenge lies in reliably finding these critical intervals. Using this new technique, we provide a tight analysis showing that FirstFit uses at most 2ω colors when all endpoints are integral, and an upper bound of $\lceil \frac{7}{3}\omega \rceil - 2$ in the general case of open and closed unit intervals, where ω denotes the optimal number of colors needed.

Keywords: Online algorithms · FirstFit algorithm · Interval coloring · Unit intervals · Open and closed intervals

1 Introduction

This work studies the *online open/closed unit-length interval coloring* problem, which is a variant of the *online interval coloring* problem. The input consists of a sequence of unit-length intervals, where each interval is either an *open* interval $(r, r+1) = \{i \mid r < i < r+1\}$ or a *closed* interval $[r, r+1] = \{i \mid r \leq i \leq r+1\}$. The intervals are released to the algorithm one by one. Once an interval is released, the online algorithm has to irrevocably assign it a color that differs from the colors assigned to any of the previously released intervals it intersects with. The aim is to minimize the colors used. In this work, we study the performance of the FirstFit algorithm for coloring the open/closed unit-length intervals.

As a fundamental problem in graph theory and scheduling, intensive research is being conducted on the online interval coloring problem. The optimal online algorithm for the general case where intervals have arbitrary lengths uses at most $3\omega - 2$ colors, where ω is the optimal number of colors [1,10,13]. Recently, interest in coloring bounded-length intervals arose, and better algorithms were found for special families of instances. For intervals with length within a fixed range between 1 and $\sigma \in \mathbb{Q}_{\geq 1}$, a $(\sigma+1)$-competitive algorithm was proposed [2], which improved the performance of online algorithms in the cases where $\sigma < 2$.

J. Matuschke and J. Verschae (Eds.): WAOA 2025, LNCS 16077, pp. 157–171, 2026.
https://doi.org/10.1007/978-3-032-06706-7_11

In comparison, the FIRSTFIT algorithm, a classical greedy algorithm, is interesting for its elegance. The FIRSTFIT algorithm is currently the best algorithm for (closed) unit-length intervals, which uses $2\omega - 1$ colors for any ω-colorable set of unit-length intervals [1,5]. For arbitrary-length intervals, the competitive ratio is known to be between 5 and 8 [9,11], but the exact value has remained an open problem for a long time.

The motivation to investigate the performance of the FIRSTFIT algorithm on the coloring of open and closed unit-length intervals is four-fold. First, FIRSTFIT is one of the most intuitive and easy-to-implement algorithms, making it appealing in practical settings. Second, analyzing its performance provides valuable insight into the nature of online problems. As a greedy strategy, FIRSTFIT always chooses the locally optimal option at each step. While this approach works well for some online problems, it can perform poorly in others. Studying its performance provides insight into the fundamental differences between problems by examining how much an adversary[1] can cause the algorithm to deviate from the offline optimal solution due to its reliance on greedy decision-making.

Third, despite its importance, the performance of FIRSTFIT is not fully understood. It is known to be exactly $(2 - \frac{1}{\omega})$-competitive for unit-length closed intervals, and at least 5-competitive in the general case. However, its behavior on intermediate cases remains unclear. The open/closed unit-length intervals case, being the simplest nontrivial instance with bounded lengths, is key to determining the exact competitive ratio of FIRSTFIT.

As a final but critical motivation, studying open/closed intervals sheds light on the impact of "ε-uncertainty". Closed unit-length intervals can be viewed as approximately $\varepsilon \approx 0$ longer than open ones. This subtle difference increases the problem's complexity, as open intervals can now be proper subsets of closed intervals–unlike in the purely closed case. FIRSTFIT has been shown to be at least 2-competitive in this setting [3], implying that the open/closed case is strictly harder for FIRSTFIT than the closed-only case. Determining the competitive ratio of FIRSTFIT in this setting quantifies the power of the uncertainty of ε-difference by measuring the extent to which the adversary can trap FIRSTFIT using this uncertainty.

Related Work. For the online interval coloring problem, Kierstead and Trotter designed an optimal online algorithm that uses at most $3\omega - 2$ colors on an ω-colorable interval graph [10]. Independently, Chrobak and Ślusarek found the same upper and lower bounds [1,13].

Recently, interest in coloring bounded-length intervals arose. Chybowska-Sokół et al. [2] studied the online interval coloring problem where the size of the intervals is within some fixed range $[1, \sigma_{\geq 1}]$. When $\sigma = \infty$, it is the previously mentioned general case, and when $\sigma = 1$, it is the unit-length case. They proposed a $(\sigma+1)$-competitive algorithm, which improves upon the general algorithm for $\sigma < 2$. Curbelo [4] then shows that, in the setting where the algorithm

[1] An online problem can be viewed as a game between the online algorithm and an adversary, where the adversary sequentially presents the input to maximize the algorithm's competitive ratio relative to the optimal solution.

receives vertices of the interval graph instead of intervals, there exists a $\sigma > 1$ such that the algorithm is at least 3-competitive.

FIRSTFIT. Despite the aforementioned results that included sophisticated algorithms, people are also interested in the performance of the naive but easy-to-implement FIRSTFIT algorithm. The exact competitive ratio of the FIRSTFIT algorithm is a long standing open problem. As early as 1976, Witsenhausen [14], and independently, in 1988, Chrobak and Ślusarek [1], proved that FIRSTFIT is at least 4 competitive. Later, Ślusarek [13] improved this bound to 4.45. The best known lower bound to date is from Kierstead et al. [9] in 2016. They prove that FIRSTFIT is at least 5-competitive.

For the upper bounds on the competitive ratio of FIRSTFIT, Kierstead [7] proved in 1988 a constant competitive ratio of 40. This result was improved to 25.72 by Kierstead and Qin [8]. Next, Pemmaraju et al. [12] showed that FIRSTFIT is at most 10-competitive. In 2008, Narayanaswamy and Subhash Babu [11] improved this analysis, and showed that FIRSTFIT uses at most 8ω colors.

The performance of the FIRSTFIT algorithm has also been of interest in special cases. Chybowska-Sokół et al. [2] studied the FIRSTFIT algorithm where the size of the intervals is within some fixed range $[1, \sigma_{\geq 1}]$. The authors show tighter bounds for various values of σ.

For unit-length closed intervals, Chrobak and Ślusarek [1], and independently, Epstein and Levy [5] proved that FIRSTFIT uses exactly $2\omega - 1$ colors.

In 2022, Biró and Curbelo proved that FIRSTFIT, on open and closed unit-length intervals, uses at least 2ω and at most $3\omega - 3$ colors [3, Theorem 6.0.1 (Biró and Curbelo (unpublished))]. Furthermore, Curbelo claimed that FIRSTFIT uses at most $3\omega - f(\omega)$ colors for a non-specified function f, which increases with ω [3]. In this paper, we improve the upper bound for this special case.

FIRSTFIT and Counting. The analysis of the FIRSTFIT algorithm is generally performed by carefully counting the intersections any interval can have. The elegance of the analysis lies in determining which properties limit the extent of overestimation in intersecting intervals. The analysis by Kierstead [7] identifies a large set of mutually intersecting intervals and then carefully counts structural properties of the graph, such as the number of overlapping intervals. Later results by Pemmaraju et al. [12] and Narayanaswamy and Subhash Babu [11] represent the overlapping structure of intervals using "columns". The columns are assigned distinct labels that encode information about the structure of interval intersections and their corresponding color assignments. The analysis then carefully counts the occurrences of each label to bound the total number of colors used by the FIRSTFIT algorithm. Finally, for unit-length (closed) intervals, the analysis carefully counts for each interval at what positions an interval could intersect it and how many of such intervals there could be [1,5].

Our Contribution. Following the traditional counting approach, we analyze the FIRSTFIT algorithm on open and closed unit-length intervals by counting how many intersections an interval can have.

Our main technical contribution is the Pivot bound, a generalization of the traditional neighborhood bound of the FirstFit algorithm (Lemma 2). Informally, this lemma shows that the color assigned to interval I by FirstFit can be bounded via another interval intersecting I that is critical in the sense that it is a relatively high-colored interval but with relatively few intersections. Our results are built on this newly-observed property of FirstFit. In a nutshell, for any interval I, our analysis identifies the critical interval intersecting I, using that the input instance consists only of unit-length open/closed intervals. It shows that the number of intersections is bounded by sophisticated counting. Although the technical details of the proofs in this paper rely on the intervals being of unit length, the aforementioned pivot bound remains applicable in broader settings.

For a special case where all intervals have integral endpoints, we prove that FirstFit is 2-competitive, which is tight as it matches the FirstFit algorithm lower bound by Biró and Curbelo [3]. Formally, for any instance that can be colored by ω colors, we show that

Theorem 1. *For open and closed unit intervals with integral endpoints, the* FirstFit *algorithm for online coloring uses at most* 2ω *colors.*

Then, for unit-length intervals with arbitrary endpoints, we prove:

Theorem 2. *For open and closed unit intervals with arbitrary endpoints, the* FirstFit *algorithm for online coloring uses at most* $\lceil \frac{7}{3}\omega \rceil - 2$ *colors.*

Paper Organization. Section 2 formally defines the problem and introduces key lemmas, the analytical framework, and core concepts used throughout the paper. Section 3 begins with the proof of 2-competitiveness (Theorem 1) of FirstFit on open/closed unit-length intervals with integral endpoints, and establishes properties that hold both in the restricted and more general case with arbitrary endpoints. Section 4 analyzes this general case and proves Theorem 2. Section 5 concludes. Most proofs from Sect. 4 appear in the full version due to space constraints.

2 Preliminaries, Key Lemmas, and the Framework

The input to our online coloring problem is an ordered set of open and closed unit-length intervals $\mathcal{I} = \{I_1, I_2, \cdots\}$, where the ordering of the set of intervals is the order in which they are revealed to the online algorithm. That is, I_j is revealed earlier than I_k if $j < k$. Each interval $I_j \in \mathcal{I}$ is either an *open* unit-length interval $(r_j, r_j + 1)$ or a *closed* unit-length interval $[r_j, r_j + 1]$ for some real number r_j. In either case, we say r_j and $r_j + 1$ are the *endpoints* of the interval. Note that the value of r_j is unrelated to the order of I_j in $\mathcal{I}$. Namely, r_j can be smaller than, equal to, or larger than r_k when $j < k$. We ignore the suffix of intervals when the reveal ordering is not necessarily to be emphasized.

Two intervals are referred to as *twins* when they are identical in terms of the exact location. That is, for interval $I = [r, r+1]$, all intervals $I' = [r, r+1]$ are its twins. Similarly, for interval $I = (r, r+1)$, all intervals $I' = (r, r+1)$ are its twins. However, an interval $(r, r+1)$ is not a twin of $[r, r+1]$, and vice versa. When intervals I and I' are twins, we write $I \equiv I'$. We further denote the set of twins of I by $\mathcal{T}(I)$. Note that we define $I \notin \mathcal{T}(I)$.

A *proper coloring* of a set of intervals is a function $c : \mathcal{I} \to \mathbb{N}$, such that $c(I) \neq c(I')$ if $I \cap I' \neq \emptyset$. The algorithm aims to find a proper coloring of the intervals using the least number of colors.

We use terms from graph algorithms. The intervals in $\mathcal{I}$ form an *interval graph*, where each $I \in \mathcal{I}$ is a vertex in the interval graph, and there is an edge between the vertices corresponding to I and I' if and only if $I \cap I' \neq \emptyset$. We denote the *neighborhood* of I by $\mathcal{N}(I)$, which is the set of all the intervals intersecting I.

As the coloring depends on intersecting intervals as well as the ordering of the input, we define the *ℓ-neighborhood* of an interval as its neighborhood at the moment when I_ℓ is revealed to the algorithm. Formally, $\mathcal{N}_\ell(I_j) = \{I_k \mid I_j \cap I_k \neq \emptyset \text{ and } k < \ell\}$.[2] Naturally, $\mathcal{N}(I_j) = \mathcal{N}_{|\mathcal{I}|}(I_j) \supseteq \mathcal{N}_\ell(I_j)$ for any $\ell \in [j, |\mathcal{I}|]$. Note that $\mathcal{N}_j(I_j)$ is the set of all intervals that intersect with I_j when I_j is revealed. With slight abuse of notation we sometimes write $N_{I_j}(I_j)$ instead of $N_j(I_j)$ when it is not clear what the index of the interval is.

Interval graphs are *perfect graphs* [6], and the chromatic number of the graph is equal to the maximum clique size ω. This clique corresponds to the maximum set of mutually intersecting intervals for the corresponding intervals. We denote ω as the number of colors an optimal solution uses for the given input $\mathcal{I}$.

FIRSTFIT Algorithm. The FIRSTFIT algorithm assigns each interval the least available color, that is, using $\texttt{FF}(I_j)$ to denote the color assigned to I_j by FIRSTFIT, $\texttt{FF}(I_j) \leftarrow \min\{\mathbb{N} \setminus \{\texttt{FF}(I) \mid I \in \mathcal{N}_j(I_j)\}\}$. This greedy strategy provides an upper bound for the color of any interval: an interval's color is at most one more than the number of previously revealed intervals it intersects.

Lemma 1 (Neighborhood bound). *For any interval $I_j \in \mathcal{I}$, its color assigned by* FIRSTFIT *is at most $1 + |\mathcal{N}_j(I_j)|$.*

Pivot Interval and Ideas of Further Bonding the Color. In this work, we further generalize Lemma 1 by considering a *pivot* interval in the neighborhood of the interval I and use the pivot interval to bound the color $\texttt{FF}(I)$ more carefully. Informally, given any interval I and a pivot interval I^* that can be any interval intersecting with I, $\texttt{FF}(I)$ is at most 1 plus the number of intervals in I's neighborhood with color greater than the color of I^*, plus the color of I^*.

Lemma 2 (Pivot bound). *For any interval $I_j \in \mathcal{I}$, let I^* be any interval in $\mathcal{N}(I_j)$ and let $\mathcal{S}^* \subset \mathcal{N}(I_j)$ be the set of intervals in the neighborhood of I_j such that for all $I' \in \mathcal{S}^*$, $FF(I') > FF(I^*)$. Then, $FF(I_j) \leq FF(I^*) + |\mathcal{S}^*| + 1$.*

Proof. Assume aiming towards a contradiction that $\texttt{FF}(I_j) \geq 2 + \texttt{FF}(I^*) + |\mathcal{S}^*|$. It then follows from the pigeonhole principle that there must exist at least one color c, such that $\texttt{FF}(I^*) < c < 2 + \texttt{FF}(I^*) + |\mathcal{S}^*|$, that is not assigned to any of the intervals in $\mathcal{S}^*$. Then, $\texttt{FF}(I_j) = \min\{\mathbb{N} \setminus \{\texttt{FF}(I) \mid I \in \mathcal{N}_j(I_j)\}\} \leq c < 2 + \texttt{FF}(I^*) + |\mathcal{S}^*|$. Contradicting our assumption that $\texttt{FF}(I_j) \geq 2 + \texttt{FF}(I^*) + |\mathcal{S}^*|$. □

[2] With slight abuse of notation, we sometimes write $\mathcal{N}_{I_\ell}$ in stead of $\mathcal{N}_\ell$.

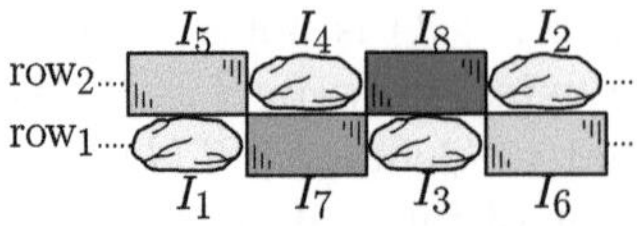

Fig. 1. Rows denote the color assigned by OPT, actual colors are assigned by FIRSTFIT.

Note that the Neighborhood bound is a special case of the Pivot bound, where the pivot is an empty set. In this extreme case, $\text{FF}(I^*) = 0$, and $\mathcal{S}^* = \mathcal{N}_{|\mathcal{I}|}(I)$.

Intuitively, for a smaller upper bound of $\text{FF}(I_j)$, by the Pivot bound, we can choose a pivot I^* with a small color $\text{FF}(I^*)$ or choose I^* with a small $|\mathcal{S}^*|$. However, we have a two-fold challenge. First, it is not trivial to find the best choice between minimizing $\text{FF}(I^*)$ or minimizing $|\mathcal{S}^*|$. Second, even when a good pivot I^* is given, it does not necessarily have non-trivial upper bounds of $\text{FF}(I^*)$ and $|\mathcal{S}^*|$. To cope with these difficulties, we construct a mechanism to balance the two choices of minimizing $\text{FF}(I^*)$ or $|\mathcal{S}^*|$.

Optimal Coloring and the Image of Walls. In the analysis of FIRSTFIT algorithm, we compare the solution of the FIRSTFIT algorithm to an *optimal offline algorithm* OPT that knows the complete input in advance. We imagine that OPT orients the intervals into a wall, row by row, using the fewest rows. More specifically, the intervals on the i-th row (from the bottom of the wall) are the intervals assigned color i by OPT. Since OPT uses precisely ω colors, the number of rows is exactly ω. Furthermore, no two intervals drawn in the same row will intersect with each other, as otherwise OPT admits no proper coloring.

To avoid confusion between the color assigned by the optimal offline algorithm and the color assigned by the FIRSTFIT algorithm, from now on, we distinguish the color of an interval by FIRSTFIT or by OPT by its *color* or its *row*. More specifically, the color of an interval is the color assigned by FIRSTFIT algorithm, and the row refers to the color assigned by OPT. An example of the distinction between colors and rows is shown in Fig. 1. Given any interval, we partition the rows according to their alignment with the interval I.

Definition 1. *For any interval $I \in \mathcal{I}$, let $\mathcal{R}_i(I)$ be the set of rows in the optimal solution OPT($\mathcal{I}$) that contain exactly i intervals intersecting I. And let $r_i(I)$ denote $|\mathcal{R}_i(I)|$.*

As a special case, we define interval I to be contained in $\mathcal{R}_0(I)$. Strictly speaking, the set $\mathcal{R}_i(I)$ contains rows instead of intervals. However, by slight abuse of notation, we say that interval $I' \in \mathcal{R}_i(I)$ if I' is in $\mathcal{N}(I)$ and placed in one of the rows in $\mathcal{R}_i(I)$ by OPT.

As we consider unit intervals, for some i the set $\mathcal{R}_i(I)$ must be empty.

Lemma 3. *Consider any $I \in \mathcal{I}$.*

(a) $\mathcal{R}_0(I)$ contains at least the row where I is, and $r_0(I) \geq 1$.
(b) If I is closed, $r_i(I) = 0$ for all $i \geq 4$.
(c) If I is open, $r_i(I) = 0$ for all $i \geq 3$.
(d) Since OPT uses ω rows, $r_0(I) + r_1(I) + r_2(I) + r_3(I) = \omega$.

By the definition of $\mathcal{R}_{\mathrm{i}}(I)$, the number of intervals that intersect with I can be upper bounded in terms of $r_{\mathrm{i}}(I)$. More specifically,

Observation 1. *For any interval $I \in \mathcal{I}$, $|\mathcal{N}_{\mathcal{I}}(I)| \leq r_1(I) + 2 \cdot r_2(I) + 3 \cdot r_3(I)$.*

Recall from Lemma 3 (b) and (c) that only closed intervals I can have non-empty $\mathcal{R}_3(I)$. Furthermore, there is only a single way for an interval to intersect three intervals in the same row. Therefore, all rows in $\mathcal{R}_3(I)$ contain three sets of intervals, where the intervals in each set are twins. Formally, given any interval $I = [r, r+1]$ with $r \in \mathbb{R}$, we partition the intervals in $\mathcal{N}(I)$ into the twins of interval I, the aligned intervals, and the misaligned intervals:

$\mathcal{T}(I)$: contains all intervals $I' \equiv [r, r+1]$, but not I itself.
$\mathcal{A}(I)$: contains all intervals $I' \equiv [r-1, r]$, $I' \equiv (r, r+1)$, and $I' \equiv [r+1, r+2]$.
$\mathcal{M}(I)$: contains all intervals in $\mathcal{N}(I) \setminus (\mathcal{T}(I) \cup \mathcal{A}(I))$.

Framework. For any interval $I \in \mathcal{I}$, we bound its color from FirstFit by using the Neighborhood bound (Lemma 1) and the Pivot bound (Lemma 2). Specifically, we pick a pivot I^* with special structural properties. Using the property, we bound the size of the corresponding $\mathcal{S}^*$. Then, we bound the color $\mathtt{FF}(I^*)$ using the Neighborhood bound. Finally, we use the Pivot bound to bound the color of I.

Practically, we first identify the "easy cases" where $\mathtt{FF}(I)$ is at most 2ω. Then, after peeling off these easy cases, we focus on the tough kernel of the analysis where there is no trivial choice for I^* such that the sum of $\mathtt{FF}(I^*)$ and $|\mathcal{S}^*|$ is small. For this tough kernel of analysis, we identify two mediocre choices for I^* for which the sum of $\mathtt{FF}(I^*)$ and $|\mathcal{S}^*|$ cannot both be large at the same time.

Small Neighborhoods. We first consider intervals I with small $r_3(I)$, and open intervals. These intervals have relatively small neighborhoods. Formally, by the Neighborhood bound and Lemma 3 (d), we can show the following:

Lemma 4. *For each interval $I \in \mathcal{I}$, if $r_3(I) < r_1(I) + 2$, then $\mathtt{FF}(I) \leq 2\omega$.*

Since any interval in $\mathcal{R}_3(I)$ must be in $\mathcal{A}(I)$, an empty $\mathcal{A}(I)$ implies $r_3(I) = 0$. Hence, we have a useful corollary:

Corollary 1. *If $\mathcal{A}(I) = \emptyset$, then $\mathtt{FF}(I) \leq 2\omega$.*

By Lemma 3 (c), open intervals cannot have too many intersections per row.[3]

Theorem 3. *For each open interval $I \in \mathcal{I}$, $\mathtt{FF}(I) \leq 2\omega - 1$.*

Closed Intervals I with Large $r_3(I)$. Intuitively, because the intervals in $\mathcal{T}(I)$ have an identical neighborhood to interval I, these intervals are tricky to deal with. This leaves us to consider the intervals in the sets $\mathcal{A}(I)$ and $\mathcal{M}(I)$ as potential pivot intervals. In the ideal scenario, by choosing the interval with the

[3] This is a generalization from results of Chrobak and Ślusarek [1] and Epstein and Levy [5], which showed for closed unit intervals, FirstFit uses exactly $2\omega - 1$ colors.

largest color in the sets $\mathcal{A}(I)$ and $\mathcal{M}(I)$ as a pivot, $\mathcal{S}^* \subseteq \mathcal{T}(I)$, making the size of the pivot set easy to bound. Using the Neighborhood bound, we bound the color of this highest-colored interval. However, it is not always possible to pick such a pivot that $\mathcal{S}^* \subseteq \mathcal{T}(I)$. In the case that $\mathcal{S}^* \not\subseteq \mathcal{T}(I)$, we look more closely at what additional intervals are part of the pivot set $\mathcal{S}^*$. Finally, we apply the Pivot bound in order to prove an upper bound on the color of interval I.

In Sects. 3 and 4, we will focus on closed intervals I with large $r_3(I)$.

3 Warm-Up: Regular Case Where Endpoints are Integral

As a starter, we consider a special case where all intervals have integral endpoints. More formally, for any interval $I \in \mathcal{I}$, $I = (i, i+1)$ for $i \in \mathbb{Z}$ if I is open, and $I = [i, i+1]$ if I is closed. In this section, we show that any $I \in \mathcal{I}$ has color $\texttt{FF}(I) \leq 2\omega$, where ω is the number of colors used by the optimal solution. Note that in the following, we focus on the closed intervals $I \in \mathcal{I}$ with $r_3(I) \geq r_1(I)+2$.

For such a set of integral-endpoints intervals, it is an important property that $\mathcal{M}(I) = \emptyset$ for any $I \in \mathcal{I}$. That is, the intervals in $\mathcal{N}(I)$ are either in $\mathcal{T}(I)$ (twins of I) or in $\mathcal{A}(I)$ (aligned with I). An example is shown in Fig. 2.

Pick a Pivot I^*. For any closed interval $I \in \mathcal{I}$ with $r_3(I) \geq r_1(I) + 2$, we first identify the interval in $\mathcal{A}(I)$ with the highest color by FirstFit.

Definition 2. (Dominating interval $D^{\mathtt{A}}(I)$ in $\mathcal{A}(I)$) *Let interval $D^{\mathtt{A}}(I)$ be the interval in $\mathcal{A}(I)$ with the highest color by* FirstFit.

In the following analysis, we bound $\texttt{FF}(I)$ by using $D^{\mathtt{A}}(I)$ as a pivot I^* and then apply the Pivot bound (Lemma 2). Note that, by Corollary 1, $D^{\mathtt{A}}(I)$ exists.

Bound the Size of $\mathcal{S}^*$. Given that we use $D^{\mathtt{A}}(I)$ as the pivot I^*, we bound the size of $\mathcal{S}^*$. Recall that $\mathcal{S}^*$ is the set of intervals in $\mathcal{N}(I)$ which are assigned higher color than $\texttt{FF}(I^*)$. Since all intervals in $\mathcal{I}$ have integral endpoints, any interval in $\mathcal{N}(I)$ is either in $\mathcal{A}(I)$ or in $\mathcal{T}(I)$. Therefore, by the definition of $D^{\mathtt{A}}(I)$, only intervals in $\mathcal{T}(I)$ are possible to contribute to $\mathcal{S}^*$. Moreover, all intervals in $\mathcal{T}(I)$ must be in the rows in $\mathcal{R}_1(I)$. We use the following definition to represent the fraction of rows in $\mathcal{R}_1(I)$ that cannot contribute to $\mathcal{S}^*$.

Definition 3. *Given any $\hat{I} \in \mathcal{R}_1(I)$, $\alpha(\hat{I}) \in [0,1]$ denotes the fraction of rows $\mathcal{R} \in \mathcal{R}_1(I)$ where $I' \in \mathcal{R} \cap \mathcal{N}(I)$ satisfies* 1) *$I' \notin \mathcal{A}(I)$ and* 2) *$\texttt{FF}(I') \leq \texttt{FF}(\hat{I})$.*

Note that we make Definition 3 more general for later usage, as $I' \in \mathcal{T}(I)$ instead of $I' \notin \mathcal{A}(I)$ for this setting would have sufficed. For the integral-endpoints case, it is sufficient to set $\hat{I}$ as $D^{\mathtt{A}}(I)$, and $\alpha(D^{\mathtt{A}}(I)) \cdot r_1(I)$ is the number of intervals in $\mathcal{T}(I)$ that are not in $\mathcal{S}^*$.

Lemma 5. *By selecting $D^{\mathtt{A}}(I)$ as a pivot I^*, $|\mathcal{S}^*| = (1 - \alpha(D^{\mathtt{A}}(I))) \cdot r_1(I)$.*

Proof. First, by the definition of $\mathcal{R}_0(I)$, I does not intersect any interval in $\mathcal{R}_0(I)$. Thus, no interval in $\mathcal{R}_0(I)$ can be in $\mathcal{S}^*$. Next, as all intervals have integral

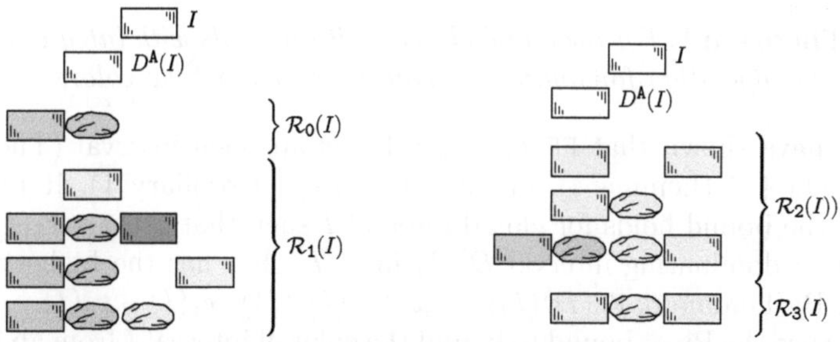

Fig. 2. All possible intersections with intervals I and $D^{\mathsf{A}}(I)$ per type of row. The intervals in $\mathcal{A}(I)$ are drawn in yellow, the intervals in $\mathcal{T}(I)$ in orange, and the remaining intervals intersecting $D^{\mathsf{A}}(I)$ are drawn in pink. (Color figure online)

endpoints, all intervals in $\mathcal{R}_2(I)$, $\mathcal{R}_3(I)$, or $\mathcal{R}_1(I)\backslash\mathcal{T}(I)$ are in the set $\mathcal{A}(I)$. Thus, all these intervals are assigned a color at most $\mathtt{FF}(D^{\mathsf{A}}(I))$ and cannot contribute to $\mathcal{S}^*$. The remaining intervals are those in $\mathcal{T}(I)$. By definition, $\alpha(D^{\mathsf{A}}(I))\cdot r_1(I)$ of these intervals have a color at most $\mathtt{FF}(D^{\mathsf{A}}(I))$. Hence, the $(1-\alpha(D^{\mathsf{A}}(I)))\cdot r_1(I)$ remaining intervals are the only intervals intersecting I that have a color greater than $\mathtt{FF}(D^{\mathsf{A}}(I))$. That is, $|\mathcal{S}^*| = (1-\alpha(D^{\mathsf{A}}(I)))\cdot r_1(I)$. □

Bound the Color $\mathtt{FF}(I^*)$**.** We bound the color of $D^{\mathsf{A}}(I)$ using the Neighborhood bound (Lemma 1). In general, apart from intervals in $\mathcal{T}(I)$, the number of intervals any interval in $\mathcal{A}(I)$, which includes $D^{\mathsf{A}}(I)$, can intersect is shown in the following observation (also see Fig. 2):

Observation 2. *Any interval in the set* $\mathcal{A}(I)$ *intersects*

(a) at most 2 *intervals per row in* $\mathcal{R}_0(I)$*,*
(b) at most 2 *intervals which are not twins with* I *per row in* $\mathcal{R}_1(I)$*,*
(c) at most 2 *intervals per row in* $\mathcal{R}_2(I)$*, and*
(d) exactly 1 *interval per row in* $\mathcal{R}_3(I)$*.*

Lemma 6. *For integral endpoints* $\mathcal{I}$*,* $\mathtt{FF}(D^{\mathsf{A}}(I)) \le 2\omega + \alpha(D^{\mathsf{A}}(I))\cdot r_1(I) - r_3(I)$*.*

Proof. By the Neighborhood bound, $\mathtt{FF}(D^{\mathsf{A}}(I)) \le 1+|\mathcal{N}_{D^{\mathsf{A}}(I)}(D^{\mathsf{A}}(I))|$. By Observation 2, there are at most $2r_0(I)+2r_1(I)+2r_2(I)+r_3(I)$ intervals in $\mathcal{A}(I)$ that are in $\mathcal{N}(D^{\mathsf{A}}(I))$. Together with the $\alpha(D^{\mathsf{A}}(I))\cdot r_1(I)$ intervals that "occupied" the smaller colors from the choice of $D^{\mathsf{A}}(I)$ at the moment when $D^{\mathsf{A}}(I)$ arrives, $|\mathcal{N}_{D^{\mathsf{A}}(I)}(D^{\mathsf{A}}(I))| \le 2r_0(I) + 2r_1(I) + \alpha(D^{\mathsf{A}}(I))\cdot r_1(I) + 2r_2(I) + r_3(I)$. Observe that we count the row that contains interval $D^{\mathsf{A}}(I)$ itself. Since $D^{\mathsf{A}}(I)$ cannot intersect any interval on this row, we should subtract at least 1 from this bound. Thus, $\mathtt{FF}(D^{\mathsf{A}}(I)) \le 1 + |\mathcal{N}_{D^{\mathsf{A}}(I)}(D^{\mathsf{A}}(I))| - 1 \le 2\omega + \alpha(D^{\mathsf{A}}(I))\cdot r_1(I) - r_3(I)$. □

Note that due to the abstraction of $\alpha(D^{\mathsf{A}}(I))$, this proof does not rely on the property of integral endpoints. Then, since Observation 2 holds also when the intervals in $\mathcal{I}$ have arbitrary endpoints, Lemma 6 implies the following:

Corollary 2. *For any* $\mathcal{I}$*,* $\mathtt{FF}(D^{\mathsf{A}}(I)) \le 2\omega + \alpha(D^{\mathsf{A}}(I))\cdot r_1(I) - r_3(I)$*.*

Proof of Theorem 1. *For open and closed unit intervals with integral endpoints, the* FIRSTFIT *algorithm for online coloring uses at most* 2ω *colors.*

Proof. We have shown that $\texttt{FF}(I) \leq 2\omega$ if I is an open interval (Theorem 3), $r_3(I) < r_1(I) + 2$ (Lemma 4), or $\mathcal{A}(I)$ is empty (Corollary 1). It remains to prove that the bound holds for closed interval I such that $r_1(I) \leq r_3(I) - 2$ and there exists a dominating interval $D^{\texttt{A}}(I)$ in $\mathcal{A}(I)$ that has the highest color by FIRSTFIT. By Lemma 6, $\texttt{FF}(D^{\texttt{A}}(I)) \leq 2\omega + \alpha(D^{\texttt{A}}(I)) \cdot r_1(I) - r_3(I)$.

We now use the Pivot bound to bound the color of interval I from above, using $D^{\texttt{A}}(I)$ as the role of I^*. By Lemma 5, $|\mathcal{S}^*| = (1 - \alpha(D^{\texttt{A}}(I))) \cdot r_1(I)$. Altogether, we have $\texttt{FF}(I) \leq \texttt{FF}(D^{\texttt{A}}(I)) + |\mathcal{S}^*| + 1 \leq 2\omega - 1$. □

4 General Case: Intervals with Arbitrary Endpoints

In this section, we consider the instance $\mathcal{I}$, where intervals have arbitrary endpoints. More specifically, by Lemma 4 and Theorem 3, we focus on $I \in \mathcal{I}$ that are closed and $r_3(I) \geq r_1(I) + 2$. Recall from Sect. 3 that we partition $\mathcal{N}(I)$ into $\mathcal{T}(I)$, $\mathcal{A}(I)$, and $\mathcal{M}(I)$.

Primarily, we pick a pivot I^* and bound its color using the Neighborhood bound. The choice of pivot should provide a plausible way to bound the size of the corresponding $\mathcal{S}^*$. However, applying the Pivot bound to bound $\texttt{FF}(I)$ in this general case is more challenging than the integral-endpoints case due to the misaligned structure of intervals. Naively applying the same technique as in the integral-endpoints case results in a ratio of 3 (see full version).

In this general case, we have another possible candidate of I^*. Symmetric to $D^{\texttt{A}}(I)$, we define interval $D^{\texttt{M}}(I)$ as the dominating interval in rows in $\mathcal{R}_2(I)$ that is misaligned with I and assigned the highest color by FIRSTFIT. Formally,

Definition 4. (Dominating interval $D^{\texttt{M}}(I)$ in $\mathcal{R}_2(I) \cap \mathcal{M}(I)$) *Let interval $D^{\texttt{M}}(I)$ be the interval in both $\mathcal{R}_2(I)$ and $\mathcal{M}(I)$ with the highest color by* FIRSTFIT.

Note that while an interval in $\mathcal{R}_1(I)$ could, in principle, dominate $\mathcal{M}(I)$, we do not consider such intervals as pivots. The reason is intuitively that the intervals that belong to both $\mathcal{R}_1(I)$ and $\mathcal{M}(I)$ are less densely packed compared to those in $\mathcal{R}_2(I), \mathcal{R}_3(I)$, or those that are in both $\mathcal{T}(I)$ and $\mathcal{R}_1(I)$. As a result, including such intervals are less advantageous for an adversary constructing a worst-case instance, since they effectively waste potential. Therefore, we focus on pivots that are both misaligned intervals and lie in rows from $\mathcal{R}_2(I)$.

A special property of the selection of $D^{\texttt{M}}(I)$ is that, since $D^{\texttt{M}}(I) \in \mathcal{M}(I)$, any row in $\mathcal{R}_0(I), \mathcal{R}_1(I)$ or $\mathcal{R}_3(I)$ cannot be in $\mathcal{R}_3(D^{\texttt{M}}(I))$. Hence, all rows in $\mathcal{R}_3(D^{\texttt{M}}(I))$ must also be in $\mathcal{R}_2(I)$ (see Fig. 3). Formally,

Lemma 7. *$\mathcal{R}_3(D^{\texttt{M}}(I)) \subseteq \mathcal{R}_2(I)$.*

In the general case, we pick one of $D^{\texttt{A}}(I)$ and $D^{\texttt{M}}(I)$ as the pivot and use the Pivot bound to bound the color $\texttt{FF}(I)$ accordingly. More specifically, we consider the following two cases.

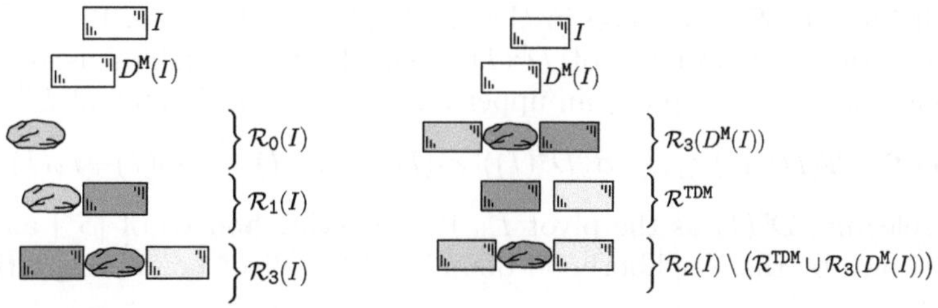

Fig. 3. An exemplary non-exhaustive list of possible rows per type of row. Yellow intervals intersect only interval I, pink intervals intersect only $D^{\mathtt{M}}(I)$, and orange intervals intersect both $D^{\mathtt{M}}(I)$ and I. (Color figure online)

Case (1): $\mathtt{FF}(D^{\mathtt{M}}(I)) \leq \mathtt{FF}(D^{\mathtt{A}}(I))$**.** In this case, the misaligned intervals in $\mathcal{R}_2(I)$ are not in $\mathcal{S}^*$, since they all have colors at most $\mathtt{FF}(D^{\mathtt{M}}(I)) \leq \mathtt{FF}(D^{\mathtt{A}}(I))$. Using $D^{\mathtt{A}}(I)$ as the pivot and closely following the case of integral-endpoints, we show:

Lemma 8. *For any interval $I \in \mathcal{I}$, if $FF(D^{A}(I)) \geq FF(D^{M}(I))$, then $FF(I) \leq 2\omega$.*

Case (2): $\mathtt{FF}(D^{\mathtt{M}}(I)) > \mathtt{FF}(D^{\mathtt{A}}(I))$. In this case, applying the Pivot bound with $I^* = D^{\mathtt{A}}(I)$ leads to a bound of $\mathtt{FF}(I) \leq 3\omega$ as shown in the naive attempt. To be precise, it is no longer true that $\mathcal{S}^* \subseteq \mathcal{R}_1(I)$, and any interval in $\mathcal{R}_2(I)$ with a color greater than $\mathtt{FF}(D^{\mathtt{A}}(I))$ is also a candidate of $\mathcal{S}^*$.

An alternative strategy is to pick the interval $D^{\mathtt{M}}(I)$ as the pivot in this case. However, in the general case, the intervals in $\mathcal{R}_1(D^{\mathtt{M}}(I))$ are not particularly structured, More precisely, we do not know which intervals also intersect with I, and the neighborhood of $D^{\mathtt{M}}(I)$ can be huge. When $D^{\mathtt{M}}(I)$ has a large number of intersections, using the Neighborhood bound to bound $\mathtt{FF}(D^{\mathtt{M}}(I))$ can lead to a high bound. This leads to a high bound of $\mathtt{FF}(I)$ by the Pivot bound. To deal with this situation, we first identify the rows containing intervals in $\mathcal{R}_1(D^{\mathtt{M}}(I))$ for which the exact positions are known, namely, the twin intervals of $D^{\mathtt{M}}(I)$.

Definition 5. *Let the set of rows in $\mathcal{R}_2(I)$ with intervals identical to $D^{M}(I)$ be: $\mathcal{R}^{TDM}(I) = \{\mathcal{R} \in \mathcal{R}_2(I) \mid \mathcal{R}$ contains at least one interval that is in $\mathcal{T}(D^{M}(I))\}$.*

In combination with the set $\mathcal{R}^{\mathtt{TDM}}(I)$ we are interested in the set $\mathcal{R}_3(D^{\mathtt{M}}(I))$. Similar to the definition of α, we define fractions β and γ of specific $\mathcal{R}_2(I)$ rows:

Definition 6. *Given $D^{M}(I) \in \mathcal{R}_2(I)$, let $\beta, \gamma \in [0,1]$ denote the fraction of rows $\mathcal{R} \in \mathcal{R}_2(I)$ where the intervals $I' \in \mathcal{R} \cap \mathcal{N}(I)$ are in $\mathcal{R}^{TDM}(I)$ and $\mathcal{R}_3(D^{M}(I))$, respectively.*

Note that $\mathcal{R}^{\mathtt{TDM}}(I) \subseteq \mathcal{R}_1(D^{\mathtt{M}}(I))$, and thus $r_3(D^{\mathtt{M}}(I)) \leq |\mathcal{R}^{\mathtt{TDM}}(I)|$ implies that $r_3(D^{\mathtt{M}}(I)) \leq r_1(D^{\mathtt{M}}(I))$. On the contrary, $D^{\mathtt{M}}(I)$ potentially has a large neighborhood if $r_3(D^{\mathtt{M}}(I)) > |\mathcal{R}^{\mathtt{TDM}}(I)|$. According to the size of $|\mathcal{R}^{\mathtt{TDM}}(I)|$, we consider the following cases **2.a** and **2.b**.

Case (2.a): $r_3(D^{\mathtt{M}}(I)) \leq |\mathcal{R}^{\mathtt{TDM}}(I)|$**.** In this case, we pick $D^{\mathtt{M}}(I)$ as the pivot I^*. By the selection of $D^{\mathtt{M}}(I)$ and $\mathtt{FF}(D^{\mathtt{M}}(I)) > \mathtt{FF}(D^{\mathtt{A}}(I))$, the only intervals that can contribute to $\mathcal{S}^*$ are those in the set $\mathcal{R}_1(I)$. Furthermore, by $|\mathcal{R}^{\mathtt{TDM}}(I)| \geq r_3(D^{\mathtt{M}}(I))$, the neighborhood of $D^{\mathtt{M}}(I)$ is small, which enables us to use the Neighborhood bound to prove an upper bound the color of interval $D^{\mathtt{M}}(I)$:

Lemma 9. $FF(D^M(I)) \leq \omega + \alpha(D^M(I)) \cdot r_1(I) - \beta \cdot r_2(I) + \gamma \cdot r_2(I) + r_2(I) + r_3(I)$.

By selecting $D^{\mathtt{M}}(I)$ as the pivot I^*, the following bounds of $|\mathcal{S}^*|$ and $\mathtt{FF}(I)$ can be shown by the Neighborhood bound and the Pivot bound respectively.

Lemma 10. *When* $FF(D^M(I)) > FF(D^A(I))$, *by selecting* $D^M(I)$ *as pivot* I^*, $|\mathcal{S}^*| = (1 - \alpha(D^M(I))) \cdot r_1(I)$.

Lemma 11. *If* $|\mathcal{R}^{TDM}(I)| \geq r_3(D^M(I))$, *then* $FF(I) \leq 2\omega$.

Case (2.b): $r_3(D^{\mathtt{M}}(I)) > |\mathcal{R}^{\mathtt{TDM}}(I)|$**.** This is the most technical part of our analysis. In this case, $D^{\mathtt{M}}(I)$ may intersect more than 2 intervals per row on average and have a large neighborhood, and using the Neighborhood bound on $\mathtt{FF}(D^{\mathtt{M}}(I))$ can lead to a large bound. We apply our framework recursively on $D^{\mathtt{M}}(I)$ by first showing that the intervals in $\mathcal{R}_3(D^{\mathtt{M}}(I))$ rows do not intersect many intervals when $r_3(D^{\mathtt{M}}(I))$ is large. Specifically, we let $D^{\mathtt{AM}}(I)$ be the highest colored interval in $\mathcal{R}_3(D^{\mathtt{M}}(I))$ that intersects with interval I.

Definition 7. (Dominating interval $D^{AM}(I)$ in $\mathcal{R}_3(D^M(I))$) *Let interval* $D^{AM}(I)$ *be the interval in* $\mathcal{R}_3(D^M(I))$ *that has the highest color by* FirstFit.

It is critical to use $D^{\mathtt{A}}(I)$, $D^{\mathtt{M}}(I)$, or $D^{\mathtt{AM}}(I)$ as the pivot according to whether $\mathtt{FF}(D^{\mathtt{AM}}(I))$ is larger than $\mathtt{FF}(D^{\mathtt{A}}(I))$.

Case (2.b.i): $\mathtt{FF}(D^{\mathtt{AM}}(I)) \geq \mathtt{FF}(D^{\mathtt{A}}(I))$**.** In this case, we apply the Pivot bound by selecting $D^{\mathtt{AM}}(I)$ as the pivot I^*. Recall that $D^{\mathtt{A}}(I)$ is the dominating interval in $\mathcal{R}_3(I)$. Observe that by the definition of interval $D^{\mathtt{M}}(I)$, $\mathtt{FF}(D^{\mathtt{AM}}(I)) < \mathtt{FF}(D^{\mathtt{M}}(I))$. Hence, analogous to parameter α, we define δ.

Definition 8. *Given any interval* $\hat{I}$, $\delta(\hat{I}) \in [0,1]$ *denotes the fraction of rows* $\mathcal{R} \in \mathcal{R}_2(I) \setminus \mathcal{R}_3(D^M(I))$ *where interval* I' *in* $\mathcal{R} \cap \mathcal{N}(I) \cap \mathcal{N}(D^M(I))$ *satisfies* $FF(I') \leq FF(\hat{I})$.

The condition $\mathtt{FF}(D^{\mathtt{AM}}(I)) \geq \mathtt{FF}(D^{\mathtt{A}}(I))$ and the selection of $D^{\mathtt{AM}}(I)$ ensure that no interval in $\mathcal{R}_3(I)$ can be in $\mathcal{S}^*$. Furthermore, by the definition of $D^{\mathtt{AM}}(I)$, no interval in $\mathcal{R}_3(D^{\mathtt{M}}(I))$ can contribute to $\mathcal{S}^*$. Therefore, $\mathcal{S}^*$ is a subset of intervals in $\mathcal{R}_1(I)$ and $\mathcal{R}_2(I) \setminus \mathcal{R}_3(D^{\mathtt{M}}(I))$.

Lemma 12. *By selecting* $D^{AM}(I)$ *as pivot* I^*, $|\mathcal{S}^*| = (1 - \alpha(D^{AM}(I))) \cdot r_1(I) + (1 - \delta(D^{AM}(I)) \cdot (1 - \gamma) \cdot 2r_2(I)$.

The definition of $\mathcal{R}_3(D^{\mathtt{M}}(I))$ and the selection of $D^{\mathtt{AM}}(I)$ guarantee that a row in the set $\mathcal{R}_3(D^{\mathtt{AM}}(I))$ must contain an interval identical to the interval $D^{\mathtt{M}}(I)$. Thus, $\mathcal{R}_3(D^{\mathtt{AM}}(I)) \subseteq \mathcal{R}^{\mathtt{TDM}}(I)$, and $r_3(D^{\mathtt{AM}}(I)) \leq |\mathcal{R}^{\mathtt{TDM}}(I)|$. Moreover, since $\mathcal{T}(D^{\mathtt{AM}}(I)) \subseteq \mathcal{R}_1(D^{\mathtt{AM}}(I))$, and every $\mathcal{R}_3(D^{\mathtt{M}}(I))$ contains a twin of $D^{\mathtt{AM}}(I)$, $r_1(D^{\mathtt{AM}}(I)) \geq r_3(D^{\mathtt{M}}(I))$. Together with the condition $r_3(D^{\mathtt{M}}(I)) > |\mathcal{R}^{\mathtt{TDM}}(I)|$, it guarantees that $r_1(D^{\mathtt{AM}}(I)) \geq r_3(D^{\mathtt{M}}(I)) > |\mathcal{R}^{\mathtt{TDM}}(I)| \geq r_3(D^{\mathtt{AM}}(I))$. Therefore, $D^{\mathtt{AM}}(I)$ intersects at most 2 intervals per row on average, and $\mathtt{FF}(D^{\mathtt{AM}}(I))$ can be effectively bounded by the Neighborhood bound.

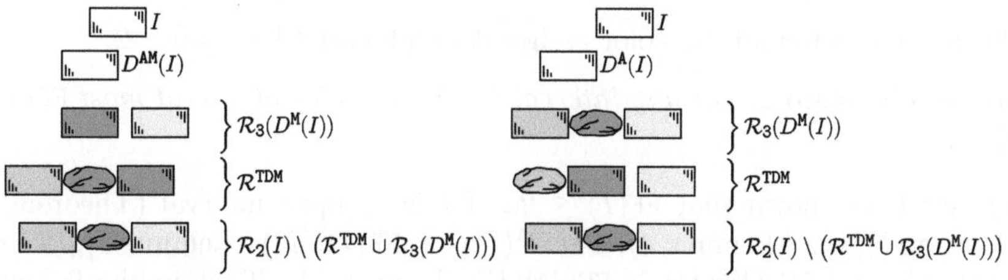

Fig. 4. An exemplary non-exhaustive list of possible rows per type of $\mathcal{R}_2(I)$. Yellow intervals intersect only interval I, pink intervals intersect only $D^{\mathtt{AM}}(I)$ (resp. $D^{\mathtt{A}}(I)$) and orange intervals intersect both $D^{\mathtt{AM}}(I)$ (resp. $D^{\mathtt{A}}(I)$) and I. (Color figure online)

Lemma 13. $FF(D^{AM}(I)) \leq r_0(I) + (1+\alpha(D^{AM}(I))) \cdot r_1(I) + \delta(D^{AM}(I)) \cdot (1-\gamma) \cdot 2r_2(I) + \beta \cdot r_2(I) + \gamma \cdot r_2(I) + 2r_3(I)$.

Lemma 14. *If* $FF(D^{AM}(I)) \geq FF(D^{A}(I))$, *then* $FF(I) \leq 2\omega$

Case (2.b.ii): $\mathtt{FF}(D^{\mathtt{AM}}(I)) < \mathtt{FF}(D^{\mathtt{A}}(I))$**.** This case is the tough kernel of the analysis. In this case, considering using $D^{\mathtt{A}}(I)$, $D^{\mathtt{M}}(I)$, or $D^{\mathtt{AM}}(I)$ as the pivot only is not sufficient to have an upper bound of $\mathtt{FF}(I)$ that is smaller than 3ω. Recall that this case is a subcase of case $\mathtt{FF}(D^{\mathtt{A}}(I)) < \mathtt{FF}(D^{\mathtt{M}}(I))$. Choosing $D^{\mathtt{A}}(I)$ as the pivot does not stop any interval in $\mathcal{R}_2(I)$ from contributing to $\mathcal{S}^*$. On the other hand, since $r_3(D^{\mathtt{M}}(I)) > |\mathcal{R}^{\mathtt{TDM}}(I)|$, $r_3(D^{\mathtt{M}}(I))$ may be much larger than $r_1(D^{\mathtt{M}}(I))$. Applying the Neighborhood bound yields a huge upper bound of $\mathtt{FF}(I^*)$. Finally, condition $\mathtt{FF}(D^{\mathtt{AM}}(I)) < \mathtt{FF}(D^{\mathtt{A}}(I))$ implies that when $D^{\mathtt{AM}}(I)$ is selected as pivot, $\mathcal{S}^*$ could contain any interval in $\mathcal{R}_3(I)$ and become too large.

To deal with this tough case, we express the upper bounds obtained by using $D^{\mathtt{A}}(I)$ or $D^{\mathtt{M}}(I)$ as the pivot in terms of $|\mathcal{R}^{\mathtt{TDM}}(I)|$ and $r_2(D^{\mathtt{M}}(I))$. When interval $D^{\mathtt{M}}(I)$ is picked as the pivot, the color $\mathtt{FF}(D^{\mathtt{M}}(I))$ and the size of $|\mathcal{S}^*|$ proven in Lemmas 9 and 10 remain true. For the case where interval $D^{\mathtt{A}}(I)$ is picked, we show the following bounds (see Fig. 4).

Lemma 15. $FF(D^{A}(I)) \leq 2r_0(I) + 2r_1(I) + \alpha(D^{A}(I)) \cdot r_1(I) + \gamma \cdot r_2(I) + r_2(I) + \delta(D^{A}(I)) \cdot (1-\gamma) \cdot r_2(I) + r_3(I)$.

Lemma 16. *When* $FF(D^{A}(I)) \geq FF(D^{AM}(I))$, *by selecting* $D^{A}(I)$ *as pivot* I^*, $|\mathcal{S}^*| \leq (1-\alpha(D^{A}(I))) \cdot r_1(I) + (1-\delta(D^{M}(I))) \cdot (1-\gamma) \cdot r_2(I) + (1-\gamma) \cdot r_2(I)$.

These formulations show that the upper bounds from applying the Pivot bound using interval $D^{\mathtt{A}}(I)$ as I^* and using interval $D^{\mathtt{M}}(I)$ as I^* form a trade-off. More specifically, the upper bound using $D^{\mathtt{A}}(I)$ is maximized when $|\mathcal{R}^{\mathtt{TDM}}(I)| = \emptyset$, while upper bound using $D^{\mathtt{M}}(I)$ is maximized when $|\mathcal{R}^{\mathtt{TDM}}(I)| = \mathcal{R}_2(I)$. Then, we use that in any circumstance, the upper bound is at most $2\omega + \frac{|\mathcal{R}^{\mathtt{TDM}}(I)|}{2}$, which is strictly less than $2\omega + \frac{r_2(I)}{2} \cdot \omega$ (Definition 5). We next prove a bound on the number of $\mathcal{R}_2(I)$ rows.

Lemma 17. *For any interval* $I \in \mathcal{I}$, *if* $r_2(I) \geq \frac{2}{3}\omega$, *then* $FF(I) \leq \frac{7}{3}\omega - 2$.

Then, we can bound the color assigned to interval I by $\lceil \frac{7}{3}\omega \rceil - 2$.

Proof of Theorem 2. *For any interval $I \in \mathcal{I}$, the color of I is at most* $\mathtt{FF}(I) \leq \lceil \frac{7}{3}\omega \rceil - 2$.

Proof. We have shown that $\mathtt{FF}(I) \leq 2\omega$ if I is an open interval (Theorem 3), $r_1(I) > r_3(I) - 2$ (Lemma 4), $\mathtt{FF}(D^{\mathtt{A}}(I)) \geq \mathtt{FF}(D^{\mathtt{M}}(I))$ (Lemma 8), $\beta \geq \gamma$ (Lemma 11) and $\mathtt{FF}(D^{\mathtt{AM}}(I)) \geq \mathtt{FF}(D^{\mathtt{A}}(I))$ (Lemma 14). Thus in the following we focus on closed intervals with $r_1(I) \leq r_3(I) - 2$, $\mathtt{FF}(D^{\mathtt{AM}}(I)) < \mathtt{FF}(D^{\mathtt{A}}(I)) < \mathtt{FF}(D^{\mathtt{M}}(I))$, and $\beta < \gamma$.

We can compute the first bound on the color of interval I, using the Pivot bound with interval $D^{\mathtt{A}}(I)$ as pivot I^*. By Lemma 15, $\mathtt{FF}(D^{\mathtt{A}}(I)) \leq 2r_0(I) + 2r_1(I) + \alpha(D^{\mathtt{A}}(I)) \cdot r_1(I) + \gamma \cdot r_2(I) + r_2(I) + \delta(D^{\mathtt{A}}(I)) \cdot (1-\gamma) \cdot r_2(I) + r_3(I)$. By Lemma 16 and taking $D^{\mathtt{A}}(I)$ as pivot I^*, $|\mathcal{S}^*| \leq (1 - \alpha(D^{\mathtt{A}}(I))) \cdot r_1(I) + (1 - \delta(D^{\mathtt{M}}(I))) \cdot (1-\gamma) \cdot r_2(I) + (1-\gamma) \cdot r_2(I)$. Thus,

$$\mathtt{FF}(I) \leq \mathtt{FF}(D^{\mathtt{A}}(I)) + |\mathcal{S}^*| + 1 \leq 2\omega + (1-\gamma) \cdot r_2(I) - 1$$

Next, we move to the other bound. By taking $D^{\mathtt{M}}(I)$ as pivot I^*, according to Lemma 9, $\mathtt{FF}(D^{\mathtt{M}}(I)) \leq \omega + \alpha(D^{\mathtt{M}}(I)) \cdot r_1(I) - \beta \cdot r_2(I) + \gamma \cdot r_2(I) + r_2(I) + r_3(I)$, and Lemma 10, $|\mathcal{S}^*| = (1 - \alpha(D^{\mathtt{M}}(I))) \cdot r_1(I)$. Then, it follows that,

$$\mathtt{FF}(I) \leq \mathtt{FF}(D^{\mathtt{M}}(I)) + |\mathcal{S}^*| + 1 \leq 2\omega + \gamma \cdot r_2(I) - 1$$

We have proven that simultaneously, the color of interval I is at most $\mathtt{FF}(I) \leq 2\omega + \gamma \cdot r_2(I) - 1$ and at most $\mathtt{FF}(I) \leq 2\omega + (1-\gamma) \cdot r_2(I) - 1$. Hence, for any value of γ the color of interval I is at most $\mathtt{FF}(I) \leq 2\omega + \frac{1}{2} \cdot r_2(I) - 1$ Then, it follows from Lemma 17 that the number of $\mathcal{R}_2(I)$ rows must be less than $r_2(I) < \frac{2}{3}\omega$, as otherwise the color of interval I is bounded by $\mathtt{FF}(I) \leq \frac{7}{3}\omega - 2$. Thus, the color of interval I is at most $\mathtt{FF}(I) < 2\omega + \frac{1}{3} \cdot \omega - 1 = \frac{7}{3}\omega - 1$. Which, by the integrality of $\mathtt{FF}(I)$, is at most $\mathtt{FF}(I) \leq \lceil \frac{7}{3}\omega \rceil - 2$ □

5 Conclusion

In this work, we develop a sophisticated counting method based on the Pivot bound and show that FirstFit uses at most 2ω colors when all open and closed unit-length intervals have integral endpoints, which matches the lower bound.

In the more general setting where the input contains open or closed unit-length intervals with any endpoints, we identify several configurations in which FirstFit still uses at most 2ω colors, again matching the lower bound. However, one configuration remains more difficult to analyze: when the dominating misaligned interval in $\mathcal{R}_2(I)$, $D^{\mathtt{M}}(I)$, receives a higher color than the dominating aligned interval $D^{\mathtt{A}}(I)$, which in turn receives a higher color than $D^{\mathtt{AM}}(I)$, that is, the dominating interval aligned with $D^{\mathtt{M}}(I)$. In this specific configuration, we were only able to reduce the upper bound on the number of colors used from $3\omega - 3$ to $\lceil \frac{7}{3}\omega \rceil - 2$. Finding a tight bound for this configuration may require new or refined analytical methods.

To improve the current upper bound for the number of colors used by the FIRSTFIT algorithm in the general setting–where arbitrary inputs are allowed–it may be of interest to first investigate the performance of the FIRSTFIT algorithm on bounded length intervals, as introduced by Chybowska-Sokól et al. [2]. The techniques and results on open/closed unit-length intervals developed in this work may serve as a first step towards investigating FIRSTFIT's performance via gradually relaxing the lengths bound.

Disclosure of Interests. The authors have no competing interests to declare that are relevant to the content of this article.

References

1. Chrobak, M., Ślusarek, M.: On some packing problem related to dynamic storage allocation. RAIRO Theor. Informatics Appl. **22**(4), 487–499 (1988). https://doi.org/10.1051/ITA/1988220404871
2. Chybowska-Sokół, J., Gutowski, G., Junosza-Szaniawski, K., Mikos, P., Polak, A.: Online coloring of short intervals. Eur. J. Comb. **118**, 103919 (2024). https://doi.org/10.1016/J.EJC.2024.103919
3. Curbelo, I.R.: The on-line width of various classes of posets. Ph. D. thesis, University of Louisville (2022). https://doi.org/10.18297/etd/3974
4. Curbelo, I.R.: Online coloring of short interval graphs and two-count interval graphs. arXiv:2412.17193 (2024)
5. Epstein, L., Levy, M.: Online Interval Coloring and Variants. In: Caires, L., Italiano, G.F., Monteiro, L., Palamidessi, C., Yung, M. (eds.) ICALP 2005. LNCS, vol. 3580, pp. 602–613. Springer, Heidelberg (2005). https://doi.org/10.1007/11523468_49
6. Golumbic, M.C.: Algorithmic Graph Theory and Perfect Graphs. Elsevier (2004)
7. Kierstead, H.A.: The linearity of first-fit coloring of interval graphs. SIAM J. Discret. Math. **1**(4), 526–530 (1988). https://doi.org/10.1137/0401048
8. Kierstead, H.A., Qin, J.: Coloring interval graphs with first-fit. Discret. Math. **144**(1–3), 47–57 (1995). https://doi.org/10.1016/0012-365X(94)00285-Q
9. Kierstead, H.A., Smith, D.A., Trotter, W.T.: First-fit coloring on interval graphs has performance ratio at least 5. Eur. J. Comb. **51**, 236–254 (2016). https://doi.org/10.1016/J.EJC.2015.05.015
10. Kierstead, H.A., Trotter, W.T.: An extremal problem in recursive combinatorics. Congr. Numer. **33**, 143–153 (1981)
11. Narayanaswamy, N.S., Babu, R.S.: A note on first-fit coloring of interval graphs. Order **25**(1), 49–53 (2008). https://doi.org/10.1007/S11083-008-9076-6
12. Pemmaraju, S.V., Raman, R., Varadarajan, K.R.: Max-coloring and online coloring with bandwidths on interval graphs. ACM Trans. Algorithms **7**(3), 35:1–35:21 (2011). https://doi.org/10.1145/1978782.1978790
13. Ślusarek, M.: A coloring algorithm for interval graphs. In: Kreczmar, A., Mirkowska, G. (eds.) MFCS 1989. LNCS, vol. 379, pp. 471–480. Springer, Heidelberg (1989). https://doi.org/10.1007/3-540-51486-4_93
14. Witsenhausen, H.S.: On woodall's interval problem. J. Comb. Theory A **21**(2), 222–229 (1976). https://doi.org/10.1016/0097-3165(76)90066-2

Buy-at-Bulk Facility Location on Trees

Shamisa Nematollahi[1] and Daniel Vaz[2](✉)

[1] Université Paris Cité, CNRS, IRIF, Paris, France
shamisa@irif.fr
[2] LIGM, Université Gustave Eiffel, CNRS, ESIEE Paris, Marne-la-Vallée, France
daniel.ramosvaz@esiee.fr

Abstract. We consider the buy-at-bulk facility location problem (BBFL), a problem combining the classic facility location problem with buy-at-bulk network design, which finds motivation in telecommunication networks. In it, we are given a graph with edge lengths, opening costs and demands for each vertex, and a monotone and subadditive capacity-cost function, and our task is to open facilities on a subset of the vertices and route the demand from each vertex to these facilities. The cost of a solution (which we want to minimize) is given by the opening costs of the chosen facilities, plus the cost on each edge, which is given by its length times the cost of providing enough capacity for the demands through the edge, given by the capacity-cost function. A common variant of the problem, the k-cable facility location problem (kCFL), considers the case where capacity is provided by buying copies of given cable types, each with a certain capacity and cost.

We study BBFL on tree instances and show, for the unit-demand and splittable variants, that the problem admits a PTAS (a $(1+\epsilon)$-approximation for any $\epsilon > 0$). We also consider kCFL in the new setting of cable-unsplittable demands, where the demand of a vertex cannot be split among multiple cables. We show that the problem is NP-hard to approximate to a factor better than 3/2 on stars, and then provide an algorithm for tree instances that outputs a solution with optimal cost, but which exceeds the capacity on each cable by a factor of $1+\epsilon$. As a consequence, we show that the problem has a 2-approximation algorithm on trees.

Keywords: Buy-at-bulk · Facility location · PTAS · Approximation algorithms

1 Introduction

The facility location problem is a classic in operations research, which concerns the placement of facilities to distribute goods or services so that the costs of delivery and installation of the facilities remain low. However, in cases such as telecommunication networks, the structure of the distribution network matters as much as the actual distances, as using common distribution trunks can vastly

J. Matuschke and J. Verschae (Eds.): WAOA 2025, LNCS 16077, pp. 172–186, 2026.
https://doi.org/10.1007/978-3-032-06706-7_12

reduce the cost of distribution. This has led to connection costs being considered from a network design perspective, giving rise to a problem formulation that more closely considers its applications.

We consider one such formulation, the *buy-at-bulk facility location (BBFL)* problem, which combines the facility location and buy-at-bulk network design problems. In it, we have *demands* on the vertices of a graph, and we want to serve these demands using *facilities*. For this purpose, we must choose some vertices of the graph on which to open facilities and further install a network with enough capacity to support connecting all of the demands to an open facility. The buy-at-bulk name comes from the fact that the network has subadditive costs: the installation cost is a unit-length cost of supporting a certain amount of demand, which is a monotone, sub-additive function of the total demand on the edge. For each edge, the cost is then given by the installation cost times the length of the edge.

Our goal in the buy-at-bulk facility location is to, given a graph with edge lengths, facility costs, vertex demands and an installation cost function, determine which facilities to open, and how to direct the demand of each vertex to an open facility, such that the total cost of open facilities and edges is minimized.

The most common version of the problem uses discrete cables to provide capacity, thus replacing the sub-additive installation cost function by the minimum cost of a configuration of cables that has capacity at least equal to the demand. To distinguish this setting from the general subadditive case, we call this version *k-cable facility location (kCFL)*. The difference is formalized by adding k cable types to the input, each with a capacity μ_i and a cost c_i. When computing the solution, for each edge with a total demand of λ, we must choose how many copies of each cable to buy, such that the total capacity is at least λ; the installation cost is then the sum of the costs of each cable multiplied by the number of copies purchased.

The buy-at-bulk facility location problem was initially introduced by Meyerson et al. [29]. They established that BBFL is a specific instance of the more general Cost-Distance problem, and thus an $O(\log n)$-approximation algorithm follows (for unsplittable demands). For kCFL, Ravi and Sinha [31] improved the approximation factor to $O(k)$. In the special case of a single cable type ($k = 1$), which they call the capacitated-cable facility location problem, they show an approximation factor of $\rho_{\mathsf{UFL}}+\rho_{\mathsf{ST}}$ in the uniform-demand version of the problem, and $2\rho_{\mathsf{UFL}} + \rho_{\mathsf{ST}}$ for non-uniform unsplittable demands, where $\rho_{\mathsf{UFL}} = 1.488$ [26] and $\rho_{\mathsf{ST}} = 1.39$ [10] denote the best-known approximation ratios for the uncapacitated facility location and Steiner tree problems respectively. kCFL (for general k) and BBFL are known to be NP-hard even on a single edge, as kCFL models the knapsack problem [27].

We focus on studying the approximability of the problem from the ground up, starting with simple instances. For this reason, this work focuses on tree graphs, and on optimizing the approximation ratio under different problem variants. Trees are simple graphs, where techniques can first be studied to obtain near-optimal approximation algorithms (PTAS), and often provide techniques that generalize to other settings. In particular, many network design and operations research problems are NP-hard on trees, with some having PTAS or good approximations, while others are hard to approximate even on trees. Furthermore, some problems are solved in general graphs by embedding to trees [15]; this technique can also be applied to BBFL to obtain an alternative $O(\log n)$-approximation, given a constant approximation on trees (see e.g. [39, Section 8.6]).

We consider the BBFL problem in the splittable demand setting, in which the demand of each vertex can be split into parts that head to different facilities, and give a PTAS for the problem, which similarly implies a PTAS for kCFL.

Theorem 1. *There is a PTAS for the BBFL problem on trees with splittable demands that runs in time* $f(\epsilon) \cdot \text{poly}(n)$.

Corollary 1. *There is a PTAS for the kCFL problem on trees with splittable demands that runs in time* $f(\epsilon) \cdot \text{poly}(n, k)$.

We then move on to studying kCFL under a setting that is, as far as we know, novel: using the formulation of the problem as buying cables, we consider the *cable-unsplittable* setting, where the demands passing through an edge must be partitioned into the cables that are installed on that edge, that is, each demand must be assigned to a single cable on each edge, and cannot be split into multiple cables. This is a natural setting when we consider that the "cables" in the problem can represent trucks or other discrete containers, and a demand, which can represent a package, is indivisible and thus confined to a single container.

We show that this problem is APX-hard, as it cannot be approximated to a factor better than 3/2, even when the input graph is a star.

Theorem 2. *kCFL is APX-hard when the input graph is a star and* $k = 1$. *In particular, it is NP-hard to approximate the problem to a factor of* $3/2 - \epsilon$ *for any* $\epsilon > 0$.

We then show that, by allowing our solution to slightly overload the cables (exceeding the capacity by a factor of $1 + \epsilon$), we can compute a solution with optimum cost:

Theorem 3. *For any* $\epsilon > 0$, *there is an algorithm for the kCFL problem on trees with cable-unsplittable demands that computes solution with optimum cost but exceeds cable capacities by a factor of at most* $1 + \epsilon$; *the algorithm runs in time* $n^{k \cdot f(\epsilon)}$.

As a consequence, we show that we can obtain a 2-approximation for the problem:

Theorem 4. *There is a 2-approximation algorithm for the kCFL problem on trees with cable-unsplittable demands that runs in time* $n^{O(k)}$.

We start by presenting the formal problem definitions, as well as the connection between BBFL and kCFL in Sect. 2. We then present the results for the splittable demand setting in Sect. 3. Finally, we present the results for the cable-unsplittable setting of kCFL in Sect. 4, which for simplicity we consider in the single-cable setting; a detailed discussion of the multiple-cable setting is deferred to the full version.

We present a short technical overview at the beginning of each section. For brevity, we defer the reader to the full version of the paper for omitted proofs.

1.1 Related Work

Buy-at-bulk connected facility location is a closely related variant of the problem where open facilities are required to be connected. In this setting, we are given an extra type of cable, called core cable, with a large enough installation cost of M and capacity $\mu = +\infty$, which is used to connect the open facilities. Friggstad et al. gave an LP-based constant factor approximation algorithm [16].

Another closely related problem is the airport and railway problem, corresponding to a variant of the problem where the solution is a forest and facilities have capacity limits (i.e., each facility can serve at most k units of demand). The AR problem has been studied both with splittable demands [1,2,33] and unsplittable demands [23] in various settings. Salavatipour et al. [33] gives an $O(\log n)$-approximation to the problem, which improves to a 2-approximation with uniform facility costs.

Buy-at-bulk costs are a ubiquitous concept in our society, and were first introduced to network design by Salman et al. [34]. Since then, different variants of buy-at-bulk network design have received considerable attention [5,6,11, 12,17,19,20,38], with an $O(\log n)$-approximation for the multi-commodity setting being the most well-known result [6,15].

There is a long line of work on the metric facility location problem, both in the capacitated [3,4,7,8,13,25,30] and uncapacitated variants [9,18,21,22, 26,28,36,37]. For the capacitated variant, the best approximation has factor 5 [7], and an improved 3-approximation exists when capacities are uniform [3]. For the uncapacitated variant, a 1.488-approximation algorithm by Li [26] nearly matches the inapproximability of 1.463 [18,37]. For trees, the problem is known to be polynomial-time solvable (see e.g. [35]).

2 Preliminaries

Throughout our work, we use standard graph notation (see e.g. [14]). We consider every tree T as having a root r, and define parent and child of a vertex v with respect to r. For a vertex v, the subtree rooted in v is denoted T_v.

For an edge set E, we denote by $\tilde{E}$ the set of arcs (directed edges) containing both directions of an edge, that is, $\tilde{E} = \bigcup_{uv \in E}\{(u,v),(v,u)\}$, and denote by $\delta^-(v)$ and $\delta^+(v)$ the incoming and outgoing arcs, respectively, of vertex v in $\tilde{E}$.

For $x, y \in \mathbb{Z}$, let $[x,y] = \{z \in \mathbb{Z} : x \leq z \leq y\}$, $[x] = [1,x]$. We use $\mathbf{1}^n(S)$ to refer to the indicator vector of $S \subseteq [n]$ in $\{0,1\}^n$, that is, a vector $x \in \{0,1\}^n$ such that $x_i = 1$ if $i \in S$ and $x_i = 0$ otherwise; we write $\mathbf{1}^n(v)$ instead of $\mathbf{1}^n(\{v\})$

and omit the superscript if clear from context. We use $\mathbf{0}^n \in \{0\}^n$ to be the all zeroes vector, and write $\mathbf{0}$ if n is clear from context. Given sets $Y \subseteq X$ and a function $f\colon X \to \mathbb{R}_{\geq 0}$, we denote $f(Y) = \sum_{y \in Y} f(y)$.

In our algorithms, we sometimes say that one solution *dominates* another to mean that the first is at least as good as the second. For instance, let (μ, c) represent a cable, where μ is the capacity and c is the cost. We might say that a cable (μ, c) dominates another cable (μ', c') if $\mu \geq \mu'$ and $c \leq c'$, as it has at least as much capacity and costs at most as much. Though the definition of "domination" depends on the context and will be introduced as needed, we always say that a solution is *non-dominated* within a set A if no different solution dominates it, and we denote the *set of non-dominated solutions* in A as nondom(A).

2.1 Problem Definitions

We consider two problems, buy-at-bulk facility location (BBFL) and k-cable facility location (kCFL), which blend buy-at-bulk network design and facility location [39, Sec. 4.5, Sec. 8.6]. This section first explains these two settings, how they combine into BBFL, and how the concept of cables leads to kCFL.

Buy-at-Bulk Network Design. Network design problems include many well-known combinatorial optimization problems, such as minimum spanning tree and TSP. At its core, the goal is to find a minimum-cost subgraph that satisfies given demand constraints, where the simplest version is to provide sufficient capacity to serve point-to-point demands.

The buy-at-bulk network design problem takes as input a graph $G = (V, E)$ with edge lengths $\ell\colon E \to \mathbb{R}_{\geq 0}$, a non-decreasing subadditive capacity cost function $\mathrm{c_{cap}}\colon \mathbb{Z}_{\geq 0} \to \mathbb{R}_{\geq 0}$ and source-sink pairs (s_i, t_i, d_i) with demand d_i. The goal is to choose an s_i-t_i-path P_i for each i, such that the total cost on the edges is minimized, where the cost on an edge depends on the total demand through it, $\lambda_e = \sum_{P_i \ni e} d_i$, and is given as the product of the length and capacity cost for the demand, for a total of $\sum_{e \in E} \ell(e) \cdot \mathrm{c_{cap}}(\lambda_e)$.

The assumption that $\mathrm{c_{cap}}$ is non-decreasing and subadditive captures the *buy-at-bulk* principle: as demand grows, the cost of installing capacity increases, but the cost per unit demand decreases. Formally, subadditivity means that $\mathrm{c_{cap}}(x + y) \leq \mathrm{c_{cap}}(x) + \mathrm{c_{cap}}(y)$ for all $x, y \in \mathbb{Z}_{\geq 0}$, reflecting the fact that it is more economical to install a single larger capacity to serve combined demand than to install multiple smaller ones separately.

In our problems, we consider *set-to-set* demands of the form (X_i, Y_i, d_i), where the solution must simultaneously support a flow of value d_i from X_i to Y_i for each demand.

Definition 1. *Let G be a graph and $\mu\colon E \to \mathbb{R}_{\geq 0}$ be edge capacities.*

A flow $g\colon \tilde{E} \to \mathbb{R}_{\geq 0}$ is a function that satisfies edge capacities, that is, $g(e) \leq \mu(e)$ for every edge $e \in E$.

The excess *of g at a vertex $v \in V(G)$ is defined as $\mathrm{ex}_g(v) := \sum_{e \in \delta^-(v)} g(e) - \sum_{e \in \delta^+(v)} g(e)$.*

*We say that a flow is an S-T-*flow, *or* from S to T, *if* $\text{ex}_f(v) \geq 0$ *for* $v \in T$, $\text{ex}_f(v) \leq 0$ *for* $v \in S$, *and* $\text{ex}_f(v) = 0$ *otherwise. Its* value *is defined as* $|g| := \sum_{v \in T} \text{ex}_f(v)$.

Facility Location. In the facility location problem, we are given a set of clients V and a set of facilities F, as well as connection costs c_{ij} of assigning a client $j \in V$ to a facility $i \in F$, and opening costs $f\colon F \to \mathbb{R}_{\geq 0}$ for each facility. The goal of this problem is to pick a subset of the facilities $I \subseteq F$ and an assignment $\sigma\colon V \to I$ such that we minimize the total opening and connecting costs, $\sum_{i \in I} f(i) + \sum_{j \in V} c_{\sigma(j)j}$.

The Buy-at-Bulk Facility Location Problem. The buy-at-bulk facility location problem (BBFL) is obtained by combining the above as follows: we start with a facility location problem on a graph G, where both facilities and clients correspond to the vertices of G; each vertex also has a demand, which must be served by one or more facilities, and thus the solution is a subgraph that supports flows from each vertex to the opened facilities with value equal to the demand; for this purpose, capacities on the edges are not fixed but can be purchased as in a buy-at-bulk setting, with an installation cost function dictating the unit-length cost of providing capacity for a certain demand.

Formally, the input is a tuple $(G, \ell, f, d, \mathrm{c_{cap}})$ with the following parts:

- a graph $G = (V, E)$ with edge lengths $\ell\colon E \to \mathbb{R}_{\geq 0}$;
- opening costs $f\colon V \to \mathbb{R}_{\geq 0}$;
- demands $d\colon V \to \mathbb{Z}_{\geq 0}$;
- a monotone and sub-additive installation cost function $\mathrm{c_{cap}}\colon \mathbb{Z}_{\geq 0} \to \mathbb{R}_{\geq 0}$.

A solution is a tuple $(I, \mathbf{g})$ consisting of:

- A set $I \subseteq V$ of facilities to open;
- For each vertex $v \in V$, a flow g_v of value $d(v)$ from $\{v\}$ to I .

The goal of the problem is to find a solution with total minimum cost, given by

$$\sum_{i \in I} f(i) + \sum_{e \in E} \ell(e) \cdot \mathrm{c_{cap}}(g'(e)),$$

where $g'(uw) := (\sum_v g_v(u, w) + g_v(w, u))$ is the *demand on the edge uw*.

It is particularly important for our algorithms that a $(1 + \epsilon)$-approximation y for $\mathrm{c_{cap}}(x)$ ($\mathrm{c_{cap}}(x) \leq \mathrm{y} \leq (1 + \epsilon)\, \mathrm{c_{cap}}(x)$) be computable in polynomial-time.

Let $(I, \{g_v\}_v)$ be a solution. We say that a demand of δ *passes* an edge uw (in the direction (u, w)) if for some $v \in V$, $g_v(u, w) = \delta$, that is, the flow of v on (u, w) is δ. Similarly, we say that a demand of δ *leaves* u if for some $v \in V$, a flow of δ leaves u, $g_v(\delta^+(u)) = \delta$. We denote by $\text{cost}(uv, \lambda) = \ell(uv) \cdot \mathrm{c_{cap}}(\lambda)$ the cost of sending λ units of demand through edge uv.

The k-Cable Facility Location Problem. The k-cable facility location problem is a special case of BBFL where capacity is obtained by buying a set of cables, each of which has a cost and a capacity. In particular, we have k available cable types, each with a capacity μ_i and cost c_i, and for each edge we buy copies of each cable so that the total capacity is sufficient to cover the demand.

In the literature, it is generally assumed that the cables satisfy economies of scale, meaning that as capacity increases, cost increases, but the cost-per-capacity ratio decreases. While this assumption is reasonable, it is not required in our algorithms.

We remark that kCFL is a special case of BBFL, as the function is monotone, as covering a larger total demand cannot be cheaper, and subadditive, as covering the sum of two demands can be done by putting together their cable sets.

An instance of kCFL is a tuple $(G, \ell, f, d, \mathbf{q})$, where:

- G, ℓ, f, d are as in BBFL;
- $\mathbf{q} = \{(\mu_1, c_1), (\mu_2, c_2), \dots, (\mu_k, c_k)\}$ has k cable types, each with capacity μ_i and cost c_i.

We denote by $Q \subseteq U^*$ a multi-set containing elements of U with arbitrary repetition, and refer to a multi-set of cables $Q \subseteq \mathbf{q}^*$ as a *bag*. The capacity of a bag Q is $\mu(Q) = \sum_{i \in Q} \mu_i$ and its cost is $c(Q) = \sum_{i \in Q} c_i$.
A solution is a tuple $(I, \mathbf{g}, \mathbf{Q})$, where:

- I and $\mathbf{g}$ are as in BBFL;
- For each edge $uw \in E$, Q_{uw} is a bag that has sufficient capacity to support the flows $\{g_v\}_v$, i.e. $\mu(Q_{uw}) \geq g'(uw) := (\sum_v g_v(u, w) + g_v(w, u))$.

The objective function, after replacing c_{cap}, is $\sum_{i \in I} f(i) + \sum_{e \in E} \ell(e) \cdot c(Q_e)$. A self-contained definition of the problem is deferred to the full version.

The lemma below shows that we can compute a $(1+\epsilon)$-approximation to the minimum-cost bag supporting a given demand λ, as required by the definition of BBFL. The proof follows standard techniques for the knapsack problem [39, Sec. 3.1] and is deferred to the full version.

Lemma 1. *Let $\{(\mu_i, c_i)\}_{i \in [k]}$ be a set of k cables and λ be a target demand. There exists an $(1+\epsilon)$-approximation algorithm to find a minimum-cost bag that has total capacity at least λ, and which runs in time* $\text{poly}(k, 1/\epsilon)$.

Variants of BBFL and kCFL. Both of the problems above can be considered in different variants, depending on whether we allow the demands to be split along different paths. The following classic variants are usually considered:

- Unit demand: $d(v) = 1$ for each $v \in V$;
- Unsplittable: g_v is supported on a single path for each $v \in V$;
- Splittable: No restrictions.

Notice that in the unit-demand case, any flow is supported on a path w.l.o.g. by the integral flow theorem, since capacities are integers [24, Corollary 8.7].

Depending on the setting, we might omit some parts of the solution if there is a clear, optimal choice. For instance, in the unit demand case when the input graph is a tree, the flow g_v is always of value 1 and determined by the facility that serves the demand of v.

Cable-Unsplittable Demands. We further introduce the *cable-unsplittable* variant of kCFL, where the demand of a single vertex, besides being supported on a single path, has to be assigned to a specific cable on each edge. In other words, for each edge e, we ask for a bag Q_e such that the demands passing through e are partitioned onto the cables of Q_e, with the total demand assigned to each cable not exceeding its capacity. We consider that cables are uni-directional, that is, the demands assigned to each cable flow in the same direction.

Formally, Q_{uw} is partitioned (according to direction) into $Q_{(u,w)}$ and $Q_{(w,u)}$, and we have $h_{(u,w)}\colon V \to Q_{(u,w)}$, $h_{(w,u)}\colon V \to Q_{(w,u)}$, such that for each $q = (c, \mu) \in Q_{uw}$, $\sum_{v:h_{(u,w)}(v)=q} g_v(u,w) \leq \mu$ and $\sum_{v:h_{(w,u)}(v)=q} g_v(w,u) \leq \mu$. We do not explicitly state the partition of Q_{uw} or the assignments $h_{(u,w)}$, $h_{(w,u)}$ as part of the solution unless needed.

3 Splittable-Demand BBFL

In this section, we show a PTAS for the BBFL problem on trees with splittable demands. The following lemmas are useful when designing dynamic programs for the problem.

Lemma 2. *In the context of splittable (or unit) demand and a monotone cable cost function, there is an optimal solution without crossings; that is, for any edge uw, demand cannot pass in the directions (u,w) and (w,u) simultaneously.*

Lemma 3. *In the setting of uncapacitated facilities, there is an optimal solution without facility-crossings; that is, no demand leaves an open facility.*

Observation 5. *Lemma 2 does not hold for unsplittable demands.*

3.1 Splittable-Demand BBFL on Trees

In this section, we design an algorithm for splittable demand BBFL on trees. Let T represent the tree with a root node r. We assume that T is binary by the lemma below.

Lemma 4. *Any tree T can be transformed into a binary tree T' without changing the optimum or the cost of solutions.*

For each vertex $v \in T$, we define total(v) as a set of pairs (λ, c), where c is the minimum cost to serve the subtree T_v with a demand constraint of λ as follows:

- If $\lambda > 0$, a demand of λ leaves node v to its parent.
- If $\lambda \leq 0$, a demand of $|\lambda|$ arrives at node v from its parent.

In either case, c includes the cost on the edge from v to the parent to support demand $|\lambda|$.

In a later step, we will round the costs to multiples of $\epsilon \cdot \mathrm{opt}/n$ and compute the minimum-cost solution using the dynamic program on the modified instance. As we will show, this results in a $(1+\epsilon)$-approximation to the problem.

Dynamic Program. The values $\mathrm{total}(v)$ are defined recursively, using the previously computed values for the children.

The base case corresponds to a leaf node w with parent p. We either send the demand to the parent without opening a facility, or we open a facility at w. Formally,

$$\mathrm{total}(w) = \{(d(w), \mathrm{cost}(pw, d(w))), (-\lambda, f(w) + \mathrm{cost}(pw, \lambda)) : \lambda \in [0, D]\}$$

For internal nodes, the DP considers two cases. If a facility is opened at v, we can serve $d(v)$ locally and allow both children (v' and v'') and the parent to send demand to v. The total cost includes the optimal costs of the subtrees $T_{v'}$ and $T_{v''}$, routing demand from children to v, facility opening cost, and routing demand from the parent to v. Formally,

$$\mathrm{total}_A(v) = \{(-\lambda, f(v) + \mathrm{cost}(pv, \lambda) + \mathrm{total}_{\min}(v') + \mathrm{total}_{\min}(v'')) : \lambda \in [0, D]\}$$

with $\mathrm{total}_{\min}(v') = \min\{c : (\lambda', c) \in \mathrm{total}(v')\}$.

If no facility is opened at v, demand cannot be served there; in this case, all flow at v must pass through it. That means the total demand leaving v toward the parent equals the sum of the flows from both children plus $d(v)$. The algorithm thus computes the minimum cost for each valid combination of these flows:

$$\begin{aligned}\mathrm{total}_B(v) = \{(\lambda, \mathrm{cost}(pv, |\lambda|) + c' + c'' :\\ (\lambda', c') \in \mathrm{total}(v'), (\lambda'', c'') \in \mathrm{total}(v''), \lambda = \lambda' + \lambda'' + d(v)\}.\end{aligned}$$

We then take $\mathrm{total}(v) = \mathrm{nondom}(\mathrm{total}_A(v) \cup \mathrm{total}_B(v))$, the non-dominated solutions among those computed.

Let OPT be an optimum solution and let $\lambda^*(v)$ be the demand in the optimum solution, leaving v to its parent or arriving from its parent as above. Let also $\mathrm{total}^*(v)$ represent the total cost of opening facilities and routing demand in P_{vw}, including the demand of value $\lambda^*(v)$ arriving/leaving from v from/to its parent.

Lemma 5. *For every* $v \in V$, $\mathrm{total}(v, \lambda^*(v))$ *exists.*

Lemma 6. *For every* $v \in V$, *let* $(\lambda^*(v), c) \in \mathrm{total}(v)$ *be the pair corresponding to the optimal solution. Then,* $c \leq c^*(v)$.

Lemma 7. *The values* total *can be calculated in time* $O(n \cdot D^2)$*, where* $D = \sum_v d_v$.

The following proof demonstrates PTAS for Theorem 1 by detailing the cost-rounding and dynamic programming approach.

Proof (Theorem 1). We design a PTAS using a combination of cost rounding and a dynamic programming algorithm on binary trees. Let opt be the cost of the optimal solution. We will do a binary search on a bound $\text{opt} \leq L \leq 2\,\text{opt}$, starting with $L = \min_v f(v)$, and doubling each time that the result of the algorithm has cost superior to $(1+\epsilon)L$.

We round the cable installation costs $\ell(e) \cdot c_{\text{cap}}(\lambda)$ to the nearest multiple of $\delta = \varepsilon \cdot L/2n$, for every edge e and demand λ. Each edge cost changes by at most δ, so the total rounding error across all n vertices is at most $n \cdot \delta = \varepsilon \cdot L/2 \leq \varepsilon \cdot \text{opt}$. Hence, rounding introduces at most an additive $\varepsilon \cdot \text{opt}$ error, and the cost of the rounded solution is at most $(1+\varepsilon) \cdot \text{opt}$.

We then use the dynamic program from Sect. 3.1, which computes, for each subtree, the minimum cost of satisfying demands under flow constraints. Lemmas 5 and 6 show that the DP correctly finds the minimum-cost solution under the rounded costs.

By Lemma 7, the DP table for each node contains $O(D)$ entries, where $D = \sum_v d(v)$, and each entry can be computed in $O(D)$ time. This results in an overall runtime of $O(n \cdot D^2)$. Since all cable costs are rounded to multiples of δ, the number of distinct cost values remains polynomial in n and $1/\varepsilon$, so the total running time is $f(\varepsilon) \cdot \text{poly}(n)$ for some function f. Thus, the algorithm returns a $(1+\varepsilon)$-approximate solution in time $f(\varepsilon) \cdot \text{poly}(n)$.

4 Cable-Unsplittable-Demand kCFL on Trees

This section addresses the cable-unsplittable demand version of kCFL, where demands on an edge cannot be split over multiple cables, and each cable can only carry demand in a single direction. Unlike the splittable-demand case handled in the previous sections, we show that cable-unsplittable kCFL is APX-hard even in very simple instances. Thus, our goal is to approach the best achievable approximation factor on trees, so as to introduce new techniques to handle this problem.

Our approach is to solve a resource-augmented variant of the problem, where we are allowed to slightly violate the capacities by a factor of $1+\epsilon$. To clarify, we allow our solution to put total demand $(1+\epsilon)\mu$ on a cable of capacity μ, but importantly, the optimum solution has total demand at most the capacity on each cable it uses. For more details on resource-augmentation settings, we refer the reader to the book chapter by Roughgarden [32].

We refer to this resource-augmented variant as RAkCFL, and after giving a PTAS for the problem (Theorem 7), we show how it can be used to give a 2-approximation to the original problem (Lemma 8). The factor of 2 is significantly larger than the factor for splittable-demand, but is close to the hardness of $3/2-\epsilon$ given in Theorem 6. We start by stating the APX-hardness result:

Theorem 6. *kCFL is APX-hard when the input graph is a star and $k = 1$. In particular, it is NP-hard to approximate the problem to a factor of $3/2 - \epsilon$ for any $\epsilon > 0$.*

The main result of this section is a PTAS for kCFL in the resource augmented setting.

Definition 2. *Let $(G, \ell, f, d, \mathbf{q})$ be an instance of the kCFL problem and $\epsilon > 0$.*

We say that $(I, \mathbf{g}, \mathbf{Q})$ is a solution to the kCFL instance with ϵ-resource augmentation *if each g_v is a flow from $\{v\}$ to I of value d_v, and for every edge uw, the flow values $\{g_v(u,w)\}$ and $\{g_v(w,u)\}$ can be partitioned into cables such that*

- *Each cable has total assigned flow at most $(1+\epsilon)\mu$, where μ is its capacity;*
- *Each flow value is assigned to a cable with at least as much capacity;*
- *The flows assigned to each cable are in the same direction.*

We refer to the kCFL problem in the ϵ-resource augmented setting as ϵ-RAkCFL, for a given $\epsilon > 0$. Before proving the main result, we show how an approximation for RAkCFL is helpful to solve kCFL. We remark that the reduction below applies to general graphs.

Lemma 8. *Let $(G, \ell, f, d, \mathbf{q})$ be an instance of the kCFL problem and $(I, \{g_v\}_v, \{Q_e\}_e)$ be an α-approximate solution with ϵ-resource augmentation for some $0 < \epsilon < 1/2$, and $\alpha \geq 1$.*

Then there is a 2α-approximate solution to the kCFL problem.

The proof follows by doubling cables that exceed the capacity.

The main result of this section is the following:

Theorem 7. *Let $(G, \ell, f, d, \mathbf{q})$ be an instance of kCFL where G is a tree, and $0 < \epsilon \leq 1/10$. There is an algorithm for RAkCFL that computes an ϵ-resource augmented solution with optimal cost in time $n^{k \cdot f(\epsilon)}$, for some computable function f.*

We prove Theorem 4 by applying Theorem 7 with $\epsilon = 1/10$, and then the transformation of Lemma 8 (which runs in polynomial time). For simplicity and ease of exposition, we present the result on instances with a single cable type ($k = 1$) and defer to the full version the details for multiple cables.

4.1 RAkCFL on Trees for a Single Cable Type ($\mathbf{k} = 1$)

Our approach is based on two main ideas: the first is to round demands so that similar values can be treated as identical and there are few values to consider; the second is to impose some structure onto the solution by grouping small demands together onto the same paths, as to reduce the combinatorial explosion of handling a large number of small demands. While the first idea contributes to underestimating the value of demands, the second causes demands to be served along sub-optimal paths, and thus both lead to resource augmentation.

Let (μ, c) be the single cable type. For each demand of at least $\epsilon\mu/2$, we round it down to the next multiple of $\epsilon^2\mu$; as we will show, these demands shrink by at most $\epsilon^2\mu$, and thus contribute to an excess of at most $2\epsilon\mu$ in the demand of each cable. Small demands must be handled differently: they are grouped into *bundles* totaling at least $\epsilon\mu/2$, with an additional leftover bundle of total demand less than $\epsilon\mu/2$. Both small demands and the leftover bundle are rounded more finely to multiples of $\epsilon^2\mu/n$, and become a (non-leftover) bundle when they are joined into demand at least $\epsilon\mu/2$. The demands in each bundle cannot be separated once joined, and must be served along the same path by the same facility. This structure can be enforced at the price of 2ϵ-resource augmentation.

We must then show that any solution to the original problem can be modified so that small demands are grouped, which overloads cables by a small fraction of the capacity. This, together with the increased demand of the original values instead of the rounding, leads to the excess of $\epsilon\mu$ on each cable, but allows us to write a dynamic programming to obtain a solution with optimum cost.

The dynamic program works by considering, for each vertex v, tuples that specify the number of large demands and bundles of each rounded size, the demand in the leftover bundle, both going up to the parent and down from the parent, as well as the minimum cost to serve these demands in the subtree G_v. Due to the transformation described above, our dynamic program assigns demand at most $(1+2\epsilon)\mu$ to each cable, so as to find the 2ϵ-resource augmented optimal solution. Converting the solution back to the original demands incurs a factor of $(1 + 2\epsilon)$, for a total demand of at most $(1 + 5\epsilon)\mu$ on each cable.

Most of the technical work in this section is to show that demands are preserved sufficiently to obtain a resource-augmented solution of optimal cost.

Rounding Demands. Let $\delta_b = \epsilon^2\mu$, $\delta_s = \epsilon^2\mu/n$ be the two rescaling factors used. For every vertex $v \in V$, we take $b_v := \lfloor d_v/\delta_b \rfloor$, $s_v = 0$ if $d_v \geq \epsilon\mu/2$ and $b_v = 0$, $s_v = \lfloor d_v/\delta_s \rfloor$ otherwise. Notice that $b_v \in \{0\} \cup [\epsilon^{-1}/2, \epsilon^{-2}]$ and $s_v \in [0, \epsilon^{-1}n/2 - 1]$, and that only one of these values is non-zero for each v.

Dynamic Program. The dynamic program will compute, for every possible set of demands rounded as indicated above, the minimum cost needed to serve the demands in the subtree G_v, by opening facilities in the subtree and paying for cable installation in the edges of G_v and the edge from v to its parent, given the demands going to or coming from the parent.

Formally, we compute a set of non-dominated solutions total(v) for each vertex $v \in V$ with parent $u \in V$, which contains tuples of the form $(a, \lambda_s, \bar{\lambda}_s, c)$, where $\lambda_s, \bar{\lambda}_s \in [0, \epsilon^{-1}n/2 - 1]$, $a \in \mathcal{A} := \{0\}^{\epsilon^{-1}/2} \times [-n, n]^{\epsilon^{-2}-\epsilon^{-1}/2}$, i.e. a has ϵ^{-2} values between $-n$ and n, but the first $\epsilon^{-1}/2$ are fixed to 0 (for convenience).

A tuple $(a, \lambda_s, \bar{\lambda}_s, c)$ corresponds to a solution where

- $a_i \in [-n, n]$ is the number of demands of size $\delta_b i$, $i \in [\epsilon^{-1}/2, \epsilon^{-2}]$ going up to u if $a_i > 0$, or coming down from u if $a_i < 0$,
- $\delta_s\lambda_s$, $\lambda_s \in [0, \epsilon^{-1}n/2 - 1]$ (resp. $\delta_s\bar{\lambda}_s$) is the total of small demands going up to (resp. coming down from) u, and

- c is the cost of serving the demands of G_v plus demands coming from the parent specified by $a, \bar{\lambda}_s$, except demands going to the parent specified by a, λ_s, including cost of opening facilities and installing cables on edges of G_v and the edge uv.

The set total(v) can be computed recursively starting from the leaves and moving up in the tree. We defer the formal recursive rules to the full version.

Acknowledgments. SN was supported by ANR project ANR-21-CE48-0016 (COMCOPT). The work of DV was supported by ANR project ANR-21-CE48-0022 (S-EX-AP-PE-AL).

Disclosure of Interests. The authors have no competing interests to declare that are relevant to the content of this article.

References

1. Adamaszek, A., Antoniadis, A., Kumar, A., Mömke, T., Approximating airports and railways. In: 35th Symposium on Theoretical Aspects of Computer Science, STACS: Caen, France, February 28 to March 3, 2018, volume 96 of LIPIcs, pp. 5:1–5:13. Schloss Dagstuhl - Leibniz-Zentrum für Informatik (2018). https://doi.org/10.4230/LIPICS.STACS.2018.5
2. Adamaszek, A., Antoniadis, A., Mömke, T., Airports and railways: Facility location meets network design. In: 33rd Symposium on Theoretical Aspects of Computer Science, STACS 2016(February), pp. 17–20: Orléans, France, volume 47 of LIPIcs, pp. 6:1–6:14. Schloss Dagstuhl - Leibniz-Zentrum für Informatik (2016). https://doi.org/10.4230/LIPICS.STACS.2016.6
3. Aggarwal, A., et al.: A 3-approximation algorithm for the facility location problem with uniform capacities. Math. Program. **141**(1–2), 527–547 (2013). https://doi.org/10.1007/S10107-012-0565-4
4. An, H.-C., Singh, M., Svensson, O.: LP-based algorithms for capacitated facility location. SIAM J. Comput. **46**(1), 272–306 (2017)
5. Andrews, J.H.: Relevant empirical testing research: challenges and responses. ACM SIGSOFT Softw. Eng. Notes **29**(5), 1–4 (2004)
6. Awerbuch, B., Azar, Y.: Buy-at-bulk network design. In: 38th Annual Symposium on Foundations of Computer Science, FOCS 1997, Miami Beach, Florida, USA, 19–22 October 1997, pp. 542–547. IEEE Computer Society (1997)
7. Bansal, M., Garg, N., Gupta, N.: A 5-approximation for capacitated facility location. In: Epstein, L., Ferragina, P. (eds.) ESA 2012. LNCS, vol. 7501, pp. 133–144. Springer, Heidelberg (2012). https://doi.org/10.1007/978-3-642-33090-2_13
8. Behsaz, B., Salavatipour, M.R., Svitkina, Z.: New approximation algorithms for the unsplittable capacitated facility location problem. Algorithmica **75**(1), 53–83 (2016). https://doi.org/10.1007/S00453-015-0012-Z
9. Byrka, J.: An optimal bifactor approximation algorithm for the metric uncapacitated facility location problem. In: Charikar, M., Jansen, K., Reingold, O., Rolim, J.D.P. (eds.) APPROX/RANDOM -2007. LNCS, vol. 4627, pp. 29–43. Springer, Heidelberg (2007). https://doi.org/10.1007/978-3-540-74208-1_3
10. Byrka, J., Grandoni, F., Rothvoss, T., Sanità, L.: Steiner tree approximation via iterative randomized rounding. J. ACM **60**(1) (2013). https://doi.org/10.1145/2432622.2432628

11. Chekuri, C., Hajiaghayi, M.T., Kortsarz, G., Salavatipour, M.R.: Approximation algorithms for nonuniform buy-at-bulk network design. SIAM J. Comput. **39**(5), 1772–1798 (2010). https://doi.org/10.1137/090750317
12. Chekuri, C., Khanna, S., Naor, J.: A deterministic algorithm for the cost-distance problem. In: Proceedings of the Twelfth Annual Symposium on Discrete Algorithms, Washington, DC, USA, 7–9 January 2001, pp. 232–233. ACM/SIAM (2001). http://dl.acm.org/citation.cfm?id=365411.365452
13. Chudak, F.A., Williamson, D.P.: Improved approximation algorithms for capacitated facility location problems. Math. Program. **102**(2), 207–222 (2005)
14. Diestel, R.: Graph Theory. Graduate Texts in Mathematics, 4th edn., vol. 173. Springer, Heidelberg (2012). https://doi.org/10.1007/978-3-662-70107-2
15. Fakcharoenphol, J., Rao, S., Talwar, K.: A tight bound on approximating arbitrary metrics by tree metrics. J. Comput. Syst. Sci. **69**(3), 485–497 (2004)
16. Friggstad, Z., Rezapour, M., Salavatipour, M.R., Soto, J.A.: LP-based approximation algorithms for facility location in buy-at-bulk network design. Algorithmica **81**(3), 1075–1095 (2019)
17. Grandoni, F., Italiano, G.F.: Algorithms and constraint programming. In: Benhamou, F. (ed.) CP 2006. LNCS, vol. 4204, pp. 2–14. Springer, Heidelberg (2006). https://doi.org/10.1007/11889205_2
18. Guha, S., Khuller, S.: Approximation algorithms for connected dominating sets. Algorithmica **20**(4), 374–387 (1998)
19. Guha, S., Meyerson, A., Munagala, K.: Improved algorithms for fault tolerant facility location. In: Proceedings of the Twelfth Annual Symposium on Discrete Algorithms, Washington, DC, USA, 7–9 January 2001, pp. 636–641. ACM/SIAM (2001). http://dl.acm.org/citation.cfm?id=365411.365554
20. Gupta, A., Kumar, A., Rastogi, R.: Exploring the trade-off between label size and stack depth in MPLS routing. In: IEEE INFOCOM 2003, The 22nd Annual Joint Conference of the IEEE Computer and Communications Societies, San Franciso, CA, USA, March 30-April 3 2003. IEEE Computer Society (2003)
21. Jain, K., Mahdian, M., Markakis, E., Saberi, A., Vazirani, V.V.: Greedy facility location algorithms analyzed using dual fitting with factor-revealing LP. J. ACM **50**(6), 795–824 (2003)
22. Jain, S., Menzel, W., Stephan, F.: Classes with easily learnable subclasses. In: Cesa-Bianchi, N., Numao, M., Reischuk, R. (eds.) ALT 2002. LNCS (LNAI), vol. 2533, pp. 218–232. Springer, Heidelberg (2002). https://doi.org/10.1007/3-540-36169-3_19
23. Jowhari, H., Nematollahi, S.: Airports and railways with unsplittable demand. Inf. Process. Lett. **188**, 106538 (2025)
24. Korte, B., Vygen, J.: Combinatorial Optimization: Theory and Algorithms. Algorithms and Combinatorics, Springer, Heidelberg (2018)
25. Korupolu, M.R., Plaxton, C.G., Rajaraman, R.: Analysis of a local search heuristic for facility location problems. J. Algorithms **37**(1), 146–188 (2000)
26. Li, S.: A 1.488 approximation algorithm for the uncapacitated facility location problem. Inf. Comput. **222**, 45–58 (2013)
27. Lueker, G.S.: Two NP-Complete Problems in Nonnegative Integer Programming. Princeton University. Department of Electrical Engineering (1975)
28. Mahdian, M., Ye, Y., Zhang, J.: Approximation algorithms for metric facility location problems. SIAM J. Comput. **36**(2), 411–432 (2006)

29. Meyerson, A., Munagala, K., Plotkin, S.A.: Cost-distance: two metric network design. In: 41st Annual Symposium on Foundations of Computer Science, FOCS 2000, Redondo Beach, California, USA, 12–14 November 2000, pp. 624–630. IEEE Computer Society (2000)
30. Pál, M., Tardos, É., Wexler, T.: Facility location with nonuniform hard capacities. In: 42nd Annual Symposium on Foundations of Computer Science, FOCS 2001, Las Vegas, Nevada, USA, 14–17 October 2001, pp. 329–338. IEEE Computer Society (2001). https://doi.org/10.1109/SFCS.2001.959907
31. Ravi, R., Sinha II, A.: Approximating k-cuts via network strength. In: Proceedings of the Thirteenth Annual ACM-SIAM Symposium on Discrete Algorithms, San Francisco, CA, USA, 6–8 January 2002, pp. 621–622. ACM/SIAM (2002). http://dl.acm.org/citation.cfm?id=545381.545463
32. Roughgarden, T.: Resource augmentation. In: Roughgarden, T. (ed.) Beyond the Worst-Case Analysis of Algorithms, pp. 72–92. Cambridge University Press (2020). https://doi.org/10.1017/9781108637435.006
33. Salavatipour, M.R., Tian, L.: Approximation algorithms for the airport and railway problem. In: 19th Scandinavian Symposium and Workshops on Algorithm Theory, SWAT 2024(June), pp. 12–14, Helsinki, Finland, volume 294 of LIPIcs, pp. 40:1–40:16. Schloss Dagstuhl - Leibniz-Zentrum für Informatik (2024). https://doi.org/10.4230/LIPICS.SWAT.2024.40
34. Salman, F.S., Cheriyan, J., Ravi, R., Subramanian, S.: Buy-at-bulk network design: approximating the single-sink edge installation problem. In: Proceedings of the Eighth Annual ACM-SIAM Symposium on Discrete Algorithms, New Orleans, Louisiana, USA, 5–7 January 1997, pp. 619–628. ACM/SIAM (1997). http://dl.acm.org/citation.cfm?id=314161.314397
35. Shaw, D.X.: A unified limited column generation approach for facility location problems ontrees. Ann. Oper. Res. **87**, 363–382 (1999). https://doi.org/10.1023/A:1018901523519
36. Shmoys, D.B., Tardos, É., Aardal, K.I.: Approximation algorithms for facility location problems (extended abstract). In: Leighton, F.T., Shor, P.W. (eds.) Proceedings of the Twenty-Ninth Annual ACM Symposium on the Theory of Computing, El Paso, Texas, USA, 4–6 May 1997, pp. 265–274. ACM (1997). https://doi.org/10.1145/258533.258600
37. Sviridenko, M.: An improved approximation algorithm for the metric uncapacitated facility location problem. In: Cook, W.J., Schulz, A.S. (eds.) IPCO 2002. LNCS, vol. 2337, pp. 240–257. Springer, Heidelberg (2002). https://doi.org/10.1007/3-540-47867-1_18
38. Talwar, K.: the single-sink buy-at-bulk LP has constant integrality gap. In: Cook, W.J., Schulz, A.S. (eds.) IPCO 2002. LNCS, vol. 2337, pp. 475–486. Springer, Heidelberg (2002). https://doi.org/10.1007/3-540-47867-1_33
39. Williamson, D.P., Shmoys, D.B.: The Design of Approximation Algorithms. Cambridge University Press (2011). http://www.cambridge.org/de/knowledge/isbn/item5759340/

Improved Online Sorting

Jubayer Nirjhor(✉) and Nicole Wein

University of Michigan, Ann Arbor, MI 48109, USA
{nirjhor,nswein}@umich.edu

Abstract. We study the online sorting problem, where n real numbers arrive in an online fashion, and the algorithm must immediately place each number into an array of size $(1+\varepsilon)n$ before seeing the next number. After all n numbers are placed into the array, the cost is defined as the sum over the absolute differences of all $n-1$ pairs of adjacent numbers in the array, ignoring empty array cells. Aamand, Abrahamsen, Beretta, and Kleist introduced the problem and obtained a deterministic algorithm with cost $2^{O\left(\sqrt{\log n \cdot \log\log n + \log \varepsilon^{-1}}\right)}$, and a lower bound of $\Omega(\log n / \log\log n)$ for deterministic algorithms. We obtain a deterministic4 algorithm with quasi-polylogarithmic cost $\left(\varepsilon^{-1}\log n\right)^{O(\log\log n)}$. Concurrent and independent work by Azar, Panigrahi, and Vardi achieves polylogarithmic cost $O(\varepsilon^{-1}\log^2 n)$.

Keywords: Online algorithms · Sorting · Data structures

1 Introduction

In the *online sorting problem*, n real numbers arrive in an online fashion, and the algorithm must immediately place each number into an array of size $(1+\varepsilon)n$ before seeing the next number (so that each array cell contains at most one number). After all n numbers are placed into the array, the *cost* is defined as the sum over the absolute difference of all $n-1$ pairs of adjacent numbers in the array, ignoring empty array cells. Formally, let A be the array of size $(1+\varepsilon)n$ where we place input numbers, indexed starting from 1, and let $1 \le i_1 \le i_2 \le \cdots \le i_n \le (1+\varepsilon)n$ be the subsequence of n array indices that are nonempty at the end (an input number was placed there). Then we define the cost as $\sum_{k=1}^{n-1} |A[i_k] - A[i_{k+1}]|$.

It is evident that the minimum achievable cost occurs when the numbers are perfectly sorted. This natural problem was introduced by Aamand, Abrahamsen, Beretta, and Kleist [13] in the context of online geometric packing problems.

It is convenient to normalize the problem to assume that the n numbers come from the range $[0,1]$ and contain the numbers 0 and 1, because then the cost of the optimal offline algorithm is precisely 1, so the competitive ratio of an online algorithm is equal to its cost. We use this convention unless otherwise specified.

For the case of $\varepsilon = 0$ (where the array is of size n, and thus completely full at the end), an asymptotically tight competitive ratio of $\Theta(\sqrt{n})$ is known, where

J. Matuschke and J. Verschae (Eds.): WAOA 2025, LNCS 16077, pp. 187–197, 2026.
https://doi.org/10.1007/978-3-032-06706-7_13

the upper bound is a deterministic algorithm [13], and the lower bound holds even for randomized algorithms [1,13].

For the general setting of $\varepsilon > 0$, Aamand, Abrahamsen, Beretta, and Kleist [13] obtained a deterministic algorithm with cost of $2^{O\left(\sqrt{\log n \cdot \log\log n + \log \varepsilon^{-1}}\right)}$. They also obtained, for any constant ε, a lower bound of $\Omega(\log n / \log\log n)$ for deterministic algorithms (there is no known lower bound for randomized algorithms). This leaves a large gap between upper and lower bounds, and our work shrinks the gap.

1.1 Our Result

We obtain an algorithm with *quasi-polylogarithmic* competitive ratio:

Theorem 1. *Given a positive integer n and a real number $\varepsilon \in (0, 3]$, there is a deterministic algorithm for online sorting with n numbers, an array of size $(1+\varepsilon)n$, and competitive ratio $\left(\varepsilon^{-1} \log n\right)^{O(\log\log n)}$.*

Remark. Ignoring the ε dependence for simplicity, our competitive ratio can also be written as $2^{O\left((\log\log n)^2\right)}$, where the exponent $O((\log\log n)^2)$ is an exponential improvement over the exponent $O(\sqrt{\log n \cdot \log\log n})$ of prior work. Further improvement of the exponent from $O((\log\log n)^2)$ to $O(\log\log n)$ would make the final cost polylogarithmic.

Concurrent and Independent Work. In independent work, Azar, Panigrahi, and Vardi achieve a polylogarithmic bound [8] using techniques orthogonal to ours. Specifically, for any $\varepsilon \geq \Omega(\log n/n)$, they give a deterministic algorithm with competitive ratio $O(\varepsilon^{-1} \log^2 n)$. Additionally, they provide a trade-off for sparser arrays: For any $\gamma \in [O(1), O(\log^2 n)]$, for an array with γn space they give a deterministic algorithm with competitive ratio $O(\gamma^{-1} \log^2 n)$.

1.2 Related Work

Several variants of online sorting have been studied, including the stochastic [1, 9] and high-dimensional [1] versions. Online sorting is also related to geometric packing problems: lower bounds for online sorting have been used to provide lower bounds for online packing of convex polygons [13].

Online sorting is also related to other online array maintenance problems such as *list labeling* [2–4,11], *matching on the line* [5–7,10], and *B-tree leaf splitting* [12].

1.3 Our Techniques

We use the techniques of [13] as a starting point. First we will provide a high-level outline of their algorithm, and then describe the additional ideas behind our algorithm.

Algorithm of [13]. Divide the interval $[0, 1]$ into contiguous disjoint subintervals, and divide the array into contiguous disjoint subarrays called *boxes* (for this outline, we will not specify how many subintervals and boxes). For any real number $x \in [0, 1]$, let $I(x)$ be the subinterval containing x.

Note that after all n numbers are inserted, a $1/(1+\varepsilon)$ fraction of the array cells will be occupied. When the first number x_0 arrives, place x_0 into the first box and *label* the box with $I(x_0)$. In general, when a number x arrives, if there is a box labeled with $I(x)$ that is below the *fullness threshold* of $1/(1+\varepsilon)$ (i.e. a $< 1/(1+\varepsilon)$ fraction of array cells in the box are occupied), place x into that box. Otherwise, place x into the first completely empty box and label that box $I(x)$. The parameters (such as the size of each box, the fullness threshold of each box, and the number of subintervals) are set so that there always exists a box to place x.

After placing x into a box, the algorithm still needs to determine which array cell within the box to place x. This is done recursively. In order to ultimately achieve fullness $1/(1+\varepsilon)$, the recursive instance within each box must run the algorithm with fullness threshold $> 1/(1+\varepsilon)$. This is because for any box, some of its sub-boxes may be almost completely empty (e.g. with only one occupied array cell if no numbers from that particular subinterval ever arrive again in the input sequence), and to compensate, the other boxes must be fuller, hence a higher fullness threshold. Eventually, at the bottom recursive level, the fullness threshold becomes nearly 1, and the known $\Theta(\sqrt{n})$-cost algorithm for the $\varepsilon = 0$ case is applied. Ultimately, the value of ε that determines the fullness threshold on each recursive level is set to increase linearly as you go up the recursive levels.

Our Algorithm. Our improved algorithm has two key ideas:

Idea 1: The first key idea is to use *another layer of recursion* to decide in which box to place each number. In the algorithm of [13], when they label a box, they simply choose the first available box. Instead, our algorithm treats the "array of boxes" as its own recursive instance of the online sorting problem. In other words, when a number x arrives, the first step is the same as the algorithm of [13]: if there is a box labeled $I(x)$ that is below the fullness threshold, place x into that box choosing the specific array cell recursively. If there is no such box, our algorithm deviates from theirs: we recursively run the algorithm on the array of boxes to determine which empty box to place x, and then choose the specific array cell within the box with a second recursive call.

Idea 2: The second key idea is to increase ε *exponentially* instead of linearly as we proceed up the recursive levels. This is necessary because, for technical reasons, linear scaling of ε clashes with the implementation of Idea 1. The reason exponential scaling works better than linear is roughly as follows. In order to treat the array of boxes as a recursive instance of online sorting with $\varepsilon > 0$, some of the boxes need to be completely empty.

Intuitively, varying ε exponentially allows us to designate more empty boxes in the top layers of recursion, which makes the recursive algorithm on the array of boxes have lower cost.

2 Proof of Theorem 1

Recall Theorem 1:

Theorem 1. *Given a positive integer* n *and a real number* $\varepsilon \in (0, 3]$, *there is a deterministic algorithm for online sorting with* n *numbers, an array of size* $(1+\varepsilon)n$, *and competitive ratio* $\left(\varepsilon^{-1}\log n\right)^{O(\log\log n)}$.

The bulk of the proof will be in proving the following lemma:

Lemma 1. *Consider positive integers* n, k, *and real numbers* $\delta \in (0, \frac{1}{2})$, $[\alpha, \alpha+\beta) \subseteq [0, 1]$ *such that* $2^{k+1}\delta \leq 3$ *is satisfied, and* δ, k *may depend on* n. *Then, there exists an algorithm that solves the online sorting problem on an array of size* $(1+2^{k+1}\delta)n$, *over any sequence of* n *real numbers from* $[\alpha, \alpha+\beta)$, *achieving a cost of* $\beta n^{1/\omega_k}\delta^{-O(k)}$. *Here, the integer sequence* $\{\omega_i\}_{i\in\mathbb{Z}}$ *is defined recursively by* $\omega_i = \omega_{i-1} + \omega_{i-4}$ *for* $i \geq 2$, *and* $\omega_i = 2$ *for* $-\infty < i \leq 1$.

Before proving Lemma 1, we first show that Lemma 1 indeed implies Theorem 1.

Proof (of Theorem 1, assuming Lemma 1). First, we look at the recurrence $\omega_i = \omega_{i-1} + \omega_{i-4}$ for $i \geq 2$ starting with $\omega_i = 2$ for $-\infty < i \leq 1$. It's a linear recurrence with characteristic polynomial $x^4 = x^3 + 1$. We note that, among the four complex roots of this polynomial, only the root $x \approx 1.38$ satisfies $|x| \geq 1$. So for every other root x, the value $x^n \to 0$ as $n \to \infty$. From this, we can deduce that $\omega_n \propto 1.38^n$ as $n \to \infty$.

We apply Lemma 1 choosing parameters $\alpha := 0$, $\beta := 1$, $k := \lfloor \log_{1.38}\log n \rfloor$, and $\delta := \frac{\varepsilon}{2^{k+1}}$, since they satisfy $2^{k+1}\delta = \varepsilon \leq 3$ (by the assumption on the value of ε). It uses $(1+2^{k+1}\delta)n = (1+\varepsilon)n$ memory cells and yields a cost of

$$\begin{aligned} n^{1/\omega_k}\delta^{-O(k)} &= 2^{O\left(\frac{\log n}{\omega_k} + k\log\frac{2^{k+1}}{\varepsilon}\right)} \\ &= 2^{O\left(\frac{\log n}{1.38^k} + k^2 + k\log\varepsilon^{-1}\right)} \\ &= (\log n)^{O\left(\log\log n + \log\varepsilon^{-1}\right)} \end{aligned}$$

obtained by plugging in our chosen parameters. Using the identity $a^{\log b} = b^{\log a}$, we can infer that $(\log n)^{O\left(\log\varepsilon^{-1}\right)} = \left(\varepsilon^{-1}\right)^{O(\log\log n)}$. So the cost above is $\left(\varepsilon^{-1}\log n\right)^{O(\log\log n)}$.

With this, it remains to prove Lemma 1.

2.1 Proof of Lemma 1

We divide the proof into three sections. First, we present the algorithm itself. Afterwards, we prove that the proposed algorithm correctly solves the online sorting problem. Lastly, we prove that the algorithm indeed achieves a cost of $\beta n^{1/\omega_k}\delta^{-O(k)}$.

The Algorithm. We will describe the algorithm both in words and in pseudocode (Algorithm 1).

Algorithm Setup and Definitions. Define $\mathcal{A}$ as the algorithm described in [13] for online sorting with n numbers on an array of size n, that achieves the asymptotically tight competitive ratio of $\Theta(\sqrt{n})$. First off, we note that the algorithm $\mathcal{A}$ works even if the array size is bigger than n, as we could simply ignore the extra space. For a positive integer k, we will recursively define SORTER_k as our algorithm that solves the online sorting problem with n numbers on an array of size $(1+2^{k+1}\delta)n$. As k gets larger, so does the amount of empty space in the array. Intuitively, this provides us with more flexibility, allowing for a smaller cost. We further define $\text{SORTER}_k := \mathcal{A}$ for all integers $k \leq 1$.

We design the algorithm SORTER_k for $k \geq 2$, by recursively using SORTER_j for $j < k$ as subroutines. We will divide the array into ℓ disjoint contiguous subarrays that we call *boxes*, each of *width* w (i.e. each box has w array cells). Each box can receive at most $n' < w$ elements of the input sequence. When we receive a particular element of the input sequence, we will invoke an instance of our algorithm to recursively determine the box where the element will go. We will also invoke an instance of our algorithm to recursively determine the exact cell in that box where the element should be placed. We define $n' := \left\lfloor \frac{2^{k-1}\delta}{1+2^k\delta} n^{\omega_{k-1}/\omega_k} \right\rfloor$ (the number of elements that go into a certain box), and $w := \lfloor (1+2^k\delta)n' \rfloor$ (the width of each box). Then the total number of boxes that completely fit into the array is $\ell := \left\lfloor \frac{(1+2^{k+1}\delta)n}{w} \right\rfloor$ (if the length of the array $(1+2^{k+1}\delta)n$ is not divisible by w, then we ignore the leftover space). For an integer $1 \leq i \leq \ell$, we define the box B_i as the subarray spanning the array indices from $(i-1)w$ to $iw-1$. We say that a box B_i is *full* if it contains exactly n' elements of the input sequence. Intuitively, we call it full because we've reached the maximum number of elements we allow inserting into this particular subarray, while keeping enough empty space to recursively invoke our algorithm.

We will also divide the interval of real numbers $[\alpha, \alpha+\beta)$ into $b := \lfloor n^{\omega_{k-4}/\omega_k} \rfloor$ disjoint contiguous subintervals. Each box will be designated a particular subinterval of real numbers, so that only elements from that subinterval will be placed into the box. To determine which subinterval should be assigned to a particular box, we invoke our algorithm recursively. Let the algorithm BOXSORTER be a fixed instance of SORTER_{k-4} on an array of size ℓ. Intuitively, the i-th array cell here corresponds to the i-th box. We specifically use SORTER_{k-4} here because, as we'll see in the analysis, this instance is enough to guarantee that all elements of the input sequence will find an array slot following our algorithm (allowing more empty space through, say, SORTER_{k-3} or higher, does not work in our analysis). The input to BOXSORTER will be a subsequence of the original input sequence – whenever we encounter an input from a subinterval that isn't assigned to a not-full box, we will append that element to the input sequence of BOXSORTER. Then we'll insert the element into the box corresponding to the array cell where BOXSORTER places this element. In the analysis, we will show that an instance of SORTER_{k-4} does indeed work as our BOXSORTER algorithm.

We define the *pointer function* $p : [b] \longrightarrow \mathbb{Z}$ and initialize $p(i) = -1$ for each $i \in [b]$. This function maps each of the b subintervals to one of the ℓ boxes (with -1 indicating that no box has been assigned to it yet).

Final Description of Algorithm. Given a real number $x \in [\alpha, \alpha + \beta)$, let $i \in [b]$ be the subinterval that x falls into, that is, i is the unique integer satisfying $(i-1)\beta \leq (x-\alpha)b < i\beta$. If $p(i) \neq -1$ and $B_{p(i)}$ is not full (meaning it has received strictly fewer than n' elements), we place x into $B_{p(i)}$. Otherwise, we send x to BOXSORTER as an input and assign $p(i)$ to the index in $[\ell]$ where x is placed by BOXSORTER, then finally we place x into $B_{p(i)}$. To place x into an array cell inside the box $B_{p(i)}$, we run SORTER_{k-1} on the array defined by the box and send x as an input.

To summarize the recursive structure of the algorithm, SORTER_k spawns one instance (that we call BOXSORTER) of SORTER_{k-4} on an array of size ℓ, whose input values come from an interval of length β. It also spawns ℓ instances of SORTER_{k-1}, each on an array of size w and receiving at most n' elements, whose input values come from an interval of length $\frac{\beta}{b}$ (since we divide the original inter-

Algorithm 1. SORTER_k

Require: $k \in \mathbb{Z}$; array of size $(1 + 2^{k+1}\delta)n$; input interval $[\alpha, \alpha + \beta)$

if $k \leq 1$ **then**
  return $\mathcal{A}$ ▷ baseline online sorter from [13]
$n' \leftarrow \left\lfloor \frac{2^{k-1}\delta}{1+2^k\delta} n^{\omega_{k-1}/\omega_k} \right\rfloor$ ▷ fullness of each box
$w \leftarrow \lfloor (1 + 2^k\delta)\, n' \rfloor$ ▷ size of each box
$\ell \leftarrow \left\lfloor \frac{(1+2^{k+1}\delta)n}{w} \right\rfloor$ ▷ total number of boxes
$b \leftarrow \lfloor n^{\omega_{k-4}/\omega_k} \rfloor$ ▷ number of subintervals
$p[1..b] \leftarrow -1$ ▷ pointer: subinterval $\rightarrow$ box
$count[1..\ell] \leftarrow 0$ ▷ occupancy of each box
instantiate BOXSORTER $\leftarrow \text{SORTER}_{k-4}$ on array of size ℓ
for $j \leftarrow 1$ **to** ℓ **do**
  instantiate INBOXSORTER$[j] \leftarrow \text{SORTER}_{k-1}$ on array of size w
for all arriving element $x \in [\alpha, \alpha + \beta)$ **do**
  $i \leftarrow 1 + \left\lfloor \frac{x-\alpha}{\beta} \cdot b \right\rfloor$ ▷ subinterval index in $[b]$
  if $p[i] \neq -1$ **and** $count[p[i]] < n'$ **then**
    $j \leftarrow p[i]$ ▷ reuse assigned not-full box
  else
    $j \leftarrow$ BOXSORTER(x) ▷ pick a box index $j \in [\ell]$
    $p[i] \leftarrow j$
  INBOXSORTER$[j]$(x) ▷ place x inside B_j
  $count[j] \leftarrow count[j] + 1$

val $[\alpha, \alpha + \beta)$ of length β into b subintervals of equal size). The input sequences for all these instances are subsequences of the given input sequence.

Proof of Correctness. In this section, we prove, by induction on k, that our algorithm SORTER_k successfully places the given input sequence of n numbers into the array of size $(1+2^{k+1}\delta)n$. The base case of $k = 1$ is immediate, since we define $\text{SORTER}_1 := \mathcal{A}$ as the algorithm described in [13] to solve online sorting with n numbers on an array of size n (our array size is bigger than n).

By the inductive hypothesis, for a positive integer $k \geq 2$, SORTER_{k-1} can place n' numbers into an array of size $w := \lfloor (1 + 2^k\delta)n' \rfloor$. We observe that the only possible ways for us to not successfully insert an element are: either if BOXSORTER has reached the capacity (defined by a fraction of its size) and does not have a cell (corresponding to a box) to assign, or if we cannot find a cell inside the box assigned to the subinterval of the input element. But finding the cell within a particular box is handled recursively, and from the inductive hypothesis, we do know that SORTER_{k-1} can place n' elements in each box. So we only need to show that the input sequence to BOXSORTER is small enough for it to never run out of cells (corresponding to boxes) to assign. Recall that we use an instance of SORTER_{k-4} as our BOXSORTER algorithm, so we will show that the number of elements sent to BOXSORTER respects the fraction of occupied cells required by SORTER_{k-4}. The proof follows immediately from the claim below, as we will see.

Claim. Let $\mathcal{S}$ be the sequence of numbers we send to BOXSORTER as input. Then $|\mathcal{S}| \leq \frac{\ell}{1+2^{k-3}\delta}$.

Proof. Consider an element x from the original input sequence that we include in $\mathcal{S}$. Depending on the type of box where x is placed, we consider two cases.

1. *The box is full at the end.* Since a full box receives exactly n' elements, the number of full boxes is at most $\frac{n}{n'}$. So the number of such elements in $\mathcal{S}$ is also at most $\frac{n}{n'}$ (since $\mathcal{S}$ only contains at most one element from each box).
2. *The box is not full at the end.* Since each non-full box is assigned a different subinterval, the number of such boxes at the end is at most b (the number of subintervals), and so is the number of such elements in $\mathcal{S}$.

We conclude that $|\mathcal{S}| \leq \frac{n}{n'} + b$. Next we note that

$$bw \leq n^{\omega_{k-4}/\omega_k} \cdot (1 + 2^k\delta)n' \leq n^{\omega_{k-4}/\omega_k} \cdot (1 + 2^k\delta)\frac{2^{k-1}\delta}{1 + 2^k\delta} n^{\omega_{k-1}/\omega_k} = 2^{k-1}\delta n$$

using $\omega_k = \omega_{k-1} + \omega_{k-4}$. This bound helps us compute:

$$\begin{aligned}\frac{|\mathcal{S}|(1+2^{k-3}\delta)}{\ell} &\le \frac{(\frac{n}{n'}+b)(1+2^{k-3}\delta)}{\frac{(1+2^{k+1}\delta)n}{w}} \\ &= \frac{(\frac{n}{n'}w+bw)(1+2^{k-3}\delta)}{(1+2^{k+1}\delta)n} \\ &\le \frac{(\frac{n}{n'}(1+2^k\delta)n' + 2^{k-1}\delta n)(1+2^{k-3}\delta)}{(1+2^{k+1}\delta)n} \\ &= \frac{(1+3\cdot 2^{k-1}\delta)(1+2^{k-3}\delta)}{(1+2^{k+1}\delta)}.\end{aligned}$$

When $2^{k+1}\delta \le 3$, this final quantity is ≤ 1, which finishes the proof.

As a final remark, we note that we require $\varepsilon = 2^{k+1}\delta \le 3$ for the final inequality of ≤ 1 to hold. This is why the algorithm is only defined for $\varepsilon \in (0,3]$.

This claim implies that the array size ℓ used by BoxSorter satisfies $\ell \ge (1+2^{k-3}\delta)|\mathcal{S}|$. Since Sorter_{k-4} can, by the inductive hypothesis, solve online sorting with $|\mathcal{S}|$ numbers on an array of size $(1+2^{k-3}\delta)|\mathcal{S}|$, this means that ℓ is large enough as an array size to insert $|\mathcal{S}|$ elements using Sorter_{k-4}. So it correctly works as our BoxSorter algorithm.

This concludes the proof that the algorithm Sorter_k can place n numbers on an array of size $(1+2^{k+1}\delta)n$.

Cost Analysis. We are left to bound the total cost, for which we use induction. By $\text{cost}^{(k)}(r_1,\ldots,r_n)$, we denote the cost incurred by algorithm Sorter_k when facing the sequence of real numbers $r_1,\ldots,r_n$. We prove that there exists $C>0$ such that $\text{cost}^{(k)}(r_1,\ldots,r_n) \le \beta n^{1/\omega_k}\delta^{-C(k+1)}$ for any sequence of numbers $r_1,\ldots,r_n$ with $r_i \in [\alpha, \alpha+\beta)$. For $k=1$, the algorithm from [13] implies that Sorter_1 places an input sequence $r_1,\ldots,r_n \in [\alpha,\alpha+\beta)$ with a cost of $\text{cost}^{(1)}(r_1,\ldots,r_n) = 18\beta\sqrt{n}$, and we can choose C accordingly.

Now suppose $k>1$, and let $\{B_1,\ldots,B_\ell\}$ be the set of boxes in the array. First we bound the number ℓ of boxes, by plugging in the parameters $w = \lfloor (1+2^k\delta)n' \rfloor$, $n' = \left\lfloor \frac{2^{k-1}\delta}{1+2^k\delta} n^{\omega_{k-1}/\omega_k} \right\rfloor$, and $b = \lfloor n^{\omega_{k-4}/\omega_k} \rfloor$ into the following

$$\ell \le \frac{(1+2^{k+1}\delta)n}{w} = \frac{(1+2^{k+1}\delta)n}{\lfloor (1+2^k\delta)n' \rfloor} \le 2\frac{(1+2^{k+1}\delta)n}{(1+2^k\delta)n'} \le \frac{4n}{n'} = \frac{4n}{\left\lfloor \frac{2^{k-1}\delta}{1+2^k\delta} n^{\omega_{k-1}/\omega_k} \right\rfloor}$$

$$(\text{since } \omega_k = \omega_{k-1} + \omega_{k-4})$$

$$\begin{aligned}&\le \frac{8n}{\frac{2^{k-1}\delta}{1+2^k\delta} n^{\omega_{k-1}/\omega_k}} = \frac{8}{\frac{2^{k-1}\delta}{1+2^k\delta}} n^{(\omega_k-\omega_{k-1})/\omega_k} = \frac{1+2^k\delta}{2^{k-4}\delta} n^{\omega_{k-4}/\omega_k} \\ &\le 2\frac{1+2^k\delta}{2^{k-4}\delta}\left\lfloor n^{\omega_{k-4}/\omega_k} \right\rfloor = \frac{1+2^k\delta}{2^{k-5}\delta} b.\end{aligned}$$

where we repeatedly use the fact that $x \le 2\lfloor x \rfloor$ for all $x \ge 1$.

We compute the following two costs separately, which add up to the total cost:

- **Cost inside boxes.** Here, we focus on the cost generated by the elements inside a certain box and sum these up across all boxes. We can think of our algorithm as partitioning the sequence $r_1, r_2, \ldots, r_n$ into subsequences according to the box in which each element is placed. For $i \in [\ell]$, denote the subsequence of length L_i placed in B_i with $y^i_1, \ldots, y^i_{L_i}$ so that we have $\{r_1, \ldots, r_n\} = \bigcup_{i=1}^{\ell} \{y^i_1, \ldots, y^i_{L_i}\}$. Moreover, $L_i \leq n'$ for each $i \in [\ell]$. Define $\alpha'_i := (i-1) \cdot \beta b^{-1}$ and $\beta'_i := \beta b^{-1}$, then for each $j \in [L_i]$ it holds $\alpha'_i \leq y^i_j < \alpha'_i + \beta'_i$. By the induction hypothesis, the cost induced by the recursive call of SORTER_{k-1} on box B_i is bounded by

$$
\begin{aligned}
\text{cost}^{(k-1)}(y^i_1, \ldots, y^i_{L_i}) &\leq \beta'_i \cdot (L_i)^{1/\omega_{k-1}} \delta^{-Ck} \qquad\qquad \left(\text{since } \tfrac{2^{k-1}\delta}{1+2^k\delta} < 1\right)\\
&\leq \beta b^{-1} \cdot (n')^{1/\omega_{k-1}} \delta^{-Ck}\\
&\leq \beta b^{-1} \cdot \left(\frac{2^{k-1}\delta}{1+2^k\delta} n^{\omega_{k-1}/\omega_k}\right)^{1/\omega_{k-1}} \delta^{-Ck}\\
&\leq \beta b^{-1} \cdot \left(\frac{2^{k-1}\delta}{1+2^k\delta}\right)^{1/\omega_{k-1}} n^{1/\omega_k} \delta^{-Ck}\\
&\leq b^{-1} \cdot \beta n^{1/\omega_k} \delta^{-Ck}
\end{aligned}
$$

  Since there are at most ℓ boxes in total, we bound the total cost generated inside the boxes B_i as

$$
\begin{aligned}
\sum_{i=1}^{\ell} \text{cost}^{(k-1)}(y^i_1, \ldots, y^i_{L_i}) &\leq \ell b^{-1} \cdot \beta n^{1/\omega_k} \delta^{-Ck}\\
&\leq \frac{1+2^k\delta}{2^{k-5}\delta} \cdot \beta n^{1/\omega_k} \delta^{-Ck}\\
&= \left(2^{5-k}\delta^{-1} + 8\right) \cdot \beta n^{1/\omega_k} \delta^{-Ck}\\
&\leq \frac{1}{2} \beta n^{1/\omega_k} \delta^{-C(k+1)}
\end{aligned}
$$

  where the last inequality holds if we choose C large enough.
- **Cost between boxes.** Now we estimate the cost incurred by BOXSORTER which is an instance of SORTER_{k-4}. Here, we place $|S|$ elements from $[\alpha, \alpha+\beta)$. From Sect. 2.1 and the bound for ℓ above, we first note that

$$
|S| \leq \frac{\ell}{1+2^{k-3}\delta} \leq \frac{1}{2^{k-3}\delta} \frac{1+2^k\delta}{2^{k-4}\delta} n^{\frac{\omega_{k-4}}{\omega_k}} = \left(2^{7-2k}\delta^{-2} + 2^{7-k}\delta^{-1}\right) n^{\frac{\omega_{k-4}}{\omega_k}}.
$$

  We use the above bound on $|S|$ and the inductive hypothesis to bound the total cost in this case as:

$$\begin{aligned}
\mathrm{cost}^{(k-4)}(\mathcal{S}) &= \beta \cdot |\mathcal{S}|^{1/\omega_{k-4}} \delta^{-C(k-3)} \\
&\le \beta \cdot \left(\left(2^{7-2k}\delta^{-2} + 2^{7-k}\delta^{-1} \right) n^{\omega_{k-4}/\omega_k} \right)^{1/\omega_{k-4}} \delta^{-C(k-3)} \\
&= \beta \cdot \left(2^{7-2k}\delta^{-2} + 2^{7-k}\delta^{-1} \right)^{1/\omega_{k-4}} \cdot n^{1/\omega_k} \delta^{-C(k-3)} \\
&\le \beta \cdot \sqrt{2^{7-2k}\delta^{-2} + 2^{7-k}\delta^{-1}} \cdot n^{1/\omega_k} \delta^{-C(k-3)} \\
&\le \frac{1}{2} \beta n^{1/\omega_k} \delta^{-C(k+1)}
\end{aligned}$$

where the last inequality holds if we choose C large enough.

Summing up these two costs: *cost inside boxes* and *cost between boxes*, we complete the induction. This concludes the proof that the total cost incurred by SORTER_k for online sorting with n numbers on an array of size $(1 + 2^{k+1}\delta)n$ is bounded by $\beta n^{1/\omega_k} \delta^{-O(k)}$.

Acknowledgments. We would like to thank Ioana Bercea and László Kozma for introducing us to this problem at the *Algorithms and Data Structures Today* workshop at the National University of Singapore.

References

1. Abrahamsen, M., Bercea, I.O., Beretta, L., Klausen, J., Kozma, L.: Online sorting and online TSP: randomized, stochastic, and high-dimensional. In: 32nd Annual European Symposium on Algorithms (ESA), volume 308 of LIPIcs. Leibniz Int. Proc. Inform., pages Art. No. 5, 15. Schloss Dagstuhl. Leibniz-Zent. Inform., Wadern (2024.) https://doi.org/10.4230/lipics.esa.2024.5
2. Itai, A., Konheim, A.G., Rodeh, M.: A sparse table implementation of priority queues. In: Even, S., Kariv, O. (eds.) ICALP 1981. LNCS, vol. 115, pp. 417–431. Springer, Heidelberg (1981). https://doi.org/10.1007/3-540-10843-2_34
3. Bender, M.A., Conway, A., Farach-Colton, Komlós, M., Kuszmaul, W., Wein, N.: Online list labeling: breaking the $\log^2 n$ barrier. In: 2022 IEEE 63rd Annual Symposium on Foundations of Computer Science (FOCS), pp. 980–990. IEEE Computer Soc., Los Alamitos (2022)
4. Bulánek, J., Koucký, M., Saks, M.: On randomized online labeling with polynomially many labels. In: Fomin, F.V., Freivalds, R., Kwiatkowska, M., Peleg, D. (eds.) ICALP 2013, Part I. LNCS, vol. 7965, pp. 291–302. Springer, Heidelberg (2013). https://doi.org/10.1007/978-3-642-39206-1_25
5. Raghvendra, S.: Optimal analysis of an online algorithm for the bipartite matching problem on a line. In: 34th International Symposium on Computational Geometry, volume 99 of LIPIcs. Leibniz Int. Proc. Inform., pages Art. No. 67, 14. Schloss Dagstuhl. Leibniz-Zent. Inform., Wadern (2018)
6. Gupta, A., Lewi, K.: The online metric matching problem for doubling metrics. In: Czumaj, A., Mehlhorn, K., Pitts, A., Wattenhofer, R. (eds.) ICALP 2012. LNCS, vol. 7391, pp. 424–435. Springer, Heidelberg (2012). https://doi.org/10.1007/978-3-642-31594-7_36

7. Gupta, A., Guruganesh, G., Peng, B., Wajc, D.: Stochastic online metric matching. In: 46th International Colloquium on Automata, Languages, and Programming (ICALP), volume 132 of LIPIcs. Leibniz Int. Proc. Inform., pages Art. No. 67, 14. Schloss Dagstuhl. Leibniz-Zent. Inform., Wadern (2019)
8. Azar, Y., Panigrahi, D., Vardi, O.: Almost tight bounds for the online sorting problem (2025)
9. Hu, Y.: Nearly optimal bounds for stochastic online sorting. arXiv preprint abs/2508.07823 (2025)
10. Balkanski, E., Faenza, Y., Périvier, N.: The power of greedy for online minimum cost matching on the line. In: Proceedings of the 24th ACM Conference on Economics and Computation, pp. 185–205 (2023)
11. Bender, M.A., et al.: Nearly optimal list labeling. In: 2024 IEEE 65th Annual Symposium on Foundations of Computer Science (FOCS), pp. 2253–2274. IEEE Computer Soc., Los Alamitos (2024)
12. Yao, A.C.C.: On random 2 – 3 trees. Acta Informatica **9**(2), 159–170 (1977/78). https://doi.org/10.1007/bf00289075
13. Aamand, A., Abrahamsen, M., Beretta, L., Kleist, L.: Online sorting and translational packing of convex polygons. In: Proceedings of the 2023 Annual ACM-SIAM Symposium on Discrete Algorithms (SODA), pp. 1806–1833. SIAM, Philadelphia (2023). https://doi.org/10.1137/1.9781611977554.ch69

Online Algorithm for Fractional Matchings with Edge Arrivals in Graphs of Maximum Degree Three

Kanstantsin Pashkovich[1(✉)] and Thomas Snow[2]

[1] University of Waterloo, Waterloo, ON N2L 3G1, Canada
kpashkovich@uwaterloo.ca
[2] University of Toronto, Toronto, ON M5S 1A1, Canada
tsnow@cs.toronto.edu

Abstract. We study online algorithms for maximum cardinality matchings with edge arrivals in graphs of low degree. Buchbinder, Segev, and Tkach showed that no online algorithm for maximum cardinality fractional matchings can achieve a competitive ratio larger than $4/(9-\sqrt{5}) \approx 0.5914$ even for graphs of maximum degree three. The negative result of Buchbinder et al. holds even when the graph is bipartite and edges are revealed according to vertex arrivals, i.e. once a vertex arrives, all edges are revealed that include the newly arrived vertex and one of the previously arrived vertices. In this work, we complement the negative result of Buchbinder et al. by providing an online algorithm for maximum cardinality fractional matchings with a competitive ratio at least $4/(9-\sqrt{5}) \approx 0.5914$ for graphs of maximum degree three. We also demonstrate that no online algorithm for maximum cardinality integral matchings can have the competitive guarantee 0.5807, establishing a gap between integral and fractional matchings for graphs of maximum degree three. Note that the work of Buchbinder et al. shows that for graphs of maximum degree two, there is no such gap between fractional and integral matchings, because for both of them the best achievable competitive ratio is 2/3. Also, our results demonstrate that for graphs of maximum degree three best possible competitive ratios for fractional matchings are the same in the vertex arrival and in the edge arrival models.

Keywords: Online algorithms · Matchings · Bounded-degree graphs

1 Introduction

Matchings constitute an extensively studied area of mathematics and theoretical computer science with various practical applications. Indeed, matchings arise in different areas of our everyday lives: job placements for students, assigning riders to drivers on a ride-sharing platform, providing advertisement spots, etc. Some of these matchings have an online nature since the edges or vertices in the underlying graph appear at certain timepoints and are available only in a specific time frame.

J. Matuschke and J. Verschae (Eds.): WAOA 2025, LNCS 16077, pp. 198–212, 2026.
https://doi.org/10.1007/978-3-032-06706-7_14

In this work, we study maximum cardinality matchings in an adversarial edge arrival model. In this model, at every timepoint a new edge arrives. In the integral matching case, upon the arrival of a new edge, we need to immediately and irrevocably decide whether to include this new edge in our current matching. In the fractional matching case, we need to irrevocably select a value for each new edge such that for every vertex, the sum of values on incident edges is always at most one. To make these decisions, we rely on online algorithms. To estimate the performance of an online algorithm, we select as a benchmark the cardinality of a maximum matching in the already "arrived" graph.

In this paper, we focus mainly on the adversarial edge arrival model in graphs of maximum degree three. We determine the best competitive ratio of online algorithms for fractional matchings in these graphs. To do this, we provide an online algorithm that achieves the guarantee $4/(9-\sqrt{5}) \approx 0.5914$ on these graphs, where $4/(9-\sqrt{5}) \approx 0.5914$ equals the corresponding upper bound obtained in [3]. Due to the construction in [3], for graphs of maximum degree three, the best possible competitive ratio remains the same regardless of whether one considers general or bipartite graphs, and whether one considers the vertex arrival or edge arrival models. Additionally, we show that the guarantee $4/(9-\sqrt{5}) \approx 0.5914$ cannot be achieved on graphs of maximum degree four; we also show that the above guarantee cannot be achieved for integral matchings in general graphs of maximum degree three.

In general, we know that the vertex arrival and edge arrival models lead to different competitive ratios for fractional matchings. Indeed, the results of [17] show that for general graphs, online algorithms can achieve a competitive ratio 0.526 in the vertex arrival model. The results of [7] show that no online algorithm can achieve a guarantee larger than 0.5 in the edge arrival model, even for bipartite graphs. Thus, at a certain value of the maximum degree, the best competitive ratio for the vertex arrival order is strictly larger than the competitive ratio for the edge arrival order; our work shows that this degree should be at least four.

1.1 Our Results

In our work, we focus on online algorithms for fractional matchings in the adversarial edge arrival model. [3] established a series of results for online matchings in the edge arrival model when the underlying graph has a bounded maximum degree. In particular, Buchbinder, Segev, and Tkach [3] showed that no online algorithm for maximum cardinality fractional matchings can achieve a competitive ratio larger than $4/(9-\sqrt{5}) \approx 0.5914$ even for graphs of maximum degree three. The negative result in [3] holds even when the graph is a forest and edges are revealed according to vertex arrivals. In this work, we provide an online algorithm for maximum cardinality fractional matchings with a competitive ratio at least $4/(9-\sqrt{5}) \approx 0.5914$ for graphs of maximum degree three, thus showing that $4/(9-\sqrt{5}) \approx 0.5914$ is the best competitive ratio for graphs of degree three. So, for maximum cardinality fractional matchings, our result demonstrates that for graphs of maximum degree three, the competitive ratios are the same for the edge arrival and vertex arrival models. We also show that no online algorithm can achieve a guarantee larger than 0.5807 for integral matchings in graphs of maximum degree three. Thus, unlike for graphs of maximum degree two [3], our results establish a gap between the best achievable competitive ratios for fractional and integral matchings in graphs of maximum degree three.

Next, we show that the guarantee of $4/(9-\sqrt{5}) \approx 0.5914$ is not achievable in the graphs of maximum degree four. To show this, we provide an instance such that no online algorithm for maximum cardinality fractional matching can achieve a guarantee larger than ≈ 0.58884 on it in the edge arrival model.

Another important contribution in [3] is an elegant algorithm, the so called MinIndex Algorithm. Buchbinder et al. [3] show that MinIndex achieves the best possible guarantees both in the case of fractional and integral matchings when the maximum degree is at most two. We show that the guarantee achieved by the MinIndex algorithm is at most $5/9 \approx 0.555$ in forests with maximum degree three. Our result improves on the upper bound $4/7 \approx 0.571$ shown by Buchbinder et al. [3]. Note that Buchbinder et al. [3] showed that the competitive ratio of the MinIndex algorithm equals $5/9$ for both integral and fractional matchings on forests, but their upper bound construction involves graphs with maximum degree four.

1.2 Related Work

The seminal paper [12] studied online matchings in the setting where the graph is bipartite and the vertices in one part appear over time. Each time a vertex appears, all of its incident edges are revealed, and one needs to make an irrevocable decision on which one of these edges to include in the matching, if any. [12] provided a ranking algorithm that achieves the best possible competitive ratio of $(1-1/e)$.

In the general adversarial edge arrival model, Gamlath et al. [7] showed that no online algorithm has a competitive ratio larger than $1/2 + 1/(2d+2)$ when the maximum degree is d, even on bipartite graphs. Thus, [7] showed that no online algorithm can beat the greedy algorithm's competitive ratio $1/2$, even in bipartite graphs. These results hold for both fractional and integral matchings.

In [3], an algorithm with the competitive guarantee $2/3$ was provided for graphs with maximum degree two, which was shown to be optimal. [3] showed that no online algorithm can achieve a competitive ratio larger than $4/(9-\sqrt{5}) \approx 0.5914$ even on forests with maximum degree three in the vertex arrival model. Further upper bounds were obtained in [6,10].

In the edge arrival model for bipartite graphs, where all edges appear in s batches [13] developed an algorithm with a guarantee $1/2 + 1/(2^{s+2}-2)$ for both integral and fractional matchings, where s is the number of batches. For $s = 2$, the competitive ratio becomes $2/3$, and it is also optimal. [9] developed an online algorithm with a competitive ratio larger than the competitive ratio of the greedy algorithm for bipartite graphs and random uniform edge arrival orders. Online stochastic matchings with oblivious adversarial edge arrival order in bipartite graphs were studied in [8]. In [8], an algorithm was developed that achieves a guarantee of 0.503 in the above stochastic model, and they complement this result with an upper bound of 2/3 on any achievable guarantee.

There was an extensive study of the edge arrival models under the assumption of free edge disposal, i.e. an already selected edge can be disposed of at later timepoints. For the weighted version of the problem, a deterministic algorithm with guarantee $1/(3+2\sqrt{2})$ was provided in [14]; moreover, this guarantee was shown to be optimal among deterministic algorithms [1]. Later, [6] provided a randomized algorithm with a guarantee of 0.1867 for this model. There was further progress on upper bounds for possible guarantees of randomized algorithms in this model, see [6,11].

The degree of the underlying graph was also studied in the context of online matching algorithms for rounding fractional matchings [2,4,16].

For a comprehensive overview of results on online matchings, we refer the readers to the surveys [5,15], and to a recent survey [11].

1.3 Our Techniques

Our online algorithm and its analysis demonstrate that the upper bound $4/(9-\sqrt{5}) \approx 0.5914$ from [3] is the best possible for graphs with maximum degree three. Our online algorithm is inspired by the construction from [3]. Indeed, to obtain their upper bound, [3] constructs instances such that every online algorithm with the guarantee $4/(9-\sqrt{5}) \approx 0.5914$ on them should maintain a certain fractional matching. We refer to these instances as "consistent instances". We use the structure of the fractional matchings from [3] on consistent instances as "building blocks" in our algorithm. We partition the edges from the consistent instances into two types of edges "path edges" and "spokes". Our algorithm attempts to greedily construct consistent instances from arriving edges, identifying some of the arrived edges as path edges and some as spokes. Naturally, the algorithm is not able to group all edges into consistent instances, and thus, we identify the remaining edges as "bridges".

For the path edges and spokes, our algorithm attempts to keep their values close to the values as in fractional matchings from [3]. For the bridges, we need to consider several cases to carefully assign the value of the resulting fractional matching.

Our algorithm keeps both a primal solution and a dual solution, i.e. it keeps both the values of a fractional matching and the values of a fractional vertex cover. The values of the fractional cover are used mainly for the analysis, and with the exception of spokes, the assignment of values for the fractional matching does not rely on them.

2 Our Algorithm

In this section, we first provide the intuition behind our algorithm that comes from the upper bound construction in [3]. In the full version, we state our algorithm in full detail and provide all necessary notions.

2.1 Consistent Instances

First, let us introduce a particular hard instance for graphs of maximum degree three which was constructed in [3]. A *consistent* instance with n rounds contains the following edges

1. $e_1 = v_1^l v_1^r$.
2. $e_i^l = v_{i-1}^l v_i^l$ and $e_i^r = v_{i-1}^r v_i^r$ for $i = 2, \ldots, n$.
3. $\hat{e}_i^l = v_i^l \hat{v}_i^l$ and $\hat{e}_i^r = v_i^r \hat{v}_i^r$ for $i = 1, \ldots, n-2$.

Here, first the edge e_1 arrives, then with each further $i = 2, \ldots, n$ the edges e_i^l and e_i^r arrive. After that, the edges $\hat{e}_i^l$ and $\hat{e}_i^r$ arrive for $i = 1, \ldots, n-2$. See Fig. 1 for an example of a consistent instance with $n = 4$.

Buchbinder et al. show that any online algorithm for the fractional matching problem achieves a guarantee at most $c := 4/(9-\sqrt{5}) \approx 0.5914$ on the consistent instances defined above. Since the consistent instances correspond to bipartite graphs, the same upper bound $c = 4/(9-\sqrt{5}) \approx 0.5914$ holds for the guarantee of online algorithms for integral matchings.

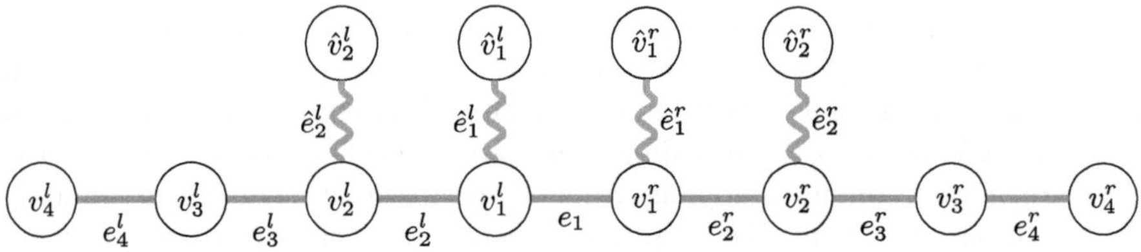

Fig. 1. Example of a consistent instance with $n = 4$ rounds.

2.2 Edge Types

Our algorithm tries to greedily construct consistent instances from the arriving edges. To do that systematically, we define three types of edges, so each edge is assigned one of these types upon its arrival:

- type 1, *path edges*.
- type 2, *spokes*.
- type 3, *bridges*.

Intuitively, the *path edges* are the edges associated with 1 and 2 in the definition of consistent instances, while *spokes* are the edges associated with 3 in the definition of consistent instances; see Fig. 1. Roughly speaking, *bridges* are the edges that run between two different consistent instances that our algorithm constructed so far. We would like to note that bridges and spokes are more tricky objects for our algorithm than the above intuition may suggest.

To keep the exposition concise, we define the function type$(\cdot)$. For each subset of edges $A \subseteq E$, the value type(A) equals (t_1, t_2, t_3), where t_1, t_2, and t_3 equal the number of edges in A of type 1, type 2, and type 3, respectively.

The most challenging case for us is to identify bridges and to assign them appropriate values. For these purposes, our algorithm is looking for special combinations of types when an edge $e = uv$ arrives. In particular our algorithms relies on the following set

$$\{\text{type}\left(\delta(u) \setminus \{e\}\right), \text{type}\left(\delta(v) \setminus \{e\}\right)\}$$

in the current graph G after the edge e arrived. If the above set lies in

$$\mathcal{B} := \{\{(1,0,0)\}, \{(1,0,0),(0,1,0)\}, \{(1,1,0),(1,0,0)\}, \{(0,2,0),(1,0,0)\}\}$$

then an arrived edge e is considered to be a bridge by our algorithm.

For the sake of exposition, we also define the function ends$(\cdot)$. For each edge $e = uv \in E$, we have ends$(uv) := \{u, v\}$.

2.3 Determining Types for the Arriving Edge

Note that our algorithm assumes that the graph has maximum degree at most three at every timepoint. Let us describe how we assign the type to a newly arrived edge $e = uv$. Here, we work with the graph G that refers to the graph after the arrival of the edge e. So, $\delta(u)$ stands for the edges incident to the vertex u in G, and deg(u) stands for the degree of u in G, etc. For each f in $E \setminus \{e\}$, the value y_f represents the value assigned to the edge f in the fractional matching constructed before the arrival of e.

Table 1 illustrates type(e) for the arriving edge $e = uv$ assigned by our algorithm. As follows from the table, the type of $e = uv$ depends on type$\left(\delta(u) \setminus \{e\}\right)$ and type$(\delta(v) \setminus \{e\})$.

There are several special cases. In Table 1 these cases are represented by enclosing the type of $e = uv$ in a box. In these special cases, both u and v have degree 3 in the graph G, so without loss of generality, in these cases, we assume

$$1 - \sum_{f \in \delta(u) \setminus \{e\}} y_f \leq 1 - \sum_{f \in \delta(v) \setminus \{e\}} y_f .$$

The empty cells in Table 1 correspond to impossible combinations of type$\left(\delta(u) \setminus \{e\}\right)$ and type$(\delta(v) \setminus \{e\})$.

Table 1. Type assignment for the newly arrived edge $e = uv$.

type($\delta(u) \setminus \{e\}$) \ type($\delta(v) \setminus \{e\}$)	(0,0,0)	(1,0,0)	(0,1,0)	(0,0,1)	(1,1,0)	(1,0,1)	(0,1,1)	(2,0,0)	(0,2,0)	(0,0,2)
(0,0,0)	1	1	1		1	2	2	2	1	
(1,0,0)	1	3	3		3	2	2	2	3	
(0,1,0)	1	3	1		1	2	2	2	1	
(0,0,1)										
(1,1,0)	1	3	1		1	$\boxed{1}$	$\boxed{1}$	$\boxed{1}$	1	
(1,0,1)	2	2	2		$\boxed{2}$	2	2	2	$\boxed{2}$	
(0,1,1)	2	2	2		$\boxed{2}$	2	2	2	$\boxed{2}$	
(2,0,0)	2	2	2		$\boxed{2}$	2	2	2	$\boxed{2}$	
(0,2,0)	1	3	1		1	$\boxed{1}$	$\boxed{1}$	$\boxed{1}$	1	
(0,0,2)										

2.4 Fractional Matching in Consistent Instances

Now that we have provided an idea of how our algorithm assigns a type to the arriving edge, let us provide a general idea of how we intend to construct a fractional matching.

The upper bound proof by Buchbinder et al. showed that for an online algorithm to achieve a competitive ratio on consistent instances defined in Sect. 2.1, the algorithm has to output a very specific fractional matching on these instances.

Let us define the values that should appear in the resulting fractional matching as per [3]. For this, let us define the following values

$$\widetilde{y}_1 := c,\ \widetilde{y}_2 := \frac{c}{2} \quad \text{and} \quad \widetilde{y}_3 := \frac{5c-2}{2}$$

and for natural n, $n \geq 4$ let us define

$$\widetilde{y}_n := \frac{(3F_n + F_{n-2} - 2)c - 2F_n + 2}{2},$$

where $\phi := \frac{1+\sqrt{5}}{2}$, $\psi := 1 - \phi$ and $F_n := \frac{\phi^n - \psi^n}{\sqrt{5}}$. So ϕ is the golden ratio, and F_n is the n-th Fibonacci number. We note the following useful property, the proof of which can be found in the full version of the paper, for all natural n we have $1 - \widetilde{y}_n - \widetilde{y}_{n+1} = c - \widetilde{y}_{n+2}$. In particular, we use $1 - \widetilde{y}_n - \widetilde{y}_{n+1}$ and $c - \widetilde{y}_{n+2}$ interchangeably.

The proof for the upper bound in [3] showed that for the algorithm to achieve a competitive ratio c on consistent instances, the algorithm needs to assign the following edge values (subject to symmetry breaking), see Sect. 2.1:

1. $e_1 = v_1^l v_1^r$ has to be assigned $\widetilde{y}_1$.
2. $e_i^l = v_{i-1}^l v_i^l$ and $e_i^r = v_{i-1}^r v_i^r$ for $i = 2, \ldots, n$ have to be assigned $\widetilde{y}_n$.
3. $\hat{e}_i^l = v_i^l \hat{v}_i^l$ and $\hat{e}_i^r = v_i^r \hat{v}_i^r$ for $i = 1, \ldots, n-2$ have to be assigned $1 - \widetilde{y}_n - \widetilde{y}_{n+1}$.

Our algorithm tries to follow these value assignments on path edges and spokes, but the presence of bridges requires us to select more nuanced assignments even on path edges and spokes. In particular, we make more careful value assignments for edges that are bridges or are incident to a bridge upon their arrival.

2.5 Algorithm

We defer the formal definition of our algorithm to the full version. The algorithm heavily relies on the primal-dual methodology.

Without loss of generality, we assume that for the arriving edge $e = uv$ that $\deg(u) \geq \deg(v)$. Our algorithm is a primal-dual algorithm. The algorithm produces a fractional matching by assigning each arrived edge $e \in E$ a non-negative value y_e such that at every timepoint and for every vertex w we have $\sum_{f\in\delta(w)} y_f$ is at most 1. Moreover, the algorithm produces a fractional vertex cover by keeping and updating nonnegative values x_w for each vertex $w \in V$. At every timepoint and for every arrived edge $e = uv$, we have that $x_u + x_v$ is at least c. Thus, at every timepoint, the values x_w, $w \in V$ scaled by $1/c$ produce a fractional vertex cover for the graph G.

Value Assignments in Consistent Instances. Thus, to prove that c is indeed the guarantee of our online algorithm, it suffices to prove that at every timepoint the sum of all y_e, $e \in E$ equals the sum of all x_w, $w \in V$. Let us provide an intuition about how the algorithm preserves this equality by updating x_w, $w \in V$. Figure 2 demonstrates the increases in the values x_w, $w \in \{v_i^l, v_{i-1}^l, v_{i+1}^l, \hat{v}_i^l\}$ after the arrivals of their incident edges in the consistent instances from Sect. 2.1. For example, after the edge $v_i^l v_{i+1}^l$ arrives, the edge $v_i^l v_{i+1}^l$ gets value $\widetilde{y}_i$ and the values x_w, $w \in \{v_i^l, v_{i+1}^l\}$ are increased by $\widetilde{y}_i - (c - \widetilde{y}_{i+1})$ and $c - \widetilde{y}_{i+1}$, respectively. We would like to emphasize that Fig. 2 depicts an ideal situation for constructing the fractional matchings and updating the fractional vertex cover. Our algorithm attempts to mimic this ideal behavior upon the arrival of path edges and spokes.

Position Indicators and Endpoints' Distinction in Consistent Instances. Even in the ideal situation depicted in Fig. 2, to assign values to path edges, our algorithm relies on determining the exact position of such edges in the consistent instance. To make sure that our algorithm has access to these positions, we keep a position indicator n_f for each path edge f in the graph. For example, in a consistent instance when $f = v_i^l v_{i+1}^l$ we have $n_f = i+1$; and when $f = v_1^l v_i^r$ we have $n_f = 1$.

Already in the consistent instances, the endpoints of the arriving edges could have different properties. For example, if an arriving edge is identified as a path edge in a consistent instance, then one of its endpoints has degree one and the

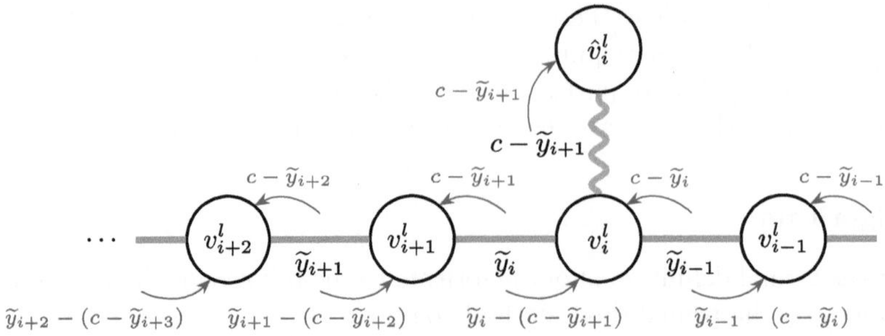

Fig. 2. The construction of the fractional matching and increases in the fractional vertex cover, when our algorithm has a consistent instance as an input.

other endpoint has degree two (immediately after this arrival). To keep track of these different properties, for some edges e, the algorithm identifies one of the endpoints as $z(e)$ and another endpoint as $w(e)$. In particular, this is crucial for the analysis of path edges and spokes, as the value assigned to such edges depends solely on $z(e)$. For example, in Fig. 2 for the path edge $f = v_i^l v_{i+1}^l$ we have $z(f) = v_i^l$ and $w(f) = v_{i+1}^l$, similarly for the path edge $f = v_{i+1}^l v_{i+2}^l$ we have $z(f) = v_{i+1}^l$ and $w(f) = v_{i+2}^l$. For the spoke $f = v_i^l \hat{v}_i^l$ in Fig. 2, we have $z(f) = v_i^l$ and $w(f) = \hat{v}_i^l$.

Partition Into Consistent Instances and Bridges. With the cover construction depicted in Fig. 2 in mind, the algorithm utilizes this construction by loosely partitioning the arriving edges into subgraphs of the consistent instance, with some exceptions; namely, bridges. The purpose of bridges is to connect, not necessarily distinct, partitions. To see how the fractional cover changes, consider, for example, the updates in the fractional cover upon the arrival of a path edge in the algorithm. These updates are identical to the ones depicted in Fig. 2 except that in certain cases we cannot use the value $\widetilde{y}_{n_e}$ from the ideal case depicted in Fig. 2 but we have to use the actually assigned value y_e. In a similar way, we can see the updates in the fractional cover upon the arrival of a spoke. We refer the reader to the full paper for specific edge assignments and cover updates for paths and spokes when the ideal values are not feasible.

Recall from Sect. 2.2 that an edge $e = uv$ is assigned to be a bridge if and only if $\{\delta(u) \setminus \{e\}, \delta(v) \setminus \{e\}\}$ is in the set $\mathcal{B}$. Consider Fig. 3, where we assume that all edges but b_1, b_2, b_3, b_4, b_5, s_1, s_2 arrive first in some specific order, and then the edges b_1, b_2, b_3, b_4, b_5 arrive, and then s_1, s_2 arrive. Now, before the arrival of b_5, both of the endpoints of b_5 are incident only to one path edge each. Thus since $\{(1,0,0)\}$ is in $\mathcal{B}$, the edge b_5 is assigned to be a bridge.

Even though it is helpful to think about our algorithm as an algorithm partitioning the arrived path edges and spokes into consistent instances, this intuition does not always extend to spokes. In particular, some of the edges are assigned to be a spoke even though they do not "naturally fit" in any consistent instance. For example, before s_1 arrives, one of its endpoints is incident to a bridge b_1

and a path edge, while the other endpoint is incident to no edges. Due to the presence of the bridge b_1, the edge s_1 is not assigned to be a path edge but to be a spoke by our algorithm. Generally, an edge becomes such a spoke when, after its arrival, precisely one of its endpoints has degree 3 and is incident to a bridge.

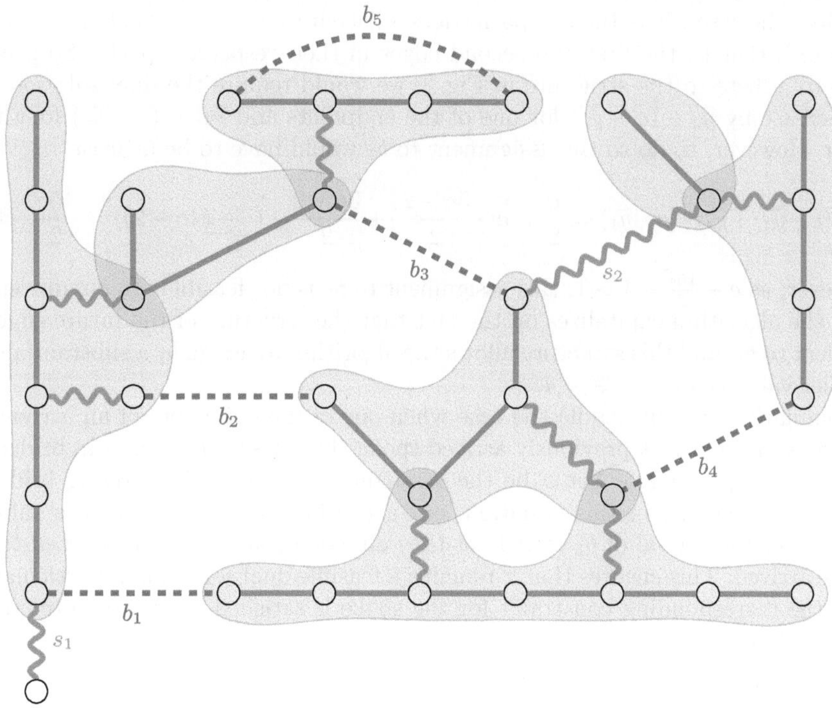

Fig. 3. An example of a potential partition indirectly maintained by our algorithm (subject to arrival order) into subgraphs of the consistent instances. Here, the straight green edges represent path edges, the wavy red edges represent spokes, and the dashed blue edges represent bridges. (Color figure online)

Value Assignments for Bridges. Bridges are divided into four classes determined by the incident edges to their endpoints, see the definition based on $\mathcal{B}$. Intuitively, an edge e is assigned a bridge when both endpoints are already in consistent instances in the current partition, and it is not clear to which consistent instance the edge e should be added. In this case, the edge e is assigned a bridge and e attempts to fulfill the role it would be given, as if it were assigned to each partition individually. For instance, in Fig. 3 as the bridge b_1 arrives it is not immediately clear which consistent instance b_1 should join. In this case, b_1 prevents the paths' "growth" in these two consistent instances beyond the endpoints of b_1. After the arrival of b_1, the algorithm needs to account for the possibility of future edges incident to b_1, and to do that the algorithm needs to update x and y appropriately. The main obstacle for finding an appropriate x and y update is the possibility of future spokes incident to b_1. Since, as explained above, b_1 prevented the paths' "growth" in two consistent instances, we might want the x update to happen as in the case where b_1 is just a new path edge in both consistent instances. However, this is not always possible. For instance, consider the case where the two path edges adjacent to b_1 are assigned $\widetilde{y}_1 = c$ and $\widetilde{y}_2 = c/2$, that is, the first and second edges in their respective paths. So if we were to adhere to the structure in Fig. 2, we would require the dual solution x to increase by $\widetilde{y}_2 - (c - \widetilde{y}_3)$ for one of the endpoints and $\widetilde{y}_3 - (c - \widetilde{y}_4)$ for the other. However, to do so the assignment to b_1 would have to be at least

$$\widetilde{y}_2 - (c - \widetilde{y}_3) + \widetilde{y}_3 - (c - \widetilde{y}_4) = \frac{c}{2} - (c - \frac{5c-2}{2}) + \frac{5c-2}{2} - (c - (4c-2)) = \frac{15c}{2} - 4\,.$$

However, as $c + \frac{15c}{2} - 4 > 1$, this assignment to b_1 is not feasible. To circumvent this, the algorithm capitalizes on the fact that the structure of the future edges incident to b_1, and this structure allows the algorithm to assign b_1 a substantially smaller value than $c + \frac{15c}{2} - 4$.

Similarly, one can handle the case when one of the endpoints of an arriving bridge is incident to a previously arrived spoke. This case can be seen in bridges b_2, b_3, and b_4 in Fig. 3. Let u be the common endpoint of the arriving bridge and existing spoke. In this situation, our algorithm guarantees that the value of x_u after the arrival of b_1 is at least the value of x_u at the moment when the spoke arrived. This ensures that x remains a feasible dual solution; in particular, that the corresponding constraint for the spoke is satisfied by x even after the arrival of b_1.

3 Upper Bound for Integral Matchings for Maximum Degree Three

In the previous sections, we showed that our algorithm achieves the guarantee c for fractional matchings. In this section, we show that the guarantee c cannot be achieved for integral matchings. In particular, in this section, we prove the following theorem.

Theorem 1. *For integral matchings in the adversarial edge arrival model, no algorithm achieves a guarantee larger than* 0.58065 *on graphs of maximum degree three.*

Let $\mathcal{ALG}$ be a randomized algorithm for integral matchings. Let γ be the competitive ratio achievable by $\mathcal{ALG}$. Consider the following graph in Fig. 4. Here, first edges to arrive are e_1^u and e_1^d followed by e_2^u and e_2^d.

The first option for the further edge arrivals is depicted in Fig. 5, where after the edges e_1^u, e_1^d, e_2^u, and e_2^d, all other edges in Fig. 5 arrive. In Fig. 5, in the boxes, there are upper bounds on the edges' probability to be included in the

Fig. 4. First edges in the considered instances.

matching by $\mathcal{ALG}$. The second option for the further edge arrivals is depicted in Fig. 6, where after the edges e_1^u, e_1^d, e_2^u and e_2^d the remaining edges depicted by solid straight lines arrive; then all curvy edges in Fig. 6 arrive; then all dashed edges arrive.

The arrival orders above allow is to set up a Linear Program to find an upper bound on γ. Solving the above linear program, we get $\gamma \leq 0.58065$ as required. We defer the details to the full version of the paper.

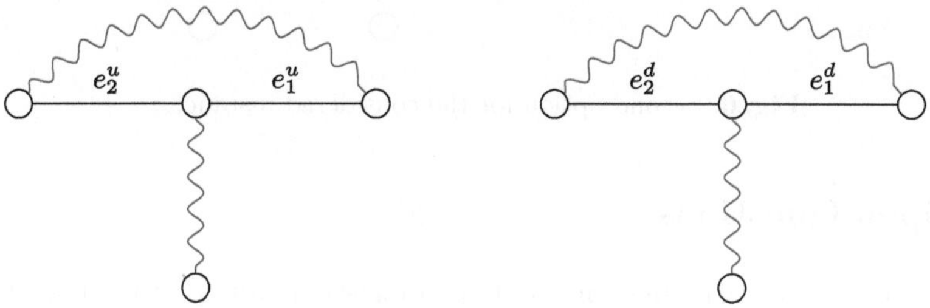

Fig. 5. First option for the considered instances.

4 Upper Bound for Fractional Matchings for Maximum Degree Four

In this section, we show that for fractional matchings, the best possible guarantee deteriorates by going from graphs with maximum degree three to maximum degree four. In particular, we show the following theorem and note that we have $0.58884 < c$. We defer the proof of Theorem 2 to the full version of the paper.

Theorem 2. *For fractional matchings in the adversarial edge arrival model, no algorithm achieves a guarantee larger than* 0.58884 *on bipartite graphs of maximum degree four.*

We note that the intent of Theorem 2 is not to optimize the bound on bipartite graphs of maximum degree four but to provide a gap on the guarantees achievable for graphs of maximum degree three and four.

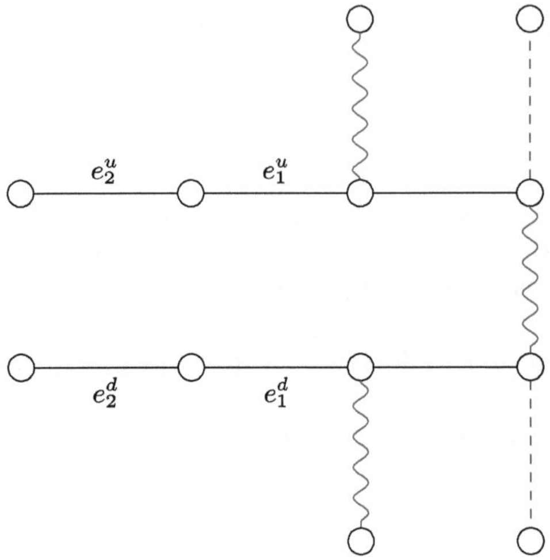

Fig. 6. Second option for the considered instances.

5 Open Questions

Let us point to further directions and open questions related to our work. The work of [3] shows that the best possible guarantee of an online algorithm equals 2/3 for graphs of maximum degree two, both for integral and fractional matchings in both vertex arrival and edge arrival models. The upper bound from [3] and our work show that the possible guarantee of an online algorithm for fractional matchings equals $c = 4/(9 - \sqrt{5}) \approx 0.5914$ for graphs of maximum degree three in both vertex arrival and edge arrival models. Our work leads to the following open question: What is the smallest value d such that online algorithms for fractional matchings achieve different best possible guarantees in vertex and edge arrival models for graphs of maximum degree d?

Note that our algorithm achieves the guarantee $c = 4/(9 - \sqrt{5}) \approx 0.5914$ for fractional matchings in both bipartite and non-bipartite graphs. Also, the work of [3] shows that c is an upper bound on the guarantee of any online algorithm for bipartite graphs. In general, for fractional matchings, the construction in [7] shows that the best possible guarantee is 1/2 for both bipartite and non-bipartite graphs in the edge arrival model. Is there d such that online algorithms for fractional matchings achieve different best possible guarantees in the edge arrival model for bipartite and non-bipartite graphs of maximum degree d? If the answer is positive, then what is the smallest such d? Apart from the maximum degree, what other parameters of the underlying graphs have a crucial role in the difference of guarantees in bipartite and non-bipartite graphs? Is it possible to obtain the guarantee c for integral matchings in the case of bipartite graphs with maximum degree three?

Acknowledgments. The research of Thomas Snow was supported by NSERC Undergraduate Student Research Award (USRA) and Math Undergraduate Research Award (MURA). The research of Kanstantsin Pashkovich was supported by NSERC Discovery Grants Program RGPIN-2020-04346. The authors are grateful to Ricardo Fukasawa for many helpful discussions. Kanstantsin Pashkovich is grateful to Laurent Poirrier for the help with encoding linear programs to obtain upper bounds.

Disclosure of Interests. The authors have no competing interests to declare that are relevant to the content of this article.

References

1. Badanidiyuru Varadaraja, A.: Buyback problem - approximate matroid intersection with cancellation costs. In: Aceto, L., Henzinger, M., Sgall, J. (eds.) ICALP 2011. LNCS, vol. 6755, pp. 379–390. Springer, Heidelberg (2011). https://doi.org/10.1007/978-3-642-22006-7_32
2. Blikstad, J., Svensson, O., Vintan, R., Wajc, D.: Online edge coloring is (nearly) as easy as offline. In: Proceedings of the 56th Annual ACM Symposium on Theory of Computing, STOC 2024, pp. 36–46. Association for Computing Machinery, New York (2024). https://doi.org/10.1145/3618260.3649741
3. Buchbinder, N., Segev, D., Tkach, Y.: Online algorithms for maximum cardinality matching with edge arrivals. Algorithmica **81**(5), 1781–1799 (2018). https://doi.org/10.1007/s00453-018-0505-7
4. Cohen, I.R., Wajc, D.: Randomized online matching in regular graphs. In: Proceedings of the Twenty-Ninth Annual ACM-SIAM Symposium on Discrete Algorithms, SODA 2018, pp. 960–979. Society for Industrial and Applied Mathematics, USA (2018). https://doi.org/10.1137/1.9781611975031.62. https://epubs.siam.org/doi/abs/10.1137/1.9781611975031.62
5. Devanur, N.R., Mehta, A.: Online matching in advertisement auctions. Online Matching-Based Market Des. 130 (2023). https://doi.org/10.1017/9781108937535
6. Epstein, L., Levin, A., Segev, D., Weimann, O.: Improved bounds for randomized preemptive online matching. Inf. Comput. **259**, 31–40 (2018). https://doi.org/10.1016/j.ic.2017.12.002. https://www.sciencedirect.com/science/article/pii/S0890540117302298
7. Gamlath, B., Kapralov, M., Maggiori, A., Svensson, O., Wajc, D.: Online matching with general arrivals. In: 2019 IEEE 60th Annual Symposium on Foundations of Computer Science (FOCS), pp. 26–37 (2019). https://doi.org/10.1109/FOCS.2019.00011
8. Gravin, N., Tang, Z.G., Wang, K.: Online Stochastic matching with edge arrivals. In: Bansal, N., Merelli, E., Worrell, J. (eds.) 48th International Colloquium on Automata, Languages, and Programming (ICALP 2021). Leibniz International Proceedings in Informatics (LIPIcs), vol. 198, pp. 74:1–74:20. Schloss Dagstuhl – Leibniz-Zentrum für Informatik, Dagstuhl, Germany (2021). https://doi.org/10.4230/LIPIcs.ICALP.2021.74. https://drops.dagstuhl.de/entities/document/10.4230/LIPIcs.ICALP.2021.74
9. Guruganesh, G.P., Singla, S.: Online matroid intersection: beating half for random arrival. In: Eisenbrand, F., Koenemann, J. (eds.) IPCO 2017. LNCS, vol. 10328, pp. 241–253. Springer, Cham (2017). https://doi.org/10.1007/978-3-319-59250-3_20

10. Huang, Z., Peng, B., Tang, Z.G., Tao, R., Wu, X., Zhang, Y.: Tight competitive ratios of classic matching algorithms in the fully online model. In: Proceedings of the 2019 Annual ACM-SIAM Symposium on Discrete Algorithms (SODA), pp. 2875–2886 (2019). https://doi.org/10.1137/1.9781611975482.178. https://epubs.siam.org/doi/abs/10.1137/1.9781611975482.178
11. Huang, Z., Tang, Z.G., Wajc, D.: Online matching: a brief survey. SIGecom Exch. **22**(1), 135–158 (2024). https://doi.org/10.1145/3699824.3699837
12. Karp, R.M., Vazirani, U.V., Vazirani, V.V.: An optimal algorithm for on-line bipartite matching. In: Proceedings of the Twenty-Second Annual ACM Symposium on Theory of Computing, STOC 1990, pp. 352–358. Association for Computing Machinery, New York (1990). https://doi.org/10.1145/100216.100262
13. Lee, E., Singla, S.: Maximum matching in the online batch-arrival model. ACM Trans. Algorithms **16**(4) (2020). https://doi.org/10.1145/3399676
14. McGregor, A.: Finding graph matchings in data streams. In: Chekuri, C., Jansen, K., Rolim, J.D.P., Trevisan, L. (eds.) APPROX/RANDOM -2005. LNCS, vol. 3624, pp. 170–181. Springer, Heidelberg (2005). https://doi.org/10.1007/11538462_15
15. Mehta, A.: Online matching and ad allocation. Found. Trends Theor. Comput. Sci. **8**(4), 265–368 (2013). https://doi.org/10.1561/0400000057
16. Wajc, D.: Matching Theory Under Uncertainty. Ph.d. thesis, Carnegie Mellon University, Pittsburgh, PA, USA (2021). https://doi.org/10.1184/R1/14428181. https://kilthub.cmu.edu/articles/thesis/Matching_Theory_Under_Uncertainty/14428181
17. Wang, Y., Wong, S.C.: Two-sided online bipartite matching and vertex cover: beating the greedy algorithm. In: Halldórsson, M.M., Iwama, K., Kobayashi, N., Speckmann, B. (eds.) ICALP 2015. LNCS, vol. 9134, pp. 1070–1081. Springer, Heidelberg (2015). https://doi.org/10.1007/978-3-662-47672-7_87

Approximation Algorithms for Graph Search Problems with Imperfect Detection

Martijn van Ee[1(✉)] and René Sitters[2]

[1] Netherlands Defence Academy, Den Helder, The Netherlands
m.v.ee.01@mindef.nl
[2] Vrije Universiteit Amsterdam, Amsterdam, The Netherlands
r.a.sitters@vu.nl

Abstract. We study graph search problems with imperfect detection. Both graph search and search with imperfect detection are well-studied subjects, but the natural combination is relatively new. In this setting, one is given an edge-weighted graph and a target is hidden at one of the vertices. One can walk through the graph to search for the hidden target and the goal is to find it as soon as possible. However, just visiting a vertex is not sufficient and a search needs to be done which takes a fixed amount of time. Each search attempt is only successful though with probability γ, given that the target is at the search location. For the hider, we consider the model where the target is hidden at random, and the adversarial model where the target is placed by an adversary. For the searcher, we consider both pathwise search and expanding search. For all cases, we obtain the first constant factor approximation guarantees.

Keywords: Approximation algorithms · Graph search · Imperfect detection

1 Introduction

Imperfect detection is a common feature in search theory [4,14–16,21], and is generally motivated by inaccuracies in sensor measurements or the inability to perfectly sweep an area. In search problems, one is usually given a set of cells, and the goal is to design a search strategy that minimizes the expected time until the target has been found. Alternatively, a common objective is the probability of finding the target, given a certain deadline. However, travel times are usually ignored in this field. On the other hand, searching with travel times is well studied in graph search problems, such as the traveling repairman problem [6,10] and the expanding search problem [11,12]. In these problems, detection is assumed to be perfect. In this article, we combine both fields, with a focus on approximation algorithms. In particular, we extend the approximability results for graph search problems with perfect detection to the case with imperfect detection.

We consider the case where the target is hidden at random, and the case where the target is hidden adversarially. If the target is hidden at random, our goal is to find an (infinite) search path that minimizes the expected time until

J. Matuschke and J. Verschae (Eds.): WAOA 2025, LNCS 16077, pp. 213–227, 2026.
https://doi.org/10.1007/978-3-032-06706-7_15

the target is found. If the target is hidden adversarially, meaning that it can observe our search strategy before choosing a vertex to hide, our goal is to find an (infinite) search path that minimizes the maximum expected time until detection over the vertices.

We consider two ways to search the graph: *pathwise search* (as is standard in routing problems) and *expanding search* [2], where we only incur the cost of an edge if we use it for the first time and any following traversal has no cost. Although, the solution here is also a path, we shall refer to this model as expanding imperfect graph search. Natural applications of searching with imperfect detection include search and rescue missions, where a local search (such as an excavation) as well as moving equipment takes substantial time. Other applications can be found in testing the state of a system by investigating its components. The reliability of a test depends on the thoroughness and precision and hence on the time invested. Moreover, switching between tests takes a setup time. This is closely related to time-critical testing problems [1].

1.1 Problem Definitions

In imperfect graph search problems, we are given a graph $G = (V, E)$ with metric distances $d(i, j)$ for $v_i, v_j \in V$, and a root $r \in V$ where the search starts. We denote $|V| = n$. We assume the searcher moves at unit speed, and that searching a vertex takes 1 time unit. The target is hidden in vertex v_i with probability p_i. The probability that the target is detected while it is at the vertex that is being searched is $\gamma \in (0, 1]$. We consider the imperfect graph search problem with either pathwise search (IGS-P) or expanding search (IGS-E).

Let C_{ij} be the arrival time at the j-th visit to vertex v_i (including the search time in v_i). In both IGS-P and IGS-E, we aim to minimize the expected time until target detection, i.e.,

$$\sum_{v_i \in V} \sum_{j=1}^{\infty} (1-\gamma)^{j-1} \gamma p_i C_{ij}.$$

We also discuss the problem of maximizing the probability of finding the target, given a deadline T. Since the case with perfect detection is known as the orienteering problem [9], we will refer to this problem as IGS-O. Here, a solution is given by, for each vertex v_i the number of times k_i it is searched, together with a path visiting all v_i with $k_i \geq 1$ such that the search is done within deadline T. The objective for IGS-O is to maximize the probability of detection, i.e.,

$$\sum_{v_i \in V} \left(1 - (1-\gamma)^{k_i}\right) p_i.$$

In imperfect graph search problems with adversarial targets, we have the same setting with the only difference that the target is not hidden at random, but instead, we assume the hider can observe our search strategy before choosing the vertex where it will hide. Hence, it chooses the vertex with the maximum expected time until detection. Therefore, we need to design a search strategy that minimizes the maximum expected time until detection over the vertices, where the expectation is taken only w.r.t. the uncertainty in detecting the target, i.e.,

$$\max_{v_i \in V} \sum_{j=1}^{\infty} (1-\gamma)^{j-1} \gamma C_{ij}.$$

We consider imperfect graph search problems with adversarial targets in both the pathwise (AIGS-P) and the expanding search paradigm (AIGS-E).

1.2 Related Work

If detection is perfect and vertex probabilities are uniform, IGS-P is known as the traveling repairman problem (TRP) or the minimum latency problem [6]. This problem is NP-hard [18], even on edge-weighted trees [19]. An initial idea of Blum et al. [6], and later improvements by Goemans and Kleinberg [10], and Chaudhuri et al. [8] have led to a 3.59-approximation for the general case. For trees, planar graphs and the Euclidean plane, a PTAS exists [20]. For non-uniform probabilities, IGS-P with perfect detection is equivalent to the *weighted* version of TRP. Here, each vertex has a non-negative weight, and the goal is to construct a path that minimizes the weighted sum of arrival times at the vertices. It was shown by Sitters [20] that weighted-TRP can be reduced to unweighted-TRP with loss of a factor $(1+\epsilon)$ [20].

For expanding search with perfect detection, the problem of minimizing the expected search time is known as the expanding search problem. This problem is also NP-hard [3]. An 8-approximation was by presented by Hermans et al. [12]. This was improved by Griesbach et al. [11] to a $(2e+\varepsilon)$-approximation, using the approach of [10].

The adversarial versions with perfect detection are well-known problems. In pathwise search, the target will hide at the last vertex that the searcher visits. Hence, we want to minimize the length of the path that visits all vertices, which is known as the path-version of the traveling salesperson problem (path TSP). For this, a $\frac{3}{2}$-approximation algorithm is known [24]. Using a reduction to TSP, one can improve this guarantee to $\frac{3}{2} - \varepsilon_0$ for some small constant ε_0 [13,22]. In expanding search, the target will also hide at the vertex that is visited last. Hence, the optimal solution is the minimum spanning tree.

The problem we consider in this article, IGS-P, was considered by Lössner and Wegener [17]. They considered the problem with non-uniform detection probabilities γ_i, for $v_i \in V$, and showed that if these are similar[1] (which includes the uniform case), the optimal solution is ultimately periodic. This means that there is an initial start-up phase, after which the solutions repeats a certain sequence of visits. There is also some work on searching in graphs with imperfect detection, where the objective is to maximize the probability of finding the target,

[1] For all $v_i, v_j \in V$, $\log \gamma_i / \log \gamma_j \in \mathbb{Q}$.

given a deadline (IGS-O) [5,7]. These articles do not consider approximation algorithms. Yu and Lin [23] consider several strategies, and evaluate their performance empirically using both objectives.

1.3 Our Contributions

In this article, we give the first approximation algorithms for imperfect graph search problems, with either random or adversarially hidden target, in either the pathwise search or the expanding search paradigm. Our first result is that IGS-P reduces, with a loss of a factor $1+\varepsilon$, to weighted-TRP. Hence, given the $(3.59+\varepsilon)$-approximation for weighted-TRP, we obtain an $(3.59+\varepsilon)$-approximation for IGS with pathwise search. We also present the consequences of this reduction for some special cases.

It turns out that roughly the same reduction also works for IGS-E. Hence, we obtain an $(2e+\varepsilon)$-approximation. As side result, we note that a similar reduction also gives the first constant factor approximation for IGS-O.

For imperfect graph search problems with adversarial targets, we present simple algorithms with constant factor approximation guarantees. For pathwise search (AIGS-P), the algorithm's solution walks back and forth along the shortest path that visits all vertices, and either searches for one unit or for $\left\lceil \frac{1}{\gamma} \right\rceil$ time units when visiting a vertex. This gives a 3.4171-approximation. For expanding search (AIGS-E), we construct the minimum spanning tree in an expanding manner, and search each vertex for one time unit. Afterwards, the solution searches the vertices in reversed order, and so on. This gives a 2.0876-approximation. Here, we note that the complexity of AIGS-E is open. It is even unclear whether the decision version of the problem is contained in NP. In Table 1 is a summary of our results, together with known approximability results for perfect detection.

Table 1. Approximability results for perfect and imperfect detection.

	Random target		Adversarial target	
	Perfect	Imperfect	Perfect	Imperfect
Pathwise search	$3.59+\varepsilon$ [8]	$3.59+\varepsilon$ (Corollary 1)	$1.5-\varepsilon_0$ [13,22]	3.4171 (Theorem 7)
Expanding search	$2e+\varepsilon$ [11]	$2e+\varepsilon$ (Corollary 2)	1	2.0876 (Theorem 8)
Orienteering	$0.5-\varepsilon$ [9]	$0.5-\epsilon$ (Corollary 3)		

2 Random Targets

The uncertainty in detection of the target adds a major difficulty to the optimization problem, which is otherwise one of the well-studied problems: the traveling repairman problem (TRP), the expanding search problem (ESP), or the orienteering problem (OP). An optimal solution might need to search some vertices multiple times before performing the first search at some other vertices. In fact, any algorithm that always searches each vertex exactly once before performing any second search at a vertex has no constant factor approximation ratio for either version. The following example makes this explicit for pathwise search and expanding search.

Given are vertices $r = v_1, v_2, \ldots, v_n$ at pairwise distance zero, except for v_n which is at a large distance $L = n^2$ from all other vertices. Let $\gamma = 0.5$ and $p_i = 1/n$ for all i. Consider the following solution: The vertices $v_1, v_2, \ldots, v_{n-1}$ are searched sequentially for $\log_2 n$ rounds before moving to v_n which is then searched $\log_2 n$ times. This path is then repeated until the target is found. The expected time until detection is $O(n + L/n) = O(n)$. Now consider a solution that searches vertices $v_1, \ldots, v_{n-1}$ at most q times before going to v_n. The probability that the target is not found before moving to v_n is at least $(1/2)^q$. Hence, the expected time until detection is at least $(1/2)^q L = \Omega((1/2)^q n^2)$.

Theorem 1. *For pathwise search and expanding search, any algorithm that searches each vertex at most q times before each vertex is searched at least once has approximation ratio* $\Omega(n(1/2)^q)$.

The above theorem implies that for an algorithm to be an $O(1)$-approximation, $\Omega(\log n)$ searches to some vertex may be needed before all vertices are searched at least once. On the other hand, as our algorithm shows, $O(\log n)$ searches in a first tour over all vertices are enough for an $O(1)$-approximation.

2.1 Pathwise Search

We show here that any instance of IGS with pathwise search (IGS-P) can be polynomially reduced to an instance of weighted-TRP with loss of an arbitrarily small factor $1+\epsilon$ in the approximation. As stated in the introduction, we assume a constant search time of 1, an arbitrary but uniform detection probability γ, and arbitrary vertex probabilities p_i.

Given an instance I of IGS-P, we copy the graph $G = (V, E)$ and distances to the weighted-TRP-instance. Then, to each vertex v_i we add a number of leaves $v_i^{(j)}$ for $j = 1, 2, \ldots, K$, all at distance 0.5 from v_i. The weight of leaf $v_i^{(j)}$ is defined as $w_{ij} = (1-\gamma)^{j-1}\gamma p_i$. Further, we add a new origin r' at distance 0.5 from the original origin. We refer to this graph as the *extended graph* and denote this TRP-instance by $I_{\text{TRP}}^{(K)}$. Let $\text{OPT}_{\text{TRP}}^{(K)}$ be its optimal value (See Fig. 1).

Lemma 1. *Let* OPT *be the optimal value of IGS-P-instance I. Then* $\text{OPT}_{TRP}^{(K)} \leq \text{OPT}$.

Proof. Let P^* be any (infinite) search path. Given P^* we shortcut it by only keeping the first K visits to each vertex. Let this be $\bar{P}$. Let C_{ij}^* ($\bar{C}_{ij}$) be the arrival time (including the search time) at the j-th visit to vertex i in P^* ($\bar{P}$). Then, for $j \leq K$ we have

$$\bar{C}_{ij} \leq C_{ij}^*.$$

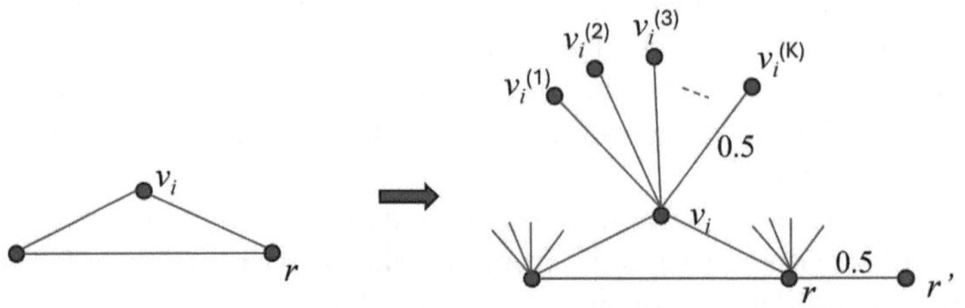

Fig. 1. Reduction of IGS-P to weighted-TRP. Vertex $v_i^{(j)}$ has weight $w_{ij} = (1-\gamma)^{j-1}\gamma p_i$ and $K = O\left(\log(n/(\epsilon p_{\min}))/\log(1/(1-\gamma))\right)$.

Now make the same path $\bar{P}$ in the TRP-graph, where the j-th visit to a vertex v_i corresponds to a visit to leaf $v_i^{(j)}$. Let P be this path and let C_{ij} be the arrival time at vertex $v_i^{(j)}$. Then

$$C_{ij} = \bar{C}_{ij} \leq C^*_{ij}.$$

Let $p_{ij} = (1-\gamma)^{j-1}\gamma p_i$ be the probability that the object is found in the j-th visit to vertex v_i. Note that this is independent of the solution and $p_{ij} = w_{ij}$. Assuming that P^* is an optimal search path we obtain that

$$\text{OPT}^{(K)}_{\text{TRP}} \leq \sum_{v_i \in V} \sum_{j=1}^{K} C_{ij} w_{ij} = \sum_{v_i \in V} \sum_{j=1}^{K} \bar{C}_{ij} p_{ij} \leq \sum_{v_i \in V} \sum_{j=1}^{K} C^*_{ij} p_{ij} \leq \text{OPT}.$$

□

Our algorithm for IGS-P is basically a reduction to TRP on an extended graph. The search path starts with a solution to the TRP-instance and then repeats an arbitrary TSP-tour until the target is found. By choosing the number of leaves K in the extended graph large enough, the probability that the target is not found on the TRP-path is small and there is no need to optimize the length of the TSP-tour.

Algorithm_IGS-P(I,K):

(1) Construct the weighted-TRP-instance $I^{(K)}_{\text{TRP}}$.
(2) Run a (weighted) TRP-algorithm to find a solution P_0 for $I^{(K)}_{\text{TRP}}$.
(3) In G, follow the search path corresponding to P_0 once and then repeat an arbitrary TSP-tour R indefinitely.

Theorem 2. *Algorithm_IGS-P returns a $(1+\epsilon)\alpha$-approximation, where α is the approximation factor for the TRP solution P_0 for the instance $I^{(K)}_{TRP}$, and $K = O\left(\log(n/(\epsilon p_{\min})/\log(1/(1-\gamma))\right)$.*

Proof. Let D be the maximum distance between any two vertices in $G = (V, E)$ (and recall that $|V| = n$) and let $p_{\min} = \min_i p_i$. Then a lower bound on the optimal expected time until detection is

$$\textsc{Opt} \geq (D+1)p_{\min}. \tag{1}$$

We shall refer to the tour formed by following P_0 and returning to r as *round* 0. Let R_0 be the length of this tour. Then

$$R_0 \leq (D+1)nK.$$

For the next rounds, in which we repeat an arbitrary TSP-tour, we use indices $h = 1, 2, \ldots$. Note that the time for one TSP-tour, including the search time is bounded by $(D+1)n$. Hence, the arrival time at any vertex v_i in round $h \geq 1$ is bounded by

$$R_0 + h(D+1)n \leq (K+h)(D+1)n. \tag{2}$$

For any vertex v_i and $h \geq 1$, the probability that the object is found in round h, given that it is in v_i, is exactly $(1-\gamma)^{K+h-1}\gamma$. Let us denote by Z_i the expected time until detection at v_i, given that the object is in v_i but not found in round 0. Then, using (2), we get

$$\begin{aligned} Z_i \leq \sum_{h\geq 1}(1-\gamma)^{h-1}\gamma(K+h)(D+1)n &= \gamma(D+1)n\sum_{h\geq 1}(1-\gamma)^{h-1}(K+h) \\ &= \gamma(D+1)n(K/\gamma + 1/\gamma^2) \\ &= (D+1)n(K+1/\gamma) \\ &\leq (K+1/\gamma)n/p_{\min}\textsc{Opt}. \end{aligned}$$

In total, the expected time until detection given that the object is not found in round 0 is

$$\sum_{v_i\in V} Z_i p_i \leq \sum_{v_i\in V} Z_i \leq (K+1/\gamma)n^2/p_{min}\textsc{Opt}.$$

Now, we choose K such that $(1-\gamma)^K\sum_{v_i\in V} Z_i p_i \leq \epsilon\textsc{Opt}$. That means, we want

$$(1-\gamma)^K(K+1/\gamma)n^2/p_{\min} \leq \epsilon,$$

which holds for

$$K = O\left(\frac{\log(n/(\epsilon p_{\min}))}{\log(1/(1-\gamma))}\right). \tag{3}$$

□

The theorem above, together with the α-approximation algorithm for TRP with $\alpha < 3.59$ from [8], and the fact that weighted-TRP can be reduced to unweighted TRP with loss of a factor $(1+\epsilon)$ [20], gives the following result.

Corollary 1. *There is a* $(3.59+\varepsilon)$*-approximation algorithm for IGS-P in any metric space.*

As shown by Sitters [20], the weighted-TRP has a PTAS for trees, planar graphs and the Euclidean plane. Any tree remains a tree after the reduction described above and the same holds for planar graphs. Hence, for these metrics, a PTAS for IGS-P follows immediately. For the Euclidean plane the argument is more subtle. The proof is omitted here due to space constraints.

2.2 Expanding Search

For expanding search, a similar reduction and algorithm applies. First, we solve a corresponding expanding search instance on an extended graph and use this as the start of the search path. For the remainder of the path we repeat an arbitrary TSP-tour, which for expanding search is just a permutation of the vertices since there is no travel cost after round zero.

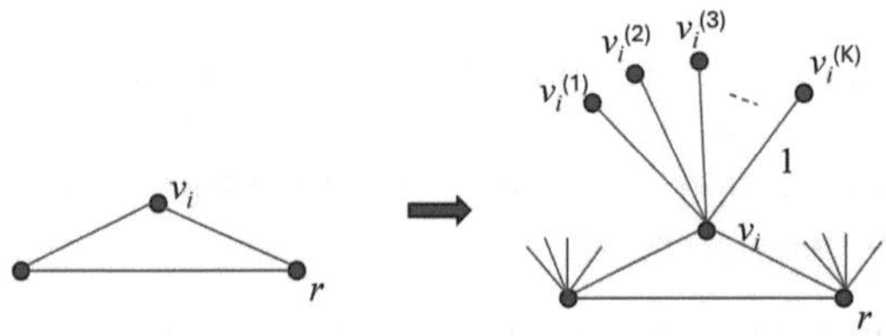

Fig. 2. Reduction of IGS-E to ESP. Vertex $v_i^{(j)}$ has weight $w_{ij} = (1-\gamma)^{j-1}\gamma p_i$ and $K = O\left(\log(n/(\epsilon p_{\min}))/\log(1/(1-\gamma))\right)$.

More precisely, we define almost the same graph as we did for pathwise search with the only difference that we keep the root r as it is and all added leaves have length 1 instead of 0.5 (Fig. 2). Note that the example given in the beginning of this section shows that we do still need $K = \Omega(\log n)$. Hence, we use K as defined in (3). The algorithm is outlined below.

Algorithm_IGS-E(I,K):

(1) Construct the weighted-ESP instance $I_{\text{ESP}}^{(K)}$.
(2) Run a (weighted) ESP algorithm to find a solution P_0 for $I_{\text{ESP}}^{(K)}$.
(3) In G, follow the search path corresponding to P_0 once and then repeat an arbitrary TSP-tour R indefinitely.

The proofs of Lemma 2 and Theorem 3 are similar to that of Lemma 1 and Theorem 2 and are omitted due to space constraints.

Lemma 2. *Let* OPT *be the optimal value of IGS-E-instance I. Then* $\text{OPT}_{ESP}^{(K)} \leq$ OPT. □

Theorem 3. *Algorithm_IGS-E returns a* $(1+\epsilon)\alpha$*-approximation, where* α *is the approximation factor for the ESP solution* P_0 *for* $I_{ESP}^{(K)}$ *and* $K = O\left(\log(n/(\epsilon p_{\min}))/\log(1/(1-\gamma))\right)$.

The theorem above, together with the $(2e+\varepsilon)$-approximation algorithm for (weighted) ESP from [11], gives the following result.

Corollary 2. *For any* $\epsilon > 0$*, there is a* $(2e+\epsilon)$*-approximation algorithm for IGS-E in any metric space.*

2.3 Maximizing the Probability

Here we show that maximizing the probability among search paths of length at most some given value T easily reduces to an orienteering problem, with loss of a factor $1+\epsilon$.

Given an instance I of IGS-O, we define an instance of the orienteering problem as shown in Fig. 1. The weight of vertex $v_i^{(j)}$ is defined as $w_{ij} = (1-\gamma)^{j-1}\gamma p_i$. We let $K = \lceil \log(\epsilon/\log(1-\gamma)\rceil$, and let $\text{OPT}_{\text{OP}}^{(K)}$ be the optimal value for this instance $I_{\text{OP}}^{(K)}$.

Lemma 3. *Let* OPT *be the optimal value of IGS-O-instance* I*. Then* $\text{OPT}_{OP}^{(K)} \geq (1-\epsilon)\text{OPT}$.

Proof. Consider any search path P^* for IGS-O of length at most T and total probability mass Z. Now we remove each j-th visit to any vertex for all $j > K$ and obtain an orienteering path P of length at most T in G. Moreover, its probability mass it at least $(1-\epsilon)Z$ since for any vertex v_i, the probability that the object is found in the first K visits to v_i, given that it is in v_i, is

$$1 - \sum_{j=1}^{K}(1-\gamma)^K > 1 - \epsilon \;\text{ for } K > \log\epsilon/\log(1-\gamma).$$

□

The algorithm, as defined below, is even simpler than the algorithms of the previous sections since there is no need for repetition. We only solve the orientering problem in the extended graph and take that as the solution for IGS-O.

Algorithm_IGS-O(I,K):

(1) Construct the weighted-OP instance $I_{\text{OP}}^{(K)}$.
(2) Run a (weighted) OP algorithm to find a solution P_0 for $I_{\text{OP}}^{(K)}$.
(3) In G, follow the search path corresponding to P_0.

Theorem 4. *Algorithm_IGS-O returns a* $(1-\epsilon)\alpha$*-approximation, where* α *is the approximation factor for the OP solution* P_0 *for* $I_{OP}^{(K)}$ *and* $K = \lceil \log \epsilon / \log(1-\gamma) \rceil$.

Proof. This follows directly from Lemma 3 and the fact that any orieentering path for $I_{\mathrm{OP}}^{(K)}$ implies a search path for I of exactly the same length and probability mass. □

Chekuri et al. [9] showed that the orienteering problem admits a $(0.5 - \epsilon)$-approximation in any metric space, which now immediately gives the following result.

Corollary 3. *There is a* $(0.5 - \epsilon)$*-approximation algorithm for IGS-O in any metric space.*

3 Adversarial Targets

In the imperfect graph search problem with adversarial targets, the goal is to minimize the maximum expected time until detection over the vertices. We again consider pathwise search and expanding search.

3.1 Pathwise Search

With pathwise search, if detection of the target is perfect, we will search every vertex exactly once, and the target will be hidden at the last vertex we search. Hence, minimizing the time until the hider is found boils down to minimizing the length of a path that visits all vertices. This is the well known path TSP problem.

If we consider imperfect detection, a natural idea is to move back and forth on the shortest path that visits all vertices. Since computing this shortest path is NP-hard, we use the $\frac{3}{2}$-approximation algorithm of [24]. We note that there is also a $\frac{3}{2} - \varepsilon_0$-approximation for some small constant ε_0 [13,22], but to simplify the analysis, we will use the approximation guarantee of $\frac{3}{2}$. In the following algorithm, we will search for one time unit each time we visit a vertex. Since we move back and forth, the endpoints of the path are searched for two time units when they are visited. We call this algorithm Single-search Back-and-forth.

Single-Search Back-and-Forth:

(1) Construct a TSP path P on V starting from r.
(2) Repeat P back and forward, where each vertex is searched for one time unit, and in each turn, the end points are searched twice.

To obtain an approximation guarantee, we use the following lower bound. Note that the time at which the last vertex that is visited for the first time has been searched is at least $\textsc{Path} + n$, where $\textsc{Path}$ is the length of the shortest path that visits all vertices, starting in r. Since all other visits occur later in time, we know

$$\text{OPT} \geq \text{PATH} + n$$

A simple analysis gives a $\frac{3}{2\gamma}$-approximation. Note that we need at most $\frac{3}{2}\text{PATH} + n$ time units to search each vertex for the first time. Also, we need at most $2(\frac{3}{2}\text{PATH} + n)$ time units to search each vertex twice. In general, we have searched each vertex for the ℓth time after at most $\ell(\frac{3}{2}\text{PATH} + n)$ time units. Hence, the expected time until detection at any vertex is at most

$$\sum_{\ell=1}^{\infty}(1-\gamma)^{\ell-1}\gamma\ell\left(\frac{3}{2}\text{PATH} + n\right) = \frac{1}{\gamma}\left(\frac{3}{2}\text{PATH} + n\right) \leq \frac{3}{2\gamma}\text{OPT}.$$

Using a more refined analysis, we can improve this guarantee. In particular, this analysis will use the fact that vertices that are visited late in the first pass, will be visited early in the second pass. The proof is omitted due to space constraints.

Theorem 5. *For any $\gamma \in (0,1]$, the Single-search Back-and-forth algorithm gives a $g(\gamma)$-approximation for AIGS-P, where*

$$g(\gamma) = \frac{3}{2}\left(\frac{\gamma^2 - 2\gamma + 2}{2\gamma - \gamma^2}\right).$$

For large values of γ, this gives satisfying results, e.g., for $\gamma \geq \frac{1}{2}$, we have a $\frac{5}{2}$-approximation. For $\gamma = 1$, we match the $\frac{3}{2}$-approximation that is known for perfect detection. However, for small values of γ, Single-search Back-and-forth is arbitrarily bad. Intuitively, too much time is spent on traveling when only one unit of time is used for searching at each visit.

An alternative algorithm, which we will call Multi-search Back-and-forth, will search for more time units when visiting a vertex. In particular, it will search a vertex at any visit for $\left\lceil \frac{1}{\gamma} \right\rceil$ time units. To analyze this algorithm, we need an additional lower bound that is concerned with the search time.

Multi-search Back-and-Forth:

(1) Construct a TSP path P on V starting from r.
(2) Repeat P back and forward where in each visit to a vertex it is searched $\left\lceil \frac{1}{\gamma} \right\rceil$ times except for the turning points (endpoints of P) which are searched for twice this amount.

Lemma 4. *For the AIGS-P, we have, for any $\gamma \in (0,1]$,*

$$\text{OPT} \geq \frac{2-\gamma}{2} \cdot \frac{n}{\gamma}.$$

Proof. First, we ignore the travel times. Consider the optimal solution, which clearly gives a lower bound on the original optimal value. We note that the value, which is the maximum expected search time over all vertices, is at least the average expected search time of this optimal solution. Moreover, the average expected search time of this optimal solution is at least the average expected search time of the solution that minimizes the average expected search time over the vertices. Minimizing the average expected search time is easy to solve.

Note that we can search a vertex at every time moment in $\mathbb{N} = \{1, 2, \ldots\}$. We assign weights to these moments in time, where n moments get weight γ, n moments get weight $(1-\gamma)\gamma$, and so on. To minimize the average expected search time over the vertices, it is clearly optimal to assign weight γ to $\{1, \ldots, n\}$, weight $(1-\gamma)\gamma$ to $\{n+1, \ldots, 2n\}$, and so on. The average expected search time of this solution equals

$$\begin{aligned}
&\gamma \cdot \frac{n+1}{2} + (1-\gamma)\gamma \cdot \frac{3n+1}{2} + (1-\gamma)^2\gamma \cdot \frac{5n+1}{2} + \ldots \\
=&\sum_{\ell=0}^{\infty}(1-\gamma)^\ell\gamma \cdot \left(\frac{(2\ell+1)n+1}{2}\right) = n\sum_{\ell=0}^{\infty}(1-\gamma)^\ell\gamma\ell + \frac{n+1}{2}\sum_{\ell=0}^{\infty}(1-\gamma)^\ell\gamma \\
=&n(1-\gamma)\gamma\sum_{\ell=0}^{\infty}(1-\gamma)^{\ell-1}\ell + \frac{n+1}{2}\gamma\sum_{\ell=0}^{\infty}(1-\gamma)^\ell = \frac{n(1-\gamma)\gamma}{\gamma^2} + \frac{(n+1)\gamma}{2\gamma} \\
=&\frac{n}{\gamma} - \frac{n}{2} + \frac{1}{2} \geq \frac{2-\gamma}{2} \cdot \frac{n}{\gamma}.
\end{aligned}$$

Hence, we conclude that, for any $\gamma \in (0, 1]$,

$$\text{OPT} \geq \frac{2-\gamma}{2} \cdot \frac{n}{\gamma}.$$

□

The proof of the following theorem is omitted due to space constraints.

Theorem 6. *For any $\gamma \in (0, 1]$, the Multi-search Back-and-forth algorithm gives a $h(\gamma)$-approximation for AIGS-P, where*

$$h(\gamma) = \left(\frac{3(1-\gamma)^{\frac{2}{\gamma}}}{1-(1-\gamma)^{\frac{2}{\gamma}}} + \frac{3}{2} + \left(\frac{2(1-\gamma)^{\frac{2}{\gamma}}}{1-(1-\gamma)^{\frac{2}{\gamma}}} + 1 + \frac{1}{2} \cdot \frac{(1-\gamma)^{\frac{1}{\gamma}}}{1+(1-\gamma)^{\frac{1}{\gamma}}}\right)\frac{2}{2-\gamma}\right).$$

In particular, Multi-search Back-and-forth is a $\frac{7}{2}$-approximation.

We observe that the worst case guarantee for Multi-search Back-and-forth is obtained for $\gamma = 1$, whereas Single-search Back-and-forth gives a $\frac{3}{2}$-approximation in this case. Therefore, we can improve the approximation guarantee by applying both algorithms, and take the best of the two outcomes.

Theorem 7. *Taking the best solution produced by Single-search Back-and-forth and Multi-search Back-and-forth is an approximation algorithm for AIGS-P with guarantee*

$$\frac{5e^2 + e + 4}{2(e^2 - 1)} < 3.4171.$$

Proof. If we take the best of the two algorithms, the worst case occurs when γ approaches 0. This is illustrated in Fig. 3. This gives a guarantee of

$$\lim_{\gamma \to 0} h(\gamma) = \frac{5e^2 + e + 4}{2(e^2 - 1)} < 3.4171.$$

□

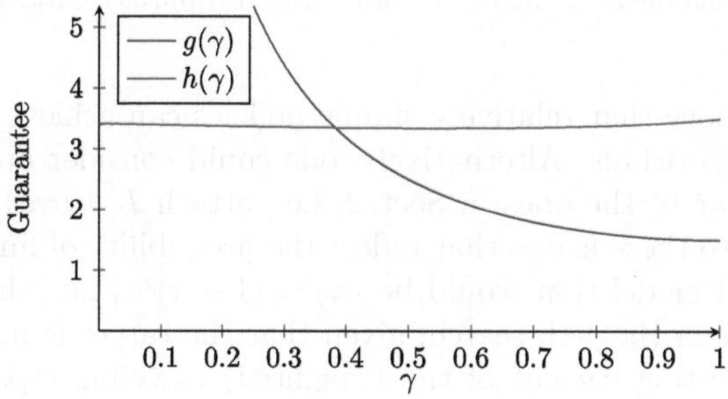

Fig. 3. Guarantee of Single-search Back-and-forth $g(\gamma)$, and guarantee of Multi-search Back-and-forth $h(\gamma)$. The minimum of the two functions is maximized when γ approaches 0.

3.2 Expanding Search

We will now consider imperfect graph search with adversarial targets and expanding search. If the detection of the target is perfect, the target will be hidden at the last vertex we visit. Hence, minimizing the time until the hider is found is equivalent to finding the minimum spanning tree, which is solvable in polynomial time.

For expanding search with imperfect detection, we use an algorithm that is similar to Single-search Back-and-forth that we used for pathwise search. We first construct a minimum spanning tree, in a way that is allowed in the expanding search paradigm, and search for one time unit each time a vertex is visited. After that, there is no more travel time. We search for another time unit at each of the vertices in reversed order and repeat this indefinitely. We will call this algorithm Single-search MST.

Single-Search MST:

(1) Compute the MST. Starting in r, use edges of the MST to visit vertices, where each edge should be connected to the already used edges. Denote P for the resulting path.
(2) Repeat P back and forward, where each vertex is searched for one time unit, and in each turn, the endpoints are searched twice.

To analyze this algorithm, note that OPT $\geq$ MST $+ n$, since this is the earliest moment in time at which the last vertex is searched for the first time. Moreover, we still have

$$\text{OPT} \geq \frac{2-\gamma}{2} \cdot \frac{n}{\gamma},$$

since the proof of Lemma 4 holds when all distances are equal to 0. The proof is omitted due to space constraints.

Theorem 8. *Single-search MST is an 2.0876-approximation algorithm for AIGS-E.*

We showed above that relatively simple policies can achieve constant factor approximation algorithms. Alternatively, one could consider applying a reduction that is similar to the ones in Sect. 2, i.e., attach K leaves to each vertex, and add weights to these leaves that reflect the probability of finding the target. In the adversarial model that would be $w_{ij} = (1-\gamma)^j\gamma$, i.e., the probability of finding the target in the j-th search, given that the target is in vertex v_i. This leads to an interesting variant of the (weighted) traveling repairman problem where the leaves of each vertex are seen as a group and the goal is to find a path that minimizes the maximum over the groups of the (weighted) total arrival time of the vertices in a group.

More formally, we have the following problem: Given is a graph $G = (V, E)$ with root $r \in V$, vertex weights w_i, and metric distances $d(i, j)$ for $v_i, v_j \in V$. We are also given a partition $V_1, \ldots, V_\alpha$ of V into α groups. The goal is to find a path that minimizes the maximum weighted *group completion time*, i.e.,

$$\max_{\beta \in \{1,\ldots,\alpha\}} \sum_{v_i \in V_\beta} w_i C_i.$$

This is a natural and interesting problem on its own, that can also be motivated by fairness considerations. Essentially, we want that the groups observe a similar average arrival time, which is usually related to the level of service. We also observe that it generalizes both TRP ($\alpha = 1$) and path TSP ($\alpha = n$). We are not aware of any prior work on this *fair*-TRP problem.

References

1. Agnetis, A., Hermans, B., Leus, R., Rostami, S.: Time-critical testing and search problems. Eur. J. Oper. Res. **296**(2), 440–452 (2022)
2. Alpern, S., Lidbetter, T.: Mining coal or finding terrorists: the expanding search paradigm. Oper. Res. **61**(2), 265–279 (2013)
3. Averbakh, I., Pereira, J.: The flowtime network construction problem. IIE Trans. **44**(8), 681–694 (2012)
4. Benkoski, S.J., Monticino, M.G., Weisinger, J.R.: A survey of the search theory literature. Naval Res. Logist. (NRL) **38**(4), 469–494 (1991)
5. Berger, J., Lo, N., Noel, M.: Exact solution for search-and-rescue path planning. Int. J. Comput. Commun. Eng. **2**(3), 266 (2013)

6. Blum, A., Chalasani, P., Coppersmith, D., Pulleyblank, B., Raghavan, P., Sudan, M.: The minimum latency problem. In: Proceedings of the Twenty-Sixth Annual ACM Symposium on Theory of Computing, pp. 163–171 (1994)
7. Chakraborty, N., Kasthurirangan, P.N., Mitchell, J.S., Nguyen, L., Perk, M.: Provable methods for searching with an imperfect sensor. arXiv preprint arXiv:2410.06069 (2024)
8. Chaudhuri, K., Godfrey, B., Rao, S., Talwar, K.: Paths, trees, and minimum latency tours. In: 44th Annual IEEE Symposium on Foundations of Computer Science, Proceedings, pp. 36–45. IEEE (2003)
9. Chekuri, C., Korula, N., Pál, M.: Improved algorithms for orienteering and related problems. ACM Trans. Algorithms **8**(3), 1–27 (2012)
10. Goemans, M., Kleinberg, J.: An improved approximation ratio for the minimum latency problem. Math. Program. **82**(1), 111–124 (1998)
11. Griesbach, S.M., Hommelsheim, F., Klimm, M., Schewior, K.: Improved approximation algorithms for the expanding search problem. In: 31st Annual European Symposium on Algorithms (ESA 2023), pp. 54–1. Schloss Dagstuhl–Leibniz-Zentrum für Informatik (2023)
12. Hermans, B., Leus, R., Matuschke, J.: Exact and approximation algorithms for the expanding search problem. INFORMS J. Comput. **34**(1), 281–296 (2022)
13. Karlin, A.R., Klein, N., Gharan, S.O.: A (slightly) improved approximation algorithm for metric TSP. In: Proceedings of the 53rd Annual ACM SIGACT Symposium on Theory of Computing, pp. 32–45 (2021)
14. Koopman, B.O.: The theory of search. I. Kinematic bases. Oper. Res. **4**(3), 324–346 (1956)
15. Koopman, B.O.: The theory of search. II. Target detection. Oper. Res. **4**(5), 503–531 (1956)
16. Koopman, B.O.: The theory of search. III. The optimum distribution of searching effort. Oper. Res. **5**(5), 613–626 (1957)
17. Lössner, U., Wegener, I.: Discrete sequential search with positive switch cost. Math. Oper. Res. **7**(3), 426–440 (1982)
18. Sahni, S., Gonzalez, T.: P-complete approximation problems. J. ACM (JACM) **23**(3), 555–565 (1976)
19. Sitters, R.: The minimum latency problem is NP-hard for weighted trees. In: Cook, W.J., Schulz, A.S. (eds.) IPCO 2002. LNCS, vol. 2337, pp. 230–239. Springer, Heidelberg (2002). https://doi.org/10.1007/3-540-47867-1_17
20. Sitters, R.: Polynomial time approximation schemes for the traveling repairman and other minimum latency problems. SIAM J. Comput. **50**(5), 1580–1602 (2021)
21. Stone, L.D.: Theory of optimal search, vol. 118. Elsevier (1976)
22. Traub, V., Vygen, J., Zenklusen, R.: Reducing path TSP to TSP. In: Proceedings of the 52nd Annual ACM SIGACT Symposium on Theory of Computing, pp. 14–27 (2020)
23. Yu, L., Lin, D.: Bayesian-based search decision framework and search strategy analysis in probabilistic search. Sci. Program. **2020**(1), 8865381 (2020)
24. Zenklusen, R.: A 1.5-approximation for path TSP. In: Proceedings of the Thirtieth Annual ACM-SIAM Symposium on Discrete Algorithms, pp. 1539–1549. SIAM (2019)

Correction to: Robust Scheduling on Uniform Machines
New Results Using a Relaxed Approximation Guarantee

Hauke Brinkop, David Fischer, and Klaus Jansen

Correction to:
Chapter 8 in: J. Matuschke and J. Verschae (Eds.): *Approximation and Online Algorithms*, LNCS 16077, https://doi.org/10.1007/978-3-032-06706-7_8

The book was published with a typo of chapter 08 in this book ID (667325_1_En). The submitted paper with the Funding Acknowledgement "Supported by the German Research Foundation (DFG) project “Fine-grained complexity and algorithms for scheduling and packing” (JA 612/25-1)."; whereas, according to new DFG rules, it has been updated as "Funded by the Deutsche Forschungsgemeinschaft (DFG, German Research Foundation) – Project Number 453769249".

The updated version of this chapter can be found at
https://doi.org/10.1007/978-3-032-06706-7_8

J. Matuschke and J. Verschae (Eds.): WAOA 2025, LNCS 16077, p. C1, 2026.
https://doi.org/10.1007/978-3-032-06706-7_16

Author Index

J. Matuschke and J. Verschae (Eds.): WAOA 2025, LNCS 16077, p. 229, 2026.
https://doi.org/10.1007/978-3-032-06706-7

MIX
Papier aus verantwortungsvollen Quellen
Paper from responsible sources
FSC® C105338

If you have any concerns about our products,
you can contact us on
ProductSafety@springernature.com

In case Publisher is established outside the EU,
the EU authorized representative is:
Springer Nature Customer Service Center GmbH
Europaplatz 3, 69115 Heidelberg, Germany

Printed by Libri Plureos GmbH
in Hamburg, Germany